AF738071

Our Multispecies Futures

Exeter Studies in Environmental Humanities

Past, Present and Future Econarratives

Series Editors:
Professor Peggy Karpouzou, National and Kapodistrian University of Athens, Greece
Dr Nikoleta Zampaki, University of Information, Technology and Management, Poland

Editorial Advisory Board:
Marco Armiero, Mieke Bal, Dominic Boyer, Jeff Diamanti, Vicki Kirby, Nina Lykke, Orsola Rignani, Scott Slovic, Imre Szeman, Robert T. Tally Jr.

Exeter Studies in Environmental Humanities: Past, Present and Future Econarratives provides a forum for innovative scholarship in Environmental Humanities, encompassing critical strands and interdisciplinary approaches to literary, ethical and socio-political explorations of life-forms' entanglements, reaching from antiquity to present. The series encourages and includes works that examine cultural representations of the environment and species' interplays, and which contribute alternative post-anthropocentric storytellings to the one of the Anthropocene, which has a disastrous impact on all Earth's ecosystems and even threatens the planet's existence.

This series seeks to engage with original manuscripts (monographs, edited collections and handbooks) on a great range of Environmental Humanities topics, exploring past, present and potential future narratives, and 'restorying' our relationship with nature to rethink sustainable planetary futures.

In order to submit a book proposal and discuss your project, please contact the Series Editors:

Professor Peggy Karpouzou (pkarpouzou@phil.uoa.gr)
Dr Nikoleta Zampaki (nzampaki@wsiz.edu.pl)

or Senior Commissioning Editor Becky Taylor (b.taylor@exeterpress.co.uk)

Our Multispecies Futures

Control, Care and Kinship in a Transformed World

VETRI NATHAN

UNIVERSITY
of
EXETER
PRESS

First published in 2026 by
University of Exeter Press
Reed Hall, Streatham Drive
Exeter EX4 4QR, UK

www.exeterpress.co.uk

Our Multispecies Futures: Control, Care and Kinship in a Transformed World

Copyright © 2026 Vetri Nathan

The right of Vetri Nathan to be identified as author of this work has been asserted by him in accordance with the Copyright, Designs and Patents Act 1988.

Exeter Studies in Environmental Humanities

ISSN 3049-771X Print
ISSN 3049-7728 Online

This publication is licensed, unless otherwise indicated, under the terms of the Creative Commons Attribution 4.0 International (CC BY 4.0) licence (http://creativecommons.org/licenses/by/4.0/), which permits use, sharing, adaptation, distribution, and reproduction in any medium or format, provided you give appropriate credit to the original author(s) and source, link to the Creative Commons licence, and indicate any modifications.

The images or other third-party material in this publication are covered by the publication's Creative Commons license, unless otherwise indicated in a reference to the material.

If the material is not covered by the publication's Creative Commons license and the intended use is not permitted by law or exceeds the permitted use, permission for use must be obtained directly from the copyright holder.

Despite careful editing, all information in this publication is provided without guarantee; any liability on the part of the author, the editor, or the publisher is excluded.

Published with the support of the Division of Humanities and the Multispecies Futures Lab at the University of California, Los Angeles.

A CIP catalogue record for this book is available from the British Library

https://doi.org/10.47788/XWMQ6666

ISBN 978-1-80413-252-4 Hardback
ISBN 978-1-80413-253-1 Paperback
ISBN 978-1-80413-254-8 ePub
ISBN 978-1-80413-255-5 PDF

EU Authorised Representative: Easy Access System Europe – Mustamae tee 50, 10621 Tallinn, Estonia, gpsr.requests@easproject.com

Cover image by Chuck Graham

Typeset in Adobe Caslon Pro by S4Carlisle Publishing Services, Chennai, India

Contents

Staying with the trouble means making oddkin; that is, we require each other in unexpected collaborations and combinations, in hot compost piles. We become—with each other or not at all.

Donna Haraway[1]

All is beautiful,
All is beautiful,
All is beautiful indeed.
Now the Mother Earth
And the Father Sky,
Meeting, joining one another
Helpmates ever, they.
All is beautiful,
All is beautiful,
All is beautiful, indeed.

Navajo Song and Blessing, cited by Gregory Cajete[2]

MULTISPECIES FUTURES LAB

To my parents, Janak and Shamim Nathan

For their unquenchable thirst for knowledge
And care for anyone in need.

And, to Buck,
My First-Born.
I miss your gentle and magnificent furry presence
And will always love you.

Acknowledgements

This project was supported by so many multispecies beings—humanimal and non.

I am grateful for my current habitat at the University of California, Los Angeles and to the natural beauty of the American West that never fails to inspire. The University of California is, and still remains, a system that is invested in outstanding public education as well as authentically interdisciplinary and experimental research for the common good. UCLA's startup research funds allowed me to complete this book and provide for its publication in Open Access format. For this I am thankful to the Division of Humanities, Dean Alexandra Minna Stern and my Department Chair, Todd Presner. Many colleagues believed in the potential of my research and teaching from the get-go: I am particularly thankful to Professors Erin Cooney, Vincente Diaz, Shelleen Greene, Thomas Harrison, Ursula Heise, David Kim, Wendy Perla Kurtz, Roberta Morosini, Robert Rushing, Dominic Thomas and Maite Zubiaurre for their collegiality and support of my transcultural predispositions during the writing of this book. Professors Silvia Antosa, Paolo Frasca and Charlotte Ross provided me with a pivotal experience in the project's final stages by inviting me to give a keynote talk at Christ Church, University of Oxford in April 2025. To the students in my *Multispecies Humanities* graduate seminar in early 2025—thank you for your trust, curiosity and sincere engagement with these new ways of thinking and doing research. As you can see, our experimental field trip to Wi'ma Island features prominently in the final version of this book. Our group owes thanks to Director Robyn Shea for hosting us at the California State University Channel Islands' Santa Rosa Island Research Station.

My editor Becky Taylor championed the project's core spirit and guided it through its various stages with great care and efficiency. Production Manager David Hawkins meticulously and empathetically shaped the book's final form. Series editors Peggy Karpouzou and Nikoleta Zampaki provided judicious support and reviewers' feedback stimulated further distillation of the manuscript. Earlier versions of material included in this book feature in an article, "Field Notes from the Multispecies Futures Lab: Entangled Lives, Bodies, and Ecocultural Systems," *California Italian Studies*, Volume 14, Issue 1; and a book chapter, "Edible Nation: Imagining Somalia through Cuisine," from the edited volume *Longing for the Future: Mal D'Afrique and Afro-Optimism in Perspectives on Somalia*, by permission from Routledge, Taylor and Francis Group.

I am particularly grateful to my visualization research assistant, Doğa Tekin, for our collaborative explorations together to produce the first two prototypes of "Multispecies Thick-Maps" that feature in this book. I am also most appreciative of the enthusiastic support of my current Lab Manager, Maria Larsson. Lab Research Assistant Emily Szpiro (funded by a grant from UCLA's *Humanities Technology* unit) worked with the utmost professionalism to parse citations, references, permissions and to oversee general readability. Hiju Kim and John Colin Marston provided valuable teaching support that allowed me to focus more efficiently on this project. The *Scala dei Turchi* project described in the Introduction was a result of a deeply rewarding collaboration with Professor Teresa Fiore at Montclair State University with assistance from UCLA graduate student Giacomo Frazzetta. The experts at the Wind River Tribal Reservation and the Greater Yellowstone Coalition continue to provide deep inspiration: thank you Wesley Martel, Jason and Richard Baldes, Ina Tillman, Shana and Charles Drimal, and the magnificent Buffalo for sharing your rich ecocultural systems with me.

The origins of my *Multispecies Futures Lab* at UCLA lie in my previous institution, Rutgers University, where it began the larval stages of its life as *The Cybercene Lab*. This initial work was funded by a Mellon Chair in Global Racial Justice. I am truly indebted to the extensive feedback given by Professor Michelle Stephens on a very early draft of this project in 2023, which set the tone for further generative thought and revisions. I am grateful to Professors Alessandro Vettori, Paola Gambarota and my original Lab Manager, Diletta Pasetti, for their invaluable support. I am also appreciative of the graduate and undergraduate students who engaged with "all things multispecies" in my courses.

Friends Mladen and Nicky and my niece Nadia read a draft of the manuscript to confirm that it was accessible and engaging for readers beyond academia. Joia Duskic attested to the manuscript's practical utility for graduate students and advanced undergraduate students interested in multispecies studies. This was important input precisely because a community-engaged project such as this must stay attentive to the needs of broader audiences.

My globally dispersed family continue to provide so much encouragement and care despite the vast distances that often separate us. This book is dedicated to my parents, Janak and Shamim, for incalculable reasons. Pramod, I will always cherish our outdoor adventures together as kids, especially during our summer vacations at our grandparents' home in Satara. Some of our naturalcultural activities included: climbing and hanging in trees all day, hiding and then "discovering" treasure boxes, and creating pretend "mini national parks" with our rudimentary backyard tools to "conserve endangered species." Lia, Rosaria, Papà Salvatore and Teresa, thanks for showing me immeasurable love and for feeding me constantly. My nephews and nieces—Adelina, Alice, Aman, Andrea, Antonio, Asim, Chiara, Nadia, Nico Senior, Nico Junior, and

Nova—you are my guaranteed source of company and joy. You let me hope that the next generation can improve upon the previous ones.

My own children were absolutely fundamental to my incremental conversion from a comfortably anthropocentric humanities professor to an educator deeply invested in seeking ways to live, teach and do research via a multispecies paradigm. Buck, Argo, Keera and Junior: *thank you* for your limitless enthusiasm, profound care and unbreakable bonds of kinship. I am who I am today because you welcomed me wholeheartedly as an equal member of your Siberian Husky pack, despite my embarrassingly human limitations.

And finally, to Angelo, the keystone species of my support system. You drove me across vast distances so that I could complete my research. You have provided me with a stable and loving home for two decades. This book would be inconceivable without your patience, empathy, resilience, and indomitable sense of humor.

Preface

Close Encounters of the Furry Kind

My graduate students and I were in tracking mode. We were seeking a charismatic and sometimes elusive creature.

After an hour or so, to our great delight we crossed paths with one. We stopped and observed it at a respectful distance in a hushed silence.

As their name suggests, Channel Island Foxes, a mini-descendant of the much larger mainland gray fox, are endemic to this fascinating archipelago located off the coast of Southern California. They were highly endangered by the late 1990s, numbering only in the few hundreds, but have since been brought back from the brink—a conservation success story. Unlike their much larger relatives, they weigh only around two kilograms, but make up for their small size by their very large personalities. This particular fox seemed both curious and unperturbed by our presence, ambling amongst the stubbly grassland by the trail, looking for something good to eat. Often, she'd become engrossed in her own research: she would focus intensely on slight movements or sounds within the short grasses and bushes. Perhaps in reaction to signs of life in the tangle, she would swiftly jump in the air in a graceful arc and then back down towards the source of the sound. After some additional hunt-and-play of this kind, our new canid companion proceeded to bask a little in the spring sun, amble a little more, and just generally "hang out" close to us.

As I enjoyed observing her antics—she suddenly stopped and returned my gaze with equal potency.

I felt an unexpected and stupefying jolt inside me. Questions and doubts poured forth in my mind.

As we were tracking her, was she... perhaps... tracking us back?

Who was "the researcher" and who "the researched"?

What meaning was I trying to generate as a human and a humanities professor by "observing" the fox? Was I able to credit her with sentience, intelligence, agency and culture, despite our inability to communicate via human verbal language to procure "evidence" of "higher" thinking? Through my observations, did I see a life that was fundamentally different from mine, or could I admit to kinship? How was this multispecies kin-building process also deeply relevant to relationship making, identity creation, as well as justice, care and wellbeing in human-specific contexts? What exactly moved inside me when she looked back at me? Could I be equally attentive to the hundreds of other life-forms on the island that were less "spectacular" or not as "cute" as these foxes?

In that marvelous moment, my relationship with the fox, the island we were visiting, and its whole web of multispecies inhabitants (both human and non) flipped decisively. In retrospect, those few seconds of disorientation were another foundational event for this entire project.

The questions, thoughts and emotions I felt in that moment echoed those that had already gradually transformed my research in the past decade. They were also the same curiosities and doubts that had led me to that island in the first place. However, despite this preparation, the suddenly reciprocated gaze of my new furry scholar-companion had re-fixed these questions front and center with an unexpected force and immediacy.

Long after this moment had passed, the fox's gaze remained as a creative and cautionary memory—those curious and penetrating eyes peeping through the grasses, gazing back—reminding me to be attentive to my anthropocentric biases and habits. I was thankful then, and am grateful even now, for her gentle yet forceful and most elegantly posed reminder. The fox's lesson is deeply embedded within every nook and cranny of this project.

We do not have to say goodbye to her just yet… we shall meet her again in this book. Besides, you can always flip to the front cover to see another individual of her species… gazing right back at you.

Introduction

We Need a Multispecies Paradigm Shift

This book is deeply entangled within the bodies, lives and ecosystems of its huge cast of characters. It contains stories from the lives of multispecies beings (including us, humanimals) all carried out in intimate collaboration with multi-kind assemblages, from rivers and swamps to cities, farmland, mountains and oceans.[3] Their presence and actions occur over timescales that range from the split-second, to spans more fathomable to *Homo sapiens* (minutes, hours, years, decades), to geological epochs lasting millions of years. They currently inhabit or have dwelled in diverse corners of our battered but still magnificent planet—from Los Angeles to Mogadishu, from the tropical jungles of Thailand to the sparkling Mediterranean Sea.

Fundamentally, this is a book about relationships. It investigates the usually unrecognized and fearsome power of human narratives, desires, delusions, hopes, fears and fantasies to forge the actual material relationships and destinies of multispecies lives, bodies and ecosystems. It excavates the fundamental and formative role that human mediatic practices play in shaping meanings, and therefore our relationships, with our living planet. It stitches together narratives about multispecies associations arranged in strange and "unexpected collaborations."[4]

Our Multispecies Futures presents the findings of a global research project that spans seven years. Like one of those Russian *Matryoshka* nesting dolls, you—the reader—may open up a case study and find that, inside it, there is another embedded story. Or, like a Möbius strip, you may discover a story turning itself around to connect to another seemingly unrelated story that you experienced a few chapters back. As lands burn, oceans boil, species go extinct and manufactured conflicts (local, regional, global) multiply, I seek to explore with you the many powerful ecological, cultural, psychic and ethical consequences of humans' relentless meaning-making on a multitude of lives, bodies and ecosystems. The hope is that this kind of naturalcultural analysis may become another small step towards the collective goal of recognizing and revaluing multispecies bonds of kinship.

I wanted to imagine future multispecies relationalities by considering our past and present because I sensed that, in the Anthropocene,[5] the fates of multispecies lives are controlled by the human-made relational stories. Such meaning-making through stories sometimes reflects the capacity of *Homo sapiens* to be creatively curious about complex ecocultural realities. Conversely, it also mirrors our propensity towards strategic fantasizing to avoid considering or to actively suppress multispecies ties. I began to suspect that we were actually living in a strange new sub-era of the Anthropocene, which I began to refer to as the *Cybercene*. This era is characterized by even greater transformation in the ways in which humans create, narrate, reproduce, negotiate, interpret and pay attention to stories, identities and relationships. Putting these two intuitions together made me realize that maybe the consequences of our new hybrid digital-embodied lives for meaning production, storytelling and relationality have not yet been fully recognized or appreciated. Nor have these transformations' profound ecological and multispecies consequences been considered, let alone properly analyzed.

The Cybercene is a conceptual term and conceit that I use as a frame to discuss the current state of our ecocultural and multispecies relationships and our "nature-connectedness," now increasingly distanced and mediated via digital means. Since the 1800s, and especially since what has been called the post-1945 Great Acceleration,[6] people's increasingly urbanized lives have been progressively deprived of proximity to and physical interactions with other-than-human lives, bodies and natural ecosystems. This lack of nearness, both actual and metaphorical, has resulted in a sharp decline in nature-attentiveness and, therefore, nature-connectedness.[7] Besides quantitative data on this trend, I sensed that there was also a hitherto underexplored qualitative and dynamic link between an overall decrease in "naturalcultural" connections (an extinction of embodied nature and multispecies experiences), an overall increase in digital- and screen-based practices, and our planet's accelerated ecological decline towards a sixth mass extinction.[8] The deeper I pursued this intuition, the more clues I found that pointed towards important and understudied interconnections between the major ecological, socio-cultural and personal/psychic crises of our times. Climate change, identity-based conflicts, challenges to democratic norms, greater inequality, the digital revolution, the planet's sixth mass extinction, the increase in loneliness, hate and paranoia did not seem to be separate phenomena at all.

I sensed that with the advent of the Cybercene, we have very rapidly developed what I call a *techno-normative* cultural perspective in addition to a metro-normative one.[9] This has added value to digital media representations and our resulting preference for online meaning-making doubly disconnects and insulates us from our multispecies surroundings. Techno-normativity also distances us from our own bodies, which are an all-important nexus for multisensorial, embodied and affective connections with natural and multispecies realities.[10] At first glance, our digital lives and habits may not seem to have an influence on the wellbeing of our multispecies kin. However, I hope to demonstrate in this book how ecological damage is mediated and accelerated in the Cybercene through a series of cascading effects, feedback loops and resonances that deeply tether new human media habits to multispecies destinies. Perhaps I may even convince readers that we should call the Cybercene era the "Greatest Acceleration."

I propose that two seemingly distinct phenomena—human-digital mediatic/cultural habits and natural ecological connection—are related, but that this link is often hidden because it is an *indirect ecocultural relationship*. I call it by this name since nature-culture associations seem analogous to what the ecological sciences would characterize as *indirect effects* or *higher order interactions* such as *cascading trophic effects*.[11] An example of such a cascading effect is when a plant species benefits from an increased predator population indirectly, due to the decrease in the number of herbivores. I do not intend to use *indirect ecocultural effects* as a metaphor, but as a useful locator of actual multispecies natural+cultural working relationships that I will illustrate within my case studies. In chapters 1 and 2, I also find it useful to frame the Cybercene's techno-normative attitudes via a historical exploration of past mediatic practices of ecocultural distancing and meaning-making. Doing so contextualizes our current digital practices and the ongoing "Artificial Intelligence (A.I.) revolution", not as an unprecedented ecocultural moment in human history, but as a continuation and quickening of previous colonial-mediatic ways of multispecies distancing, othering, domination and control.

Departing from these initial premonitions, my investigative journey eventually resulted in a new lab at the University of California—the *Multispecies Futures Lab*[12]—and this book. Both explore two fundamental questions: How can we imagine, create and restore harmonious and thriving multispecies "naturalcultural" communities regionally, nationally and globally? How and why is this restoration work crucial as the global digital revolution further transforms our relationships with lives, bodies and ecocultural systems?

To be successful in this expansive endeavor, I felt that it was necessary to introduce some flexibility, creativity and an openness to "making-with" (as Donna Haraway describes it in *Staying with the Trouble*) all the multispecies entities (human and nonhuman) that may share their place-specific and culturally situated knowledges and affects. I have been immensely humbled as a researcher, professor and humanimal during these years of inquiry, and am grateful for the many bodies, lives and ecosystems featured in my case studies that have serendipitously led me down unexpected paths in my research. I experienced feelings of amazement, delight, hope as well as deep despair and heartbreak in the course of my explorations. If you as a reader are moved by these accounts to weave in your own stories of multispecies connections, I would consider this book a success.

As a record of research activity, I hope that *Our Multispecies Futures* will be generative of further passionate and collaborative exploration, rather than being seen as an exhaustive or conclusive study. This is also why my new lab exists—to continue the journey, the conversations, the collaborations and the companionships with multispecies lives after this book is published. For scholars, students and other interested readers new to the Environmental Humanities (hereafter EH) and Multispecies Studies, I have included some additional clarifications on key terms within the endnotes. To those who are already familiar with these fields, I hope that occasionally feeling that you are retreading a known conceptual path will provide you with connections to your own ongoing initiatives.

I have chosen to foreground stories and let the more abstract or technical concepts and theoretical analyses successively unwrap these stories. This approach has inevitably and intentionally resulted in a style that amalgamates storytelling and research. By modifying my approach to conventional academic research writing, I aim to foster ongoing conversations within and beyond the many subaltern[13] studies on the responsibilities of researchers towards the public, on the role of research in the communities that are studied, and to show how different forms of writing could aid in this reflective practice. As the general public's (sometimes fair and sometimes unfair) perception of an "out of touch" academia grows, I am keen on producing scholarship that listens to, engages and re-engages with our communities. As researchers, are we able to report back to the multispecies communities that we have worked with in concrete ways, without jargon getting in the way of such engagement? Can we write in a way that demonstrates that methodological rigor does not exclude emotionally engaging work? The Channel Island Fox's lesson of unconscious anthropocentric bias and its many associated arrogances stayed with me as I shifted my writing towards a more communicative style. I am also grateful to the many scholars in EH and in Indigenous studies cited in this book, who have already shown that it is possible to convey excellent research while remaining committed to being understood by and engaged with a wider readership. Sweeney Windchief and Timothy San Pedro quote Musqueam First Nation Elder Vincent Slogan to emphasize the ethical responsibility of passing forward knowledge and research:

> We extend our left palm upwards, to symbolize reaching back to receive teachings from the Ancestors and those who have travelled before us. We are given the challenge and opportunity to live these teachings. We also have the responsibility to pass those teachings to others who may also be the younger generation, which is shown when we put our right palm downwards. In the circle, we join hands in respect, reverence and cooperation.[14]

There is a also a wider point at stake here: not only engagement with a non-academic readership, but the inclusion of forms of emotional, reciprocal, ecologically embedded and spiritual knowledge that complicate the traditional scientist/objectivist/empiricist biases of normative institutional and species-based knowledge.

Despite the strategic foregrounding of stories, this book exists to make some very precise theoretical and methodological claims. It provides a broad, transcultural overview of the importance of engaging in a multispecies paradigm shift in the humanities, social sciences and sciences and also within the fundamental worldview of the general public. This Introduction and Chapter 1 place this goal within the current scholarship of and general trends within EH. The book fosters global expansion of multispecies studies by analyzing case studies that go beyond a specific geographical or geopolitical area. It also opens up these explorations to conversations with a broad audience of researchers, thinkers and policymakers beyond the humanities: from urban planners and community leaders to educators, journalists, scientists, wildlife conservationists, and environmental policymakers. I further propose some new research methodologies in the environmental and digital humanities and sciences that demonstrate the pragmatic applicability of its ethical, conceptual and methodological choices, such as naturalcultural discourse analysis and multispecies thick-mapping. I sparingly coin some new terms, such as the *Cybercene*, redefine some, such as *discourse*, and bring new life to somewhat forgotten or infrequently used concepts, such as Mohandas Karamchand Gandhi's *soul-force*.

This project places Indigenous thought, broadly speaking, and Indigenous studies scholarship more specifically, as pioneering theories of immense ecocultural import—something that unfortunately still needs to be said explicitly within and beyond academia. Furthermore, it proposes that a broad coalition of "subaltern studies" of different kinds, such as postcolonial, decolonial Indigenous and women's/gender/sexuality studies, must be welcomed as *central* to resolving major questions of global ecocultural wellbeing, justice and sustainability, rather than these being simply positioned as fetishized niche explorations relevant only to the interests and destinies of those specific subaltern identity groups. It argues that these different subaltern studies *urgently need to talk to each other* and to the general population, rather than operating within their own specialized theoretical, methodological and communal echo chambers. By bringing scholars from these various fields into conversation with each other and with you, the reader, I contend that a hybrid "tentacular"[15] methodology of "standing with"[16] these thinkers and approaches can allow the multispecies posthumanities to play a far more decisive role as the epistemic and methodological "glue" connecting scientific, political and social advocates to explore effective solutions in ecocultural sustainability and resilience. Utilizing an unorthodox assemblage of multispecies scholars, disciplines and methods also requires positioning myself within these various fields and with these ethically committed scholars. I accomplish this in the first chapter by harnessing approaches proposed by three experts: Donna Haraway, Kim TallBear and the Channel Island Fox. For indeed, humbly understanding one's positionality is a key precursor to making any knowledge claim, even as it explicitly conveys this as a *partial* knowledge claim.

My *dadima* (grandmother), used to say that stories are blessings. So, let me begin with one. It is an Island story. I hope you will enjoy it!

Magnificent Visions: Multispecies Studies in a Screwed-Up World

Wi'ma (Santa Rosa Island), off the Coast of Southern California, March 2025

"Is the Island still within you?"

It's the question that I have been frequently asking both myself and my group of graduate students ever since our three-day field trip to Santa Rosa Island (Chumash name, *Wi'ma*).

As the culminating experience in my new lab-incubated *Multispecies Humanities* graduate seminar at the University of California, Los Angeles, I led our group (Fig. 0.1) on an exploratory field trip to this very special location. At five a.m. on the day of our departure by ferry, I had to call in to check if we would actually be able to leave, since the winds had picked up substantially the day before and the seas were quite rough. I had barely slept the night before due to excitement and anticipation. When I drowsily picked up my phone in the pre-dawn hours, I was relieved that the Island did want us to visit her after all; the ferry service folks said that the sea conditions were just calm enough to allow for a traverse.

Wi'ma is a magical place. It is one in a spectacular eight-island archipelago that lies off the coast of Southern California. Like all living places, its *ecocultures*[17] are formed by many combinations of multispecies entanglements.

Figure 0.1 Humanimal participants of my Multispecies *Humanities* seminar on the jetty of Wi'ma (Santa Rosa Island), March 2025. (Anatoli Ulyanov)

Luckily for us, Wi'ma generously revealed its many multispecies and multi-kind relationships during our stay. I realized that the Island could be the ideal storyteller to introduce the background/context of this project as well as the primary questions that I will explore. The concepts, questions and methods touched upon in this Introduction will be elaborated upon in Chapter 1; together, these two sections form the primary theoretical framework shared organically between this book project and the co-conceived *Multispecies Futures Lab*.

The book and lab's strategically transcultural and decolonial theoretical and methodological approaches are reflected in both the global reach of its individual projects and initiatives (current and forthcoming lab projects span the Americas, Asia, Africa, Oceania and Europe) and its attentiveness towards hitherto marginalized ecocultural knowledges. With the example of the Wi'ma ecocultural system in mind, the Introduction and first chapter also provide a succinct description of the rich possibilities provided by the wider disciplinary contexts of EH and multispecies studies to explore intersecting questions of nature, human culture (meaning-making) and identity construction (nationality, citizenship and other bio- and ecocultural identities such as race, ethnicity, gender, labor, sexuality, productivity and reproductivity). Wi'ma Island will help me illustrate and analyze how these formations of human culture and identity are co-constituted by multispecies agents and relationalities. Rather than seeking to present a hagiographic account, my intention in describing this book and my lab's initial entanglements with current ecocultural scholarship is to give readers possible moments of resonance with their own approaches and interests in transdisciplinary research, their own teaching and community-engaged exploration with our extraordinary multispecies worlds.

Before working on global multispecies concerns, I specialized in analyses of literary and cinematic discourse in Italy with regard to national, migrant and other hybrid identity configurations via some of the major theoretical works of postcolonial studies.[18] By early 2018, despite the rich and important results the general field of postcolonial Italian studies was producing, I was growing increasingly restless with the predominantly anthropocentric and textual approaches to analyzing migration cultures in this field. I felt that I could no longer ignore my primary academic interest in zoology and environmental sciences during high school and college. This background had provided me with the tools to learn and practice a special kind of attentiveness towards material relationships within natural ecosystems. For indeed, what is "ecology" if not the study of relationships between different beings? The ecological/environmental sciences are best conceptualized as "the study of the relationships between living organisms, including humans, and their physical environment; it seeks to understand the vital connections between plants and animals and the world around them."[19] My humanities-based doctoral research at Stanford University, on the other hand, made me understand how crucial human meanings, identities and values are to shaping our relationships to everyone and everything around us. And yet, human and natural processes and destinies were (and still are) mostly being studied separately, by separate departments, disciplines, jargon and people, often located at opposite ends of college campuses. Why? I became increasingly frustrated with the general lack of interest in the biological and environmental sciences in seeking a nuanced and critical understanding of "empirical, objective, evidence-based" research, and with the excessive reliance on theory, abstraction and jargon in the humanities.

By the late 1990s and early 2000s, pioneering scholars in EH, environmental history, animal, plant and fungi studies and political ecology were already addressing these same issues. The global impact of the Covid-19 virus in the early 2020s was another reminder to myself that I urgently needed to shift to a multispecies paradigm to engage with the world in more interdisciplinary ways. I became curious about research in the humanities that sought to bridge the cultural concerns of postcolonial theory with questions of ecological sustainability explored by EH, such as the innovative work of Rob Nixon and Elizabeth DeLoughrey.[20]

These various EH subfields are now producing a sprawling range of new EH, animal and multispecies studies perspectives. While I cite many studies from the broad field of EH, I do not seek to provide an encyclopedic collection of all the possible approaches in EH and the relatively young field of multispecies studies. Rather, I have picked one promising path within the discipline to address the main subject of this book: analyzing the multilayered entanglements between mediatic meaning-making and multispecies bodies, lives and ecosystems. As I became more familiar with the several schools of thought in EH, I found it curious that most EH scholarship seemed to forget that Indigenous thought has already been providing expertise in these matters for millennia.

Despite the increase in EH approaches globally, I would say that EH scholars are still relatively few in number and at the margins (or should I say, at the leading edge?) of academic disciplines and ecosystems rather than at the center. Indeed, the humanities in general find themselves increasingly in the margins of academic and cultural life, despite their profound value as a glue that can contextualize all knowledge fields. In any event, discovering all this newer EH scholarship was a relief. It felt liberating not to be alone in feeling uneasy about the way in which disciplinary boundaries still guarded knowledge production and, therefore, entire systems of science, thought and their applications in the actual world, mostly in the name of fostering methodological rigor.

To make my own small contribution to a push towards more prominence for approaches in EH, but also to break free from being locked once again into a disciplinary "ghetto" of EH, I decided sometime in early 2022 to start planning a new transdisciplinary lab while mulling over very early ideas for this book. I wanted to promote interdisciplinary research on *natureculture*—a key concept for this project because it recognizes that both nature and culture often work as a complex unified system rather than existing in separate realms.[21] I therefore began to formulate a broader transnational humanities lab-based structure between 2021 and 2022, modelled in the tradition of the "training labs" found in the pure sciences. The lab thus became a conduit for me to naturalculturally and ecoculturally analyze global multispecies-based identity configurations and mobilities.

The lab and book would not exist without each other: they have truly co-become (i.e., become what they are together).[22] Like two trees that have grown from seeds right next to each other, they are now inseparable. I mention the background of this co-becoming or, as Haraway calls it, "becoming with," because the lab provided a project-based focus that was essential for the development of the methodological approach of the book and its focus on concrete case studies from around the world. Although it requires adequate funding in order to function, a humanities lab possesses the advantage of not needing the extensive equipment and infrastructure indispensable for most science-based labs.

Indeed, the *Multispecies Futures Lab* exists with the purpose of creating the epistemic space for collaborative experimentation around a central theme.[23] In the true traditions of the study of epistemology, it asks some fundamental questions about naturalcultural knowledge production in human culture: how do we know what we know about nature and other species of life? What are the different knowledge-creation systems (such as storymaking and storytelling) that innovate or resist ideas about human–nonhuman relationships? How does power, authority and positionality influence how we know what we know? What are the material and ecosystemic consequences of knowing things the way we know them? How does this knowledge translate to forging power, labor and naturalcultural identities in certain ways rather than others?

The lab exists primarily to facilitate such conversations, but also, crucially, to go beyond abstract ecocultural philosophy. It seeks to maintain a steady flow between ideas and the applicability of truly interdisciplinary methods within place- and space-specific projects. It showcases and highlights global instances of IRL (in real life) initiatives of multispecies import to learn from these concrete examples. It aims to create and disseminate public-facing and multimedial knowledge and to incubate collaborative faculty and student projects. The establishment of this kind of creative space to play openly and safely with odd combinations of methods and concepts departs from other focused research working groups, which tend to have their discipline-bound preconceptions and orientations baked in. It is also different from simply being a website with a particular research content, as the lab is actively developing new methodological approaches within site-specific projects in ways that depart from the textual approaches typical of the humanities. The book you are currently reading complements the project-based focus of the lab by providing a more in-depth critical reflection and development of a theoretical framework and key concepts that determine the approaches of the lab's methods and projects.

On Wildfires and Oddkin: The Scope of Multispecies Studies

Just before the initial meeting of my *Multispecies Studies* graduate seminar, I glanced outside my office window facing west and saw, in the near distance, a massive plume of dark, billowing smoke. It was January 7, 2025, and I was looking at the Palisades Fire that had just ignited in Los Angeles. I couldn't immediately foresee the gravity of what that plume of smoke meant. As I sat down in the seminar room and faced a room of new and curious faces to introduce the topic of the seminar, I thought it best to begin by discussing why we should care about this odd topic of multispecies studies. I began with the words "the world is burning"—but then stopped mid-sentence to change my wording to "the world is *literally burning right now outside our windows*." Despite this ominous situation developing very rapidly in our city, I wanted us as a group to take a more careful, collaborative and humble approach to the circumstances and to the subject of environmental and cultural collapse, rather than quickly controlling our exploration to arrive at closed conclusions. We discussed how some forms of ecological crisis-thinking and acting can be problematic because real or imagined "crisis situations" have very often been used to justify the rapid deployment of injustices, undesirable concentrations of power or well-meaning but under-analyzed and short-sighted techno-fixes[24]—all in the name of restoring order, control and security. Over-analysis has also led to paralysis in times of crisis. Crisis overload on a personal level can frequently lead to other escapist mental maneuvers: nihilistic tendencies ("there is nothing I can do, the problem is too big, so I will just curl up on my couch and binge-watch a Netflix series"), self-serving cynicism ("everyone else is selfish or corrupt, so it is weak or stupid to think about others"), wild beliefs in techno-fixes ("Mars, anybody?") and doomscrolling apocalyptic end-of-world scenarios ("woe is us, let's build a doomsday bunker!").

I found it important to look for an affective and emotional alternative to these understandable but unhelpful approaches to our multispecies world. Faith in techno-fixes, self-interest or plain greed are reflected back in even more career (i.e., financially)-oriented, siloed and technocratic knowledge systems. The technocratic feature of these knowledge production systems lies in their disinterest (or feigned interest) in engaging in an honest and holistic exploratory analysis of whether their proposed technical or technocratic "innovations" and "solutions" are actually needed, sustainable or good for the commons. The "A.I. revolution" can be viewed as the current next step in these kinds of knowledge production for multispecies profit and for the uncritical deployment of innovation without full consideration of the consequences of their misuse.

Based on the affective reactions of my students in the classroom, I sensed that many of them, whether at undergraduate or doctoral level, possessed the same deep need that I had for alternative approaches to our knowledge paradigms to face the multiple ecocultural crises that they both inherit and facilitate. This was a need that went beyond professional training for a career.

I also sensed that addressing this need for an alternative approach to our messy world would facilitate crucial training on how to possibly make better choices. Were there reasonable pathways to understand our contemporary ecocultural crises without resorting to the extremes of tech-optimism and tech-paranoia? Could we bring the discussion back to the complex interface between human cultural frameworks, naturalcultural realities and tech practices?

With these compelling affective needs in my mind, I tasked our seminar group to explore some basic questions to embark on what I call a *multispecies paradigm shift*.[25] I decided to name this perspectival transformation a paradigm shift with the hope that such a move would reformat and revitalize the study of the humanities in ways not too dissimilar to how the Copernican revolution transformed the study of astronomy: it would remove the lonely centrality of *Homo sapiens* in the conception and study of culture and ask that this positionality be opened up and shared with the entire multispecies web of lively relations. By giving this shift a name, I seek to focus attention, attentiveness and urgency on a worldview that has deep roots in diverse human imaginative and cultural traditions and has been fostered by thinkers in many different moments, including in some current EH scholarship, but has nevertheless been so severely marginalized that it is barely a ghost or a whisper in most current normative epistemic frameworks. A methodological shift in how we "do" humanities is only one aspect of a multispecies paradigm shift. This transformation indicates a cultural conversion on a much broader and deeper level because it involves reformatting everyday worldviews in our daily lives to include a vast cast of critter-kin and "inanimate" assemblages of minerals, light, water and air. All of these efforts to build such an old/new perspective via a new multispecies civil rights movement would provide deep benefits to human societies locally, regionally, nationally and globally.

A multispecies paradigm shift would ask profound questions that are both overarching and specific to different places: How have we, *Homo sapiens*, used our creative storymaking and storytelling powers to imagine (and therefore devalue or value) the perspectives of our nonverbal, but very communicative critter-kin? Why do some of these stories gain immense reach and power while others are left unvalued and unheard? How do our biases towards our "mother tongues" and against "foreign languages" mirror our dismissive attitude towards many other nonhuman forms of sentience, affect, intelligence and communication? How have various Indigenous and other marginalized thought communities understood the land (and their relationship with the land) differently and thereby approached its caretaking differently than today's normative-colonial approaches? Do we really need "new" ways of thinking to set ourselves on a better path of multispecies ecocultural care, or do these models already exist, but have been ignored and devalued? How can we include the human *within* the natural, rather than thinking about conservation without the all-important reality of human–multispecies interplay? What are the foundational anthropocentric cultural epistemologies that have led to

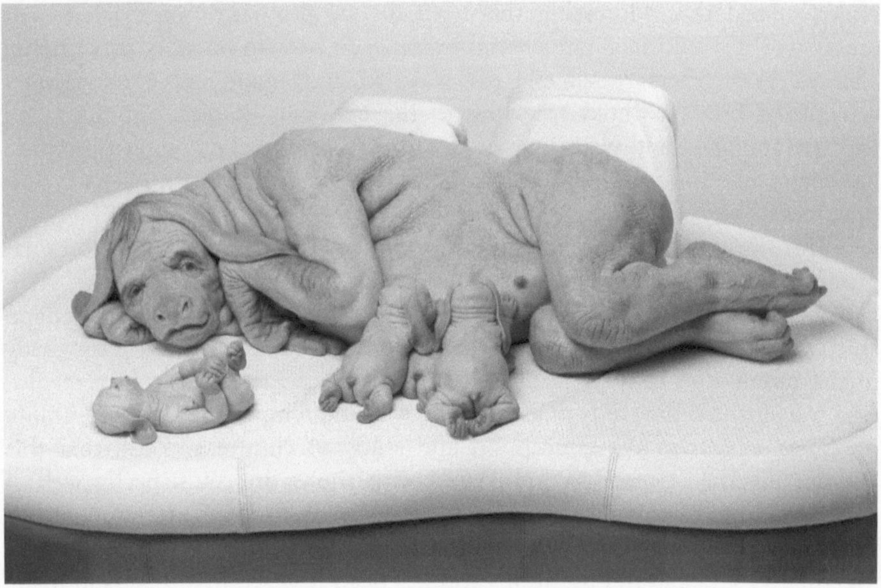

Figure 0.2 Patricia Piccinini, *The Young Family*, 2002, Silicone, fibreglass, leather, human hair, plywood. 85 × 150 × 120cm approx.

"natural" disasters? Why are these fires called "*wild*fires" and what are the affective and practical implications of this naming in how we manage them?

A striking sculptural work by Patricia Piccinini (Fig. 0.2), illustrates some of the essential questions that lie at the heart of multispecies studies approaches. The sculptural group is entitled *The Young Family*. Piccinini's sculpture is a contemporary masterpiece because it is disturbing and magnetic at the same time. It achieves this mix of affects and emotions in us because it successfully subverts our customary anthropocentric assumption that familial relations are exclusively human relations. *The Young Family* represents a unit of connected life that is neither human nor nonhuman, or is both, "glitching" our fundamental categorizing thought process that typically works to classify within the human/nonhuman duality. The young family's very ambivalent humanimality also complicates other affects and values that we associate exclusively with human kinship, such as pleasure in companionship, care, emotional vulnerability and physical frailty. The sculptural group's added power comes from the family being represented in a three-dimensional format rather than as a flat painting. Piccinini uses multispecies-derived materials to accentuate the tactile and hybrid features of bodies (leather and human hair), along with other non-animate materials that nevertheless simulate the gloss and morbidity (softness and mortality) of multispecies life. The crucial supplemental question of how we value gendered bodies arises via the obviously maternal identity of the main figure in the sculptural group—the mother nourishing her babies with milk from her body. Her gaze that meets the viewer of her body is incredibly direct and ambivalently human in its complexity—while the body from which this gaze emanates is evidently nonhuman, or some sort of chimera. Is she exhausted by her caregiving duties? Is she happy? What is her story? The artist gives us just enough information, and not too much, forcing us to ponder all these questions and to project our own biases and uncertainties onto the bodies and the narrative of this odd young family.

The epistemic, affective and cultural questions that Piccinini poses via *The Young Family* beautifully illustrate the central themes explored by the new field of multispecies studies and examined in this book: how and why thought, culture, values and spirit are considered higher in value than affect and ontology[26]—"being," living and co-becoming? What are the many profound yet ignored or repressed connections between *Homo sapiens*' existence and activity on Earth and that of this planet's other living beings (animals, plants, fungi, microorganisms)? Like Piccinini's sculptural group, multispecies studies seeks to restore the epistemic repression of the "humanity" of animals and the cultural disavowal of the animality of humans. It connects human attitudes and orientations towards multispecies lives with the cultural makings of biocultural identities such as race, gender, national identity, productivity, reproductivity and labor, just like Piccinini's young family does. It seeks to imagine, locate and revalue nonhuman forms of being, doing, sensing, feeling and practicing "cultures." Finally, it seeks to draw direct connections between all these meaning-making processes with actual material consequences in bodies, lives and ecosystems, asking how and whom do we care for, who do we consider "kin" and who do we distance and reject as the Other? Multispecies studies is not a zero-sum game: providing more attentiveness to nonhumans and humans "in unexpected collaborations and combinations" does not mean taking away attentiveness from human considerations. On the contrary, this approach could be a way to bring back an imaginative abundance to our multispecies relationships and therefore revalue human lives as well.

The humanities are "those branches of knowledge that concern themselves with human beings and their culture or with analytic and critical methods of inquiry derived from an appreciation of human values and of the unique ability of the human spirit to express itself."[27] An important branch of the humanities, namely cultural studies, studies "culture" *in its everyday forms*: it is not only concerned with products of "high" culture as commonly defined (literature, music, art, etc.) but also keenly interested in how everyday folks' beliefs, behaviors, choices, assumptions and basic cultural norms are formed, negotiated and transformed by both individuals and societies in different places and times. My graduate training lies primarily in a subset of cultural studies, literary and media studies—the study of production, interpretation and meaning-making via mediatic texts (literary, cinematic, digital, etc.). In contrast to these now-mainstream humanistic fields, multispecies (and animal/plant/fungi) studies is a subset of EH that seeks to glitch the imaginative and disciplinary limits posed by the word "human" within the humanities and the imagined human exclusivity over all forms of language, culture, identity and meaning-making. Amitav Ghosh argues in *The Great Derangement* that "the climate crisis is also a crisis of culture, and thus of the imagination."[28] We are therefore in urgent need of a major reimagination of the "human" within the humanities, and beyond. Multispecies studies seeks to eliminate the long-standing (and frankly, quite flimsy) presupposition in most normative cultures of human superiority and its corresponding claim of exclusive ownership to intangible "higher" things such as "culture, values and spirit."[29]

One way to fully open up the amazing and messy entanglements of human and nonhuman realities for far better analysis than has been done in the past is to revalue presence and being (ontology) as well as pre-meaning feelings (affect), rather than devaluing these in the favor of thinking (epistemology).[30] In EH parlance, these new foci have been called the "ontological turn" and "affective turn" respectively. However, these "turns" have been central to Indigenous worldviews for millennia and are central to Indigenous studies approaches. I will give you a quick but important methodological example of the practical efficacy and joy of including previously marginalized experts. While discussing Aboriginal worldviews, Leroy Little Bear says:

> For example, the categorizing process in many Aboriginal languages does not make use of the dichotomies either/or, black/white, saint/sinner. There is no animate/inanimate dichotomy. Everything is more or less animate. Consequently, Aboriginal languages allow for talking to trees and rocks, an allowance not accorded in English.[31]

Little Bear shows us here that innovative terminology is innovative epistemology and also innovative methodology. Reclassifying rocks as "more or less animate" allows us to engage in kin-making of a very different kind with these mineral bodies. In accordance with Little Bear's suggestion cited above, my definition of *multispecies* in this book and my lab includes *Homo sapiens*, our mortal multispecies kin (plants, animals, fungi, microbes) as well as our "more or less animate" relations, such as rocks, streams and landscapes, with which we undoubtedly share "meanings, interests and affects."[32] This approach to doing multispecies studies is not limited to investigations of mortal entities, but includes the various "more or less animate" things that sustain us all. Again, this is not "new" knowledge. As Brendan Hokowhitu eloquently states, "the material and esoteric connection of all living things (e.g., 'mauri'; all my relations) is fundamental to the Indigenous cultures that I'm aware of and really is 'Indigenous Cultures 101.'"[33] Due to this stunning amnesia in EH, I choose to engage with new materialist scholarship with some ambivalence and a few caveats. I can only ethically utilize its findings by foregrounding the Indigenous materialisms that preceded it. Hokowhitu forcefully describes how his engagement with new materialisms as an Indigenous scholar "felt like a re-witnessing of Western centricity making; a regurgitation."[34] He says that while it has value in its dismantling of the deeply held separation in Western thought between meaning and matter, "in its forgetfulness, it repossesses the 'rights' to knowledge. It forgets that Indigenous peoples were rendered incapable of cerebral rational thought and, thus, were like animals, other living things, and nature, and objects, whose sole purpose was to act as metaphors, similes and allegorical figures in re-centering the civilized rational European."[35]

Multispecies studies, via millennia-long traditions of Indigenous thought as well as recently developed EH fields such as new materialism, can bring back much-needed value and attentiveness to the "inanimate" material ecosystems (soil, air, light, water, rocks) that host this immense diversity of interconnected life, as well as to the affective bonds that tie and tether us to ourselves, each other and the world around us. Indigenous words that do not translate easily into modern European languages (such as the Hawaiian word *Āina*, which expresses the principle of land as a family relative, rather than as a resource) epistemically and affectively animate everything in an intricate web of relations and kin. Other such animating words with similar spirits include the Māori *kaitiakitanga*, Navajo *hózhó* and Inuit *ilitqusia*, to name just a few. Each of these terms contains entire ecocultural systems of multispecies thought and practice. In his seminal and now classic work *Native Science: Natural Laws of Interdependence*, Gregory Cajete explains how "most Native languages do not have a specific word for 'animals.' Rather, when animals are referred to, they are called by their specific names. The fact that there are no specific generic words for animals underlines the extent to which animals were considered to interpenetrate with human life."[36] Revaluing, adapting and caring for such already existent place- and space-specific (yet globally viable) Indigenous ecocultural concepts should become a key approach in multispecies studies. Contemporary environmental science is now catching up with these kinds of cosmologies through studies that show "unexpected collaborations and combinations"[37] between "higher" beings and "inanimate" elements: for example, studies on mycorrhizal (fungal) networks deeply complicate our categorization of soil as dead or inanimate matter. Studies show that it is actually a very fragile and complex matrix of living and nonliving co-becoming kin, rather than simply an inert resource to be plundered for extraction.

In sum, several millennia-old epistemologies already embody inclusive multispecies paradigms, contradicting dominant definitions of what "life" is, and therefore how us humans should relate to other lifeforms. In contrast to these approaches, the very definition of the *human*-ities reflects the inherent anthropocentricity of dominant colonial-based cultural perspectives. This deeply limiting human-centered attitude in turn encroaches upon all other realms of normative knowledge traditions, such as the social sciences and the biological and physical sciences, causing all kinds of conceptual, imaginative and methodological limitations. As we shall see in Chapter 2, this attitude has historically led to a lot of bad or incomplete science, oppressive cultural practices and colonial relations. Anthropocentricity is also nothing less than *the* foundational epistemic, affective and cultural reason for *Homo sapiens'* screwed-up relationship with each other and with the planet. Cultural theorists cannot continue to ignore how the many forms of cultural discrimination and bias between humans are predicated on an imaginative hierarchy between humans and the "less-than-human." Conversely, it is curious how entire structures of democratic protections and internationally negotiated laws are built upon meanings connected to fundamental "human rights."

Even when nonhumans are left out of words and definitions, multispecies meaning-making of all kinds is constantly present under the skin of so many value-based choices that affect the fragile destinies of so many lives, human and non. Consider, for example, how perpetrators of violent aggressions may be called "animals," or even how the victims of such violence may be characterized as "dehumanized" individuals. Whether explicit or implicit, multispecies imaginations and human exceptionality via a human/nonhuman duality form a master discourse, a hidden governing principle that shapes and underpins most conventional societal behaviors and norms. These considerations make it even more evident that a multispecies paradigm shift is urgently needed. The deliberate failure (or suppression) of imagination that has caused us to classify life on a linear, hierarchical scale of "inferior-to-superior" must be somehow overcome, just as Piccinini's *The Young Family* forces or inspires us to recalibrate the value of nonhuman life. Hierarchical classifications and categorizations of life consciously or unconsciously devalue and ignore *both* "inferiorized" human communities *and* other-than-human life via the nature/culture divide. Economic, cultural and political policies continue to uphold anthropocentricity as a deeply cherished paradigm, but a multispecies paradigm shift would actually better protect and conserve the vibrant and vulnerable life (human and non) that continues to exist and thrive, despite the odds.

What would a multispecies paradigm-shifted world look like in practice? I imagine that it would generate even more sound science of all kinds by banishing some of the basic anthropocentric suppositions that ground our normative knowledge-creation systems. It would be a world in which considering multispecies perspectives (human and non) becomes habitual when seeking to understand and relate to cultural, social, economic, political, psychic and spiritual phenomena. It would be a place where we seek a truly biodiverse knowledge production and affective-emotional system along with our multispecies companions. To be multispecies-oriented is to engage in *epistemic humility*—to accept that all forms of life have tremendously valuable knowledges to contribute, if only we would be more attentive to them. It is to revalue older multispecies epistemologies that have been unconsciously repressed or actively suppressed. It is to know that none of this is a zero-sum game—that giving value to multispecies perspectives doesn't mean sidestepping grave human social problems, inequalities and injustice; it actually means bringing value back to our individual and communal lives and forging more integrated solutions that will have lasting impacts. In sum, a multispecies paradigm would respond to our current profound and strategic failure of imagination to re-narrate our world in alternative ways that reinstate the importance of multispecies relationships.

When the Channel Island Fox stared back at me, I learnt yet again that *nothing exists in seclusion*. If this is true, then we are all connected by both familiar and unexpected forms of kinship. Donna Haraway calls these joyful relational identities "oddkin":

> Staying with the trouble means making oddkin; that is, we require each other in unexpected collaborations and combinations, in hot compost piles. We become—with each other or not at all.[38]

A multispecies paradigm shift in research contexts would involve developing disciplinary and methodological "oddkin": bringing together the attitudes, strengths, blind spots and preconceptions of different disciplinary voices to contribute to the bigger picture of a "thick"—complex and fluid—multispecies reality. It is to foster voices (human and non) that have so much expertise but which have been historically devalued via identity-based stereotyping. Indigenous, farming, and working-class communities, scientists whose findings reveal "inconvenient" truths, and many other individuals have been living these naturalcultural knowledges for millennia. Nonhuman life opens up unlimited possibilities to play imaginatively with similarities and differences: for example, how can we creatively bend our normal experiences of time and space, based on other species' greatly differing lifespans or sensory perceptions? The approaches generated by my own oddkin (this book and its cojoined lab)—such as "multispecies thick-mapping" and "naturalcultural discourse analysis"—all work within an assemblage of old/new knowledges seeking a multispecies paradigm shift of this kind. The hope is that this book will stimulate others to locate their own oddkin methods and pursue them, and for institutions and the public to find resonance with and support these kinds of research.

Naturalcultural Discourse Analysis: Turkish Steps and Wi'ma Island Thick-Narratives

Back on Wi'ma Island, students sought to take their own first baby steps towards relating an abstract and theoretical multispecies paradigm shift to an actual place-specific ecocultural system. Their understandable hesitations and doubts at the start of the seminar stemmed from their training in anthropocentric humanities approaches to understanding and analyzing culture—as if "culture" existed in contented isolation from our local and planetary assemblages of life! However, while initially somewhat uncertain, they ended up being supportive and even excited about experimenting with these new kinds of approaches. As a group, we were also aware of the privilege we had in being able to access resources and time to engage in these explorations. We wanted to make the best of this opportunity. After the seminar ended, I realized that another key ingredient in the students' successful experience was the moments when they were not being "productive" through academic work per se, but cultivating bonds of kinship (friendship, moral support and encouragement) amongst themselves as a group of collaborators and co-explorers.

Their story provides me with hope, even though it is just one example of a multispecies paradigm shift in a small group of engaged and curious people. Could their experience be scaled up to entire communities and cultures?

We woke up refreshed after our first night at the Santa Rosa Island Research Station, managed by the dedicated staff of the California State University Channel Islands. Most stomachs now seemed calm after having experienced the oceanic tumult of the ferry ride on the previous day. I gathered my students together in the front porch of the station, before embarking on a walk to immerse ourselves in the multispecies reality of Wi'ma. My orientation talk for the day aimed to remind us of what scholars Thom van Dooren, Eben Kirksey and Ursula Münster call "cultivating the arts of attentiveness."[39] I believe that this is a key requirement for practitioners of multispecies studies, within academia and beyond it. Anna Tsing also calls for the practicing of "passionate immersion" within the lives of both humans and nonhumans as a key methodological step towards multispecies studies research.[40] Dooren et al. declare that such immersion can

> provide "thick" accounts of the distinctive experiential worlds, modes of being, and biocultural attachments of other species. Immersive ways of knowing and being with others involve careful attention to *what matters to them*—attention to how they craft shared lives and worlds [...] In short, passionate immersion means becoming curious and so entangled, "learning to be affected" and so perhaps to understand and *care a little differently*.[41]

The challenge that I was posing to my students that morning, as we sought to experience Wi'ma Island a little differently, was: could we develop modes of transdisciplinary and qualitative analysis of how cultural and identity politics impinge upon the destinies of individuals, regions, economies and the entire *ecocultural system* of the island? How would such a perspective add to our understanding of how cultural, natural and economic configurations are mutually interdependent and co-becoming? What are the possible joyful, pragmatic and ethical implications of such a multispecies paradigm shift? Can we truly be humble enough to care for what matters *to them*? How can we admit to and openly experience *being affected* by the stories that are created and transformed via knowledge creation?

Since multispecies studies involves "passionate immersion," it includes but also seeks to go beyond the analysis of texts that represent material lives and ecocultural systems. Such indirect approaches include analyses of abstract fabulation—"speculative fictions"—or study literary representations of lives via ecocritical approaches. To study multispecies "ecologies of selves" via an inclusive, hybrid, situated and decolonial set of tools, more attention and value also need to be placed on material and ontological entanglements with cultural discourse while analyzing natureculture. This "ontological turn" in EH indicates a fundamental shift from studying how we know nature to studying the nature being itself—what exists, what is alive and how we and multispecies entities relate to each other and the world.

Stimulating and important work has been done in the field of ecocritical studies in the past two decades, but a "material turn" is an essential complement to these approaches, paying attention to the ontological rather than just the epistemic. This *and* rather than *or* approach emphasizes the direct impact of the meanings generated by global biocultural and ecocultural discourses of race, migrant status, invasive-vs-native species, citizenship, natality, productivity, etc. upon actual lives, bodies and ecosystems. Indigenous studies continues to demonstrate how deeply implicated meaning-making is with material naturalcultural systems, and vice versa. The methodological tools generated by the material turn (i.e., how meaning impinges on matter and vice versa), as expressed via Indigenous materialisms and newer traditions in EH scholarship such as new materialism, actor-network theory, object-oriented ontologies and speculative realism, have begun to chip away at the immensely destructive yet immensely durable separation between nature and culture in normative academic and cultural thought.

To carry forward these traditions of questioning the meaning/matter duality and work towards a concrete method of analysis, I intentionally began using the term *ecocultural system* rather than "landscape" or "ecosystem" or "ecology." Terminology influences methodology. The word "landscape" conceptualizes natural spaces as mere inert backdrops for human actions and considerations—a non-living matrix of mostly static materials that can be molded to serve our needs and fancies. "Ecosystem" provides a less anthropocentric conceptualization of places and spaces, but it leaves out consideration of how cultural systems entangle with and shape the material and biological destinies of these ecosystems. "Ecology" is now often used in EH and has also crept into wider cultural studies scholarship, but in a slightly confusing and non-specific way, often metaphorically representing interconnections within complex human-centric cultural systems and processes rather than actual ecological or ecocultural relationships (Black ecologies, migrant ecologies, ecologies of kindness, etc.).

By choosing to name the important spaces of Wi'ma Island as what they are—ecocultural systems—I intend to turn attention and attentiveness to the many interconnected animate beings, inanimate materials and cultural frameworks that actively *intra-act*[42] rather than simply interact, and co-become with each other. The implications of studying places and spaces ecoculturally with a multispecies paradigm shift—rather than considering them from separate ecological and cultural perspectives—are profound, for the realms of both ecological conservation and human rights, in terms of social and ecological care, sustainability and liberation.

Naturalcultural discourse analysis is the tangible result of these various ecocultural and multispecies explorations. It is the name I give to a research process that observes, investigates and narrates intra-active epistemic, ontological, material and affective relationships between naturalcultural agents in any given ecocultural system. It is a key method used throughout the rest of this book to illustrate various co-becomings of natural+cultural realities. Because the *Multispecies Futures Lab*-developed "multispecies thick-maps" are useful to easily visualize and comprehend this method, I will utilize here the *second* prototype of a multispecies thick-map, "The Ecocultures of the *Scala dei Turchi*," to introduce the workings of naturalcultural discourse analysis. I will describe multispecies thick-mapping in further detail with the help of the first prototype in Chapter 6.

The *Scala dei Turchi* digital multispecies thick-map illustrates the desirability and advantages of naturalcultural discourse analysis.[43] Born thanks to a serendipitous and fruitful collaboration with Professor Teresa Fiore (Inserra Chair in Italian Studies at Montclair State University, New Jersey), it came into being not long after I had completed research for my first prototype of such a map within the lab.[44] That first map (of the Agro Pontino region in Central Italy discussed in detail in Chapter 6) was initially built in analog form, whereas the second map went "straight to digital" and is available online as a public resource. As its name suggests, the mapped space is centered around the *Scala dei Turchi* (Turkish Stairs or Steps), a magnificent pearly white rocky geological outcrop located on the shores of the Agrigento province in Sicily—the largest island in the Mediterranean Sea. Figure 0.3 illustrates the spectacular outcrop with a rusted sign in the foreground that culturally marks this natural space simultaneously as a location of touristic value and also a piece of "Everyone's Heritage" ("*Patrimonio di tutti*"). The Scala was proposed as a location for mapping by Professor Fiore precisely because it illustrates the complex interplay of nature and culture, marine and terrestrial life, both human and non, on vastly different timescales. We aimed to produce a

Figure 0.3 The brilliant white and layered "Turkish Steps" cliffs with their rusty sign. (Vetri Nathan)

public-facing resource that could narrate how the terrestrial and marine lives and physical features of this space function as active biocultural *intra-acting* agents. The map highlights multispecies intra-activity, representing the liveliness of this ecocultural system via naturalcultural discourse analysis. To do so, the project necessarily and gladly gathered a collaborative group of experts (academic and non) from disparate disciplines, including geology, urban planning, tourism, ethnobotany, marine biology, literature and the culinary arts.[45]

The resulting multispecies thick-map demonstrates the underlying epistemic innovation of naturalcultural discourse analysis. What do I mean by this new term and method? *Cultural* discourse analysis is a method for studying how communication creates meaning within specific cultural contexts. It views media texts as a social practice of meaning-making that is deeply tethered to a community's shared beliefs, values, and norms (discourse). By analyzing how people communicate through language and media, it reveals how culture shapes behavior and interactions and how those interactions in turn shape, maintain and change culture. *Naturalcultural* discourse analysis takes cultural discourse analysis a step further. It shatters the nature/culture divide that jealously imagines cultural relationships to be exclusive to *Homo sapiens* communities. By doing this, it shows how meaning impinges on material and natural ontologies, and conversely, how in turn natural and ecological systems shape and influence meaning-making.

A scrutiny of these co-constitutive layers of meaning, material, flesh, genetics, trophic relations and affects is an important component of naturalcultural discourse analysis. However, I do not limit naturalcultural discourse analysis to investigations of meaning-matter co-constitution (in EH parlance, the "material-semiotic"), because my objective is to usefully and decisively transgress the stubborn human/nonhuman, culture/nature and animate/inanimate dualities. It is time to collapse of that age-old distinction between epistemology and ontology. I do so by redefining the term "discourse" in naturalcultural contexts to mean a biocultural and biodiverse system of shared material/ontological/ecological *and* epistemic/discursive/cultural processes and relationships within any given ecocultural community. I deploy this transgression experimentally with the intention of subverting the dominance of culture and to place it on an equal footing with the material, the ontological and the multispecies. *Cultural* discourse analysis presumes that any given cultural context possesses basic discursive parameters that have a constitutive power to shape interactions and behavior within that culture (for example, discourses that frame nature as a "resource to be exploited" lead to an embodied reality of extraction and consumption). *Ecological* analysis focuses on understanding species interactions and the reliable webs of mutualism that form the basis of stable communities in any given ecosystem. *Naturalcultural* discourse analysis presumes that not only human language and culture *but also ontologies and lived ecological relationships* between all multispecies agents possess basic parameters and equilibria in any given ecocultural system, and that these parameters co-constitute the system itself. Naturalcultural discourse analysis suggests that not only human epistemologies, but ways of being, living, relating and feeling via naturalcultural intra-action, co-becoming, feedback loops and trophic cascading effects have profound constitutive power to shape the coherence, behavior and destinies of lives, bodies and ecocultural systems.

A naturalcultural discourse analysis examines and reveals these fairly consistent patterns of relationality, interdependence and co-becomings between meaning and matter, between matter and matter and between matter and living bodies. "Stable" ecocultural systems usually have some level of established equilibria—reliable naturalcultural discursive parameters can be read, analyzed and narrated. In addition to reading these bonds of relationality, a naturalcultural discourse analysis also maps changes and transformations in all these relationships when they modulate or are disrupted or changed. These changes can be gradual, subtle and hidden, or obvious, dramatic and even catastrophic. Whatever their scale and velocity, these changes cause cascading effects throughout the ecocultural system. Naturalcultural discourse analysis can carefully investigate and map such shifts in ecocultural equilibria over time through the ever-expanding web of cascading relationships.

Enacting this kind of holistic research via a multispecies paradigm shift gives back value to practices of *being*, *feeling* and *doing* in addition to cultural meaning-making. As scholars such as Ron Broglio have discussed, creative investigations into multispecies perspectives through art, literature, music and other multi-sensorial imaginings are useful pathways to including multispecies perspectives.[46] By redefining discourse and rescuing it from the exclusive dominion of human culture, I propose naturalcultural discourse analysis as another and more radical integrative epistemic pathway to restoring visibility and value to the existing agency present within all entities, human and non.

Engaging in these kinds of analysis, visualization and narration reveals both the vibrancy and the vulnerability of complex ecocultural systems. Multispecies communities co-create, co-maintain or co-transform interdependent destinies, in beneficial and/or damaging ways. Naturalcultural discourse analysis as a method brings back attentiveness to the material, the ontological, the affective *and* the cultural and puts them in conversation with each other. Redefining discourse as shared material/ontological/ecological and epistemic/discursive/cultural processes and relationships in a given ecocultural system, usefully addresses what scholars have explored as a central obstacle faced by researchers of agency and culture: "even if we want to, how do we understand and/or represent what the non-humans feel?"[47]

Naturalcultural discourse analysis itself works both scientifically and imaginatively—but always in ways that deeply tether analysis to ecocultural realities and relationships within actual ecocultural systems. This approach thus finds a promising pathway beyond the anthropocentric question of "can the nonhuman subaltern speak?" and the implied "nature/culture" duality that undergirds this question, in order to explore deeply affecting multispecies entanglements within our immediate communities and throughout our planet. It provides new value to the "silence" of nonhuman lives—silence that is so often imposed and then used as foundational evidence upon which the entire structure of human exceptionalism is built. Naturalcultural discourse analysis demonstrates how nonhuman silence as interpreted within a deeply anthropocentric paradigm is simply an excuse for indulging in deeply self-serving Anthropocene delusions. A multispecies paradigm shift via naturalcultural discourse analysis reinterprets these silences not as an absence or empty space to be filled by de-situated knowledge, but as a *formidable tethered presence* and conduit and bridge to valuing marvelous forms of agency, vibrancy and vulnerability. By bringing back equal value to ontological and ecological activity and relationships within culture, naturalcultural discourse analysis reformats normative anthropocentric interpretations of multispecies "silence." Conversely, this interpretation also gives multispecies beings (humans and non) a "right to opacity," as Edouard Glissant calls it–i.e. the right to be respected without the need to be fully understood, categorized or possessed by others.[48]

The question of material agency is and has been a key concern of new materialist thought. As Serenella Iovino and Serpil Opperman explain:

> Their main claim is that discourses about the living world, though necessary, are per se insufficient, if separated from their broader material substratum of inanimate substances and apersonal agencies. [...] Agency assumes many forms, all of which are characterized by an important feature: they are material, and the meanings they produce influence in various ways the existence of both human and nonhuman natures. Agency, therefore, is not to be necessarily and exclusively associated with human beings and human intentionality, but it is a pervasive and inbuilt property of matter, as part and parcel of its generative dynamism. From this dynamism, reality emerges as an intertwined flux of material and discursive forces, rather than as a complex of hierarchically organized individual players.[49]

While new materialist concepts are very useful for my work in multispecies studies, I would like to keep on reiterating that attentiveness to all forms of agency, human and non, as visualized by naturalcultural discourse analysis is not "new" thought—new materialist thought is but a new iteration of foundational cosmologies that have formed the basis of Indigenous science for millennia. Also, its originality actually stops short of the profoundly innovative way in which Indigenous thought supersedes the stubborn living/nonliving divide that still underpins much new materialist theory. For example, we have already seen how Leroy Little Bear's formulation of all things being "more or less animate" brilliantly vanquishes the stubborn conceptual divide between "living flesh" and "matter," thus opening up the world to caring a little differently for both animate and inanimate naturalcultural protagonists. A rock, a river, a mountain, or a bird or beast may be ascribed agency via such an Aboriginal worldview, because these entities are intimately *related to* other living agents. *Co-constitutive kinship relations* are the key basis for ascribing agency in such a caring epistemic choice. With these important yet strategically repressed histories always in mind, the *Scala dei Turchi* prototype multispecies thick-map employs naturalcultural discourse analysis without ever forgetting that my method draws equally from my own postcolonial and subaltern origins, from my resulting ambivalent positionality within Western academia, from often-ignored, ridiculed or demonized Indigenous materialisms and cosmologies and from current research trends in EH and new materialism.

Naturalcultural discourse analysis adapts innovative approaches and concepts in multispecies studies into a proper method in order to analyze naturalcultural realities. Multispecies thick-mapping turns this method into a concrete visualization product. Indeed, the *Scala dei Turchi* multispecies map deliberately frustrates the normative separation of natural and cultural agential players by mixing these two aspects of this ecoculturally complex location. Figure 0.4 shows a screenshot of the overview image of the resulting map, showing its twelve "layers" tethered by twelve pins on the map. Each pin features an investigation and story that intra-acts with the other pins as the website visitor scrolls between pins, causing echoes and reverberations between scrambled biological, geological, archeological, economic and cultural chronicles. Each "pin" on the ArcGIS-based map becomes a kind of portal into a unique perspective on multispecies relationships that are deliberately read not in their own isolation, but in juxtaposition to other adjacent stories. Pin 2 (*The Scala in a Few Words…: Stories in Legends and Literature*) narrates legends about the origins of the cliffs' name to embed this geological feature within a deep cultural web of meanings, identities and discourses. Pin 3 (*The*

Figure 0.4 Overview of the "Ecocultures of the *Scala dei Turchi* Multispecies Thick-Map." (Vetri Nathan and Teresa Fiore)

Mediterranean's "Coral Reef": The Story of Posidonia) describes how the immense outcrop and the surrounding sand dune communities are protected from the sea's immense forces by the seagrass *Posidonia oceanica*. This endemic seagrass community is, in turn, given a stable footing in the bedrock by a diminutive but mighty date mussel (*Lithophaga lithophaga*) that bores into the rocky strata to provide a firm footing for Posidonia roots. The introductory frame of the project demonstrates how a naturalcultural discourse analysis of the Scala reveals the process of "co-becoming," as the cliffs co-become with the Posidonia, which in turn co-becomes with the date mussel:

> One curious pattern you will notice emerging in all these stories is that they showcase examples of ecological and cultural *co-becoming*. [...] This concept explains how we do not exist in isolation from our surroundings, nor do we simply interact, but *become who we are through relationships*, for example: the Posidonia seagrass's wellbeing shows deep connections to a mollusc, the area's sand dunes are born and sustained in their infancy thanks to a plant (*Cakile marittimo*), the step-like formations of the Scala are a result of Sun/Earth cycles, and the Roman Villa's mosaics are related to the local Sulphur economy.[50]

The strategic juxtaposition and layering of stories of this single Sicilian location illustrates naturalcultural *sympoiesis* by visualizing naturalcultural discourse analysis to a broad audience. Sympoiesis (from the Greek roots "sym," signifying together, and "poesis," making or creation) means "making-with" and describes the collective process of creating complex systems together—in this case, the way in which natural and cultural systems are co-becoming, in mutually beneficial or damaging ways. It is a concept popularized by Donna Haraway that emphasizes interdependence and co-creation among different entities, as opposed to "autopoiesis," or self-making. Sympoiesis highlights how various organisms, objects and systems interact to form new, emergent wholes. Naturalcultural discourse analysis works to tease out these active and formative relationships between natural and cultural agents. A sympoietically layered assemblage of stories in this thick-map illustrates seagrass "reefs" growing underwater near the beaches, the many species of plants that make up the *macchia mediterranea* ecosystem still surviving in patches, the birth of baby sand dunes, migrants crossings the Mediterranean, fish and crustacean species that end up on the dinner table, a World War II bunker, as well as literary, cinematic and mediatic representations of the cliffs. In pins 4 and 6, Professor Fiore (with research assistant Giacomo Frazzetta) analyzes how the cliffs have been a setting for the creation of divergent mediatic stories of Sicily, but also used as a deterritorialized backdrop that represses any geographical and cultural specificity for profit. Pin 11 challenges our human lifetime- oriented scales to discuss sulphur-mining operations since ancient Roman times. Pins 5 and 7 then travel to even more distant geological periods spanning millennia, thus giving visitors a palpable feeling of experiencing "thick time."[51] An experience of multispecies thick-time can generate so much wonder and humility simultaneously: for example, pin 4 in the *Scala* map narrates the *Messinian Salinity Crisis* of the Miocene epoch. In the Miocene, 5.5 million years ago, the Straits of Gibraltar closed shut, causing the Mediterranean to dry up almost completely. The map then connects that seemingly distant reality to the contemporary economic activity of the ITALKALI salt mine, one of the largest in Europe and an important source of revenue.

In addition to thick time, naturalcultural analysis further employs epistemic humility via a multispecies paradigm by engaging in *multiscalar* perspectives, zooming from sizes that range across a few microns to distances that span light-years. For example, pin 6 (*The Golden Spike: A Dance of the Earth and Sun*) broadens the vision of an already immense geological monument to a phenomenon of an almost unfathomable scale. This pin explains to site visitors how the very complex relationship, lasting almost 20,000 years, between our planet and the Sun, ultimately resulted in the iconic regular layering of plankton (*Foraminifera*) material that the Scala is so famous for today.[52]

By dramatically superimposing and making visible these naturalcultural entanglements in a multiscalar and thick-time context, this second multispecies thick-map prototype provides a startling example of multispecies co-becoming to students, scholars, local communities and tourists alike. It is important to note that a well-executed multispecies thick-map does not simply become a site of aggregation of different perspectives and knowledges (although that in itself is desirable), but through careful superimposition, analysis and storytelling, it turns this knowledge into something new. By evidencing and narrating *relationality* between oddkin natural+cultural agents, it disorients and glitches normative anthropocentric fantasies and delusions of living in insulated isolation. The hope is that, in some small way, this kind of storytelling and visualization of naturalcultural co-becomings may contribute towards better care and protection of this precious ecocultural space for future generations.

Naturalcultural discourse analysis such as that visualized in the Scala map only works well when researchers sincerely engage in epistemic humility, via both rigorous interdisciplinarity and the inclusion of non-anthropocentric ontologies and epistemologies. The researcher's ethical attentiveness towards and passionate immersion within the material, ontological and cultural realities of all agential elements in a given ecocultural system allows them to observe, and then communicate to others, the deeply complex relationships of intra-action and co-becoming. To do this is to humbly understand how healthy ecocultural systems are what Cajete calls "natural democracies"[53] in action: these include direct and indirect ecological relationships, as well as cascading trophic effects between several seemingly isolated or fixed agents, all resulting in complex and fluid intra-active situations and relationships.

The work, then, of a naturalcultural analyst involves deep commitment to training in new ways of looking and listening, but it is also wonderfully liberating: humanists can allow themselves to observe, understand and incorporate biological and ecological relationships within their research, while ecologists and other scientists can allow themselves to show their love and positionality in more subjective and personal ways. All naturalcultural analysts, including myself, would then also do well to remember their responsibility to pass on this knowledge to their communities via impactful communication.

To further introduce this "odd" method of naturalcultural discourse analysis, let's travel from the Mediterranean Sea back to Wi'ma Island in the Pacific. When our intrepid graduate seminar group had left the sheltered jetty of Ventura Harbor on our ferry to Wi'ma, we were immediately met by some curious dolphins leaping out of the massive waves to side-eye us and then disappear as quickly as they appeared, like sleek and marvelous muscular grey ghosts. Did we dream them up or were they actually there? They seemed to ferry us towards an affective and epistemic sea-change and, indeed, towards a different paradigm. The bumpy and almost four-hour traverse from Ventura Harbor to Santa Rosa Island truly gave us an experiential understanding of how the Islands' ecocultural system was and is firstly defined by both connection to and isolation from the mainland. Many of the ferry's passengers, including some in our group, felt the ocean's power in our guts, so much so that some lined up by the edge of the boat to heave up their breakfasts. We passed an abandoned oil rig, now overtaken under the water by a thriving ecosystem of marine life.[54] We then said hello to *Anyapax* (Anacapa) Island, a razor-thin line of towering cliffs that looked like the broken teeth of a drowned giant jutting out of the ocean. Despite the distraction of sea-sickness for many, it was amazing to consider that until the end of the last Ice Age, Wi'ma was actually connected to all the other three Northern Channel Islands (San Miguel, Santa Cruz and Anacapa) to form one giant island: *Santa Rosae*. Much of what was under our boat today was dry land in another era. At that moment in planetary history, when massive ice caps meant that sea levels were much lower, only one narrow but deep undersea canyon separated Wi'ma from the California mainland. This separation was slender enough to allow animals such as the massive Columbian Mammoth to swim across around 250–150,000 years ago, eventually evolving into a miniaturized island version that stood only about five feet up to its shoulder, around 17% of the size of its gargantuan ancestor.

The Channel Islands Pygmy Mammoth's story is just one of many multispecies examples of intra-active co-becoming wrought by the very complex ocean and land connections (and disconnections) created by changing sea levels over millions of years. The frolicking dolphins that guided us and one grey whale's lingering "tail salute" reminded this humanimal researcher to shed his terracentric (land-based) habits, perspectives and prejudices and consider what lies beneath the ocean's surface. Unlike many islands lying off continental shelves, the Channel Islands are divided from the mainland by

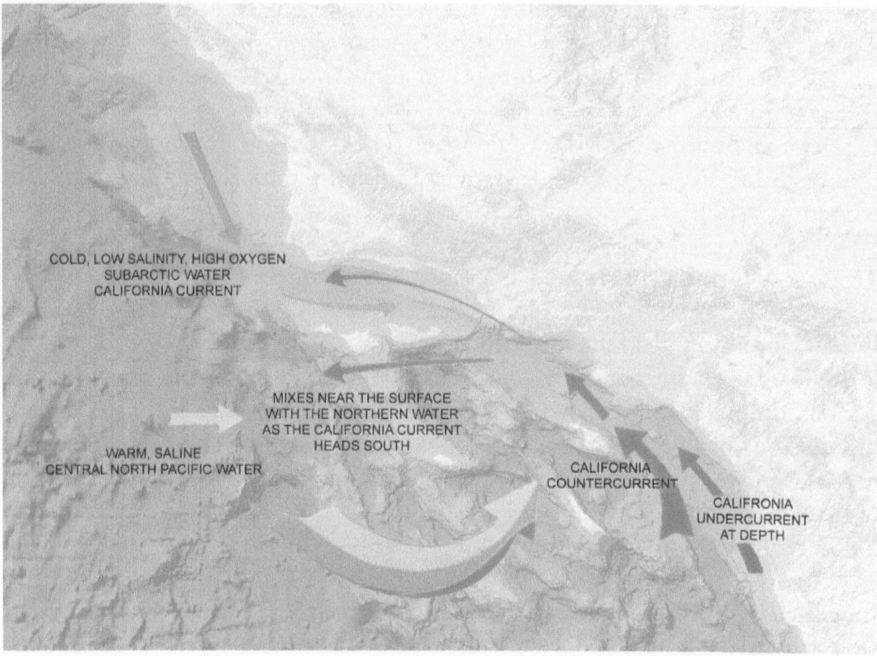

Figure 0.5 Ocean currents around Channel Islands National Park, including the California Current bringing cold water in from the north, and the Southern California Counter-current bringing warm water up from the south. (https://www.nps.gov/chis/learn/nature/currents-upwelling.htm)

very steep changes in elevation above and below sea level, due to the mega jamming-together of the Pacific and North American tectonic plates. The archipelago is also divided from and connected to the mainland by two immense oceanic currents. Indeed, as you can see in Fig. 0.5, Wi'ma finds itself in a unique "transition zone" of around a hundred kilometers where two major Pacific underwater "rivers" meet and mingle, causing an explosion of biodiversity unique to the Pacific coast of North America. The cold, low-salinity and high-oxygen water of the California Current bathes the islands further away from the coast, *Tuqan* (San Miguel) and Wi'ma, travelling all the way from British Columbia to Baja California. Prevailing winds from the northwest on the West Coast of North America play an intra-active role in this planetary dance of forces: the winds cause *Ekman Spirals*—an upwelling phenomenon caused by the rotation of the Earth that causes the Coriolis effect. The winds parallel to the coast push water away, with deeper water having to rise to take its place.[55] Ekman spirals cause the deep, cold, nutrient-rich water of the California Current to rise closer to the surface, providing abundant nutrients to the whole ecocultural system of the West Coast, including Wi'ma and Tuqan. Common California Current (cold-water) species found off Wi'ma include Black Rockfish, Sunflower Sea Stars and Red Abalone.[56] Without the winds, there would be no nutrient-rich upwelling and none of the abundant life that it supports. With this sense of newly found attentiveness to the intricacies of the oceanic movements at Wi'ma, I felt that all life supported by the Channel Island ecocultural system could ultimately be understood via naturalcultural discourse to be the offspring of many different scales of intra-action: from "small" changes in undersea topography up to the very rotation of the planet and the energy exchanges fueled by unequal heating by the Sun at the equator versus the poles. Relating to these mind-boggling multiscalar relationships in an embodied way made me feel small, yet made my heart feel so large.

Wi'ma is special, in that it is bathed not only by the California Current at its shores but also by the warm-water Southern California Counter-Current, which eddies in this specific location on its way up the coast from Baja California. The eddy is provoked by the blocking position of the Channel Islands, which causes a marine traffic jam of epic proportions that would be the envy of any Los Angeles freeway. The result of this blocking is that the Counter-Current is forced to rise to the surface in the California Bight and mix with the cold California Current approaching from the north. The milder waters around Wi'ma support warmer-water species, such as the ostentatiously orange Garibaldi fish, the California Spiny Lobster and the California Moray Eel.

The nutrient-rich Goldilocks-worthy soup produced by these mixing waters has also enabled the ocean's true "rainforests," the Giant Kelp forests, to thrive in the region. Giant Kelp (*Macrocystis pyrifera*) are a spectacular species that look like gigantic seaweed but are actually stramenopiles or heterokonts—multicellular eukaryotic protists that are neither plant nor animal nor fungus. Protists are abundant yet relatively poorly understood. Along with fungi, their in-between classification (not unlike the members of Piccinini's young family) connects them to both plant and animal kingdom characteristics. Immense kelp strands up to forty-five meters in length can grow in dense forests. Swaying gently in the currents of Wi'ma, these forests generously provide food, shelter and community to host an immense chain of kelp-kin: from Halfmoon, Blacksmith, Pacific Sardine, Kelp Bass and California Sheephead to Sheep Crab, Warty Sea Cucumber and—until it was hunted to extinction in this area for its fur—the utterly charming California Sea Otter. In addition, chains of cascading connections supported by these Kelp-inhabitants nourish passing species such as Yellowtail, Barracuda, and Bonito as well as marine mammals like California Sea Lions and Harbor Seals.[57] Cascading relationships of care abound everywhere. For example, the productive broth of the Channel Islands' mixing waters sustain phytoplankton, which in turn nurture krill, that in turn attract massive cetaceans—Grey Whales, Humpback and Right Whales, and the world's largest animal, the Blue Whale—to come and stay for a while and enjoy the buffet.

The natural fully merges with the cultural. Even the island's Chumash name is born from a cascading effect originating in oceanic flows: "Wi'ma" means redwood or driftwood, and in the past, the California Current would transport immense logs of Coast Redwood trees (*Sequoia sempervirens*) from Northern and Central California. Thanks to the eddy around the California Bight, they would arrive at Wi'ma Island's shores as driftwood, saltwater-seasoned and ready for use by the Chumash to make their canoes, or *tomol*, "one of the greatest technological achievements of any North American Indian people" (Fig. 0.6 illustrates another similarly advanced watercraft built by the Northwest Coast Indigenous community). A summary of the step-by-step instructions to make these complex works of "artisanal intelligence"[58] shows the centrality of naturalcultural intra-action in the making of a complex oceanic and archipelagic culture:

> Canoe makers selected only the best logs, which were straight grained, strong and free from knots. Using bone wedges and hammerstones they split planks from these logs and fitted them together edge-to-edge. This process involved some bending of steamed boards and some shaping with stone, bone and shell tools. The canoe makers glued each course of planks in place with a mixture of tar and pine pitch [...] Next they drilled holes near the edges of the planks and lashed them together with plant-fiber string. They built the canoe slowly, one row at a time. Finally they caulked the seams with tule pith and tar-and-pine pitch mixture. They painted the canoe with a mixture of red ochre and pine pitch as sealant. [...] canoe paddle blades were also made of redwood, attached to a shaft of ironwood (*Lyonothamnus*).[59]

Figure 0.6 A dugout canoe from the Pacific Northwest. (Wikimedia Commons)

Figure 0.7 Chumash rock art made with red ochre and charcoal, *Alaxuluxen, or the Chumash Painted Cave State Park.* (Carol M. Highsmith)

Redwood transported and cured by the ocean formed the first intra-active multispecies piece in the puzzle of making one of these amazing watercrafts. Even this abbreviated version of the *tomol*-making process allows us to count no less than five or six other species of life directly involved in the creation of a *tomol*, like the inner soft core of the tule stem, the North American Bullrush (genus *Schoenoplectus*), used for caulking *tomols* but also for weaving and other uses. Natural tar from pits is itself an intra-active "thick-time" product made of the dead bodies of millions of creatures that lived eons ago, decayed by bacteria and pressed by massive geological forces to be transformed into the thick ooze that still seeps out of the ground naturally, most famously perhaps at the La Brea Tar Pits in downtown Los Angeles. We also have Ironwood, a genus of trees in the Rose family that contains one species, the Catalina Ironwood, endemic to the Channel Islands. Red Ochre, also called bloodstone or hematite, is a reddish-brown mineral pigment also used to create some of the fascinating art in locations such as *Alaxuluxen*, or the Chumash Painted Cave State Park (Fig. 0.7). These spectacular naturalcultural artworks may variously represent *tomols*, or "canoes taking the souls of the dead to *Shimilaqsha*, the after world," or "celestial beings, such as the sun, moon, and the stars and were painted during ceremonies in order to maintain balance in the celestial world."[60] The beauty and complexity of the *tomol* and the painted rock art in *Alaxuluxen* shows how, before the arrival of European settlers, the Chumash and their ancestors had led far more sustainable coexistent lives on the island for millennia, caring for and making kin with their multispecies and multi-kind companions. Narrating the *tomol* via naturalcultural discourse analysis opens up this watercraft to a new vision: that of an immense community of multispecies knowledge, respect, kinship and care.

Ecologists are increasingly undertaking studies that show how profoundly constitutive indirect effects, like the ones we observed in Wi'ma Island, can be. For example, researchers at the Institute of Organismic and Molecular Evolution at Johannes Gutenberg University Mainz have proved that species can influence one another via a chain of environmental effects in ways that are powerful enough to permanently alter a species' genetic makeup.[61] These scientific studies demonstrate something that Indigenous peoples (for millennia) and EH scholars such as Karen Barad (for some years now) have been exploring: that ecological and naturalcultural relationships are intricate, complex, sometimes expected, sometimes unexpected, but always deeply situated, embodied and full of all kinds of intra-action via many indirect and direct effects such as cascading consequences and positive or negative feedback loops. The methods developed in the *Multispecies Futures Lab* make visible these deep indirect effects.

Another example of cascading ecological looping and webbing relationships on Wi'ma is its unexpected "cloud forests." Upwelling and the cold temperatures of the ocean surface along the Pacific coast of California form a distinct marine layer of fog, so well-known and felt during San Francisco's chilly summers. Before the arrival of European settlers, this marine layer used to support a unique cloud forest on the very tops of Wi'ma and Santa Cruz Islands. The trees of these forests act as sieves that harvest heavy marine layer fog, allowing it to drip to the ground as rain, which in turn nurtures chaparral shrubs at the base of these trees. The favor is returned by the chaparral plants, which help create a "deep, moist, and shady leaf litter bed so that acorns and other seeds can sprout and grow. It is the chaparral plants that form the backbone of this relationship, by allowing bigger trees to establish themselves, helping those trees to reproduce and spread, and protecting those trees from eroding soil."[62]

Many of the Channel Islands' complex terrestrial ecosystemic intra-active and co-becoming relationships, including the cloud forests, were devastated by the arrival of European, Mexican and American settlers, who introduced non-native grazing animals such as sheep, feral pigs, cattle, elk, deer and horses. "Culture" and "nature" were shown to be not all that separate after all. Unfortunately, the cloud forests are not recovering from this apocalypse, as the delicate chain of multispecies life that supports and cares for the bigger trees has been disrupted, perhaps permanently.

Another fascinating story-layer in our naturalcultural discourse analysis of Wi'ma Island connects pre-colonial and even prehistoric humans to the other layers of multispecies co-becomings of Wi'ma: the discovery of the "Arlington Man" or the "Arlington Springs Man." Bones were discovered in 1959 by

Phil C. Orr, curator of anthropology and paleontology at the Santa Barbara Museum of Natural History (I could not help but note the cultural and scientific implications of the curious mix in Orr's two job titles). The two femur bones were found thirty-seven feet below the current land surface, within sediment layers dating to the Pleistocene. Even without the technology of carbon-dating, Orr recognized the potential significance of finding human remains at that depth and brought them to the Museum of Natural History for long-term storage.[63] In 1987, a team under the supervision of John R. Johnson, successor to Orr, made an astounding discovery with the development of resin filtration methods to date bone collagenous protein—the bones were 13,000 years old! This meant that they were the oldest reliably dated human remains in North America, a finding of great ecocultural and even political significance.

Arlington Springs Man's presence potentially debunks the land-based theories of the arrival of humans in North America during the last Ice Age via the Aleutian land-bridge between Eurasia and North America. With Arlington Springs Man, Wi'ma glitched a land-based migration paradigm with an oceanic one. This also presented possible complications to some contemporary Native American perspectives of being descendants of the very first peoples to arrive in North America. Indeed, a *Los Angeles Times* report states that "many Native American groups strongly object to the theory that others got here first. In some cases, including one major one in the Northwest, tribes have successfully invoked the 1990 Native American Graves Protection and Repatriation Act to force researchers to return old skeletons for reburial before they can be tested."[64] Discursive elements gleaned from my own symptomatic readings[65] of media articles reporting the discovery of Arlington Springs Man's remains contain many of these biocultural tensions, including descriptions of the man as a representative of the first "conquest" of North America and obsessive questions about its racial makeup.

Another curious strand in the story of Arlington Springs Man's ecocultural meaning-making is that, for a span of about thirteen years, the Arlington Springs Man was thought to be a woman. Around 1993, visual study of the bones determined that they were probably those of a woman, and the name was changed to Arlington Woman, or Arlington Springs Woman. Talk show host Jay Leno joked about this gender change, saying something like "Here's another woman lying about her age—she said she wasn't a day over 10,000 years."[66] In 2006, this assessment was reversed yet again by the anthropologist who had initially provided the female designation. He had assigned a female identity when another researcher studied one of the femur bones and thought it was too large, "in a masculine kind of way."[67] Innuendoes about bone size and masculinity apart, the indeterminate gender story of the Wi'ma thigh bones is another element that reveals the many epistemic, discursive and affective ways in which naturalcultural realities are merged, interpreted and narrated. The many questions of whether the "very first" human remains in all of North America were connected directly to questions of biocultural import (such as what lineage relationships existed to present-day Native communities, whether these remains were male or female, or what "race" they were), became questions of wider biocultural and ecocultural significance. The fascination with the remains gained valences of many kinds because they narrated a primal origin story. As Frantz Fanon would say, the "seen" (the bones, being the first text) created the "scene" (the various stories the bones generate being the second text). The story's value stemmed from a keen interest in exploring differing biocultural discourses of kinship across time; depending on who you asked, the Arlington Man/Woman was a symbol that represented and provided material "proof" of the earliest conquest, innovation, race relations, gender roles or Indigenous presence. These differing discursive needs reflected contemporary attitudes and concerns, leading the bones to be narrated in different ways for their meaning to be "fixed" strategically within the ecocultural history of the United States and the continent of North America.

In her influential work examining the intersections of culture and geology, Kathryn Yusoff argues that the very language and practice of geology grew within a general colonial-cultural imagination of Black and Indigenous lives as the inferior Other: "Through the emplacement of Blackness in paleontological orders, Black and indigenous subjectivity is consigned to deep time[68] rather than the political present. Geography is stolen through temporal displacement."[69] The Arlington Man/Woman's storymaking and storytelling case shows how rich the naturalcultural study of the "thick" co-becomings can be and how matter (flesh, bones, cartilage, genetics) can become a mediatic text which is then interpreted and negotiated via discursive, affective means—in this case, the formation of biocultural explorations around gender, but with an accompanying fascination with constructions of racial identity. Via this story, naturalcultural knowledge creation (and from there, identity creation and stratification) can be seen as a very complex, subjective, creative and messy process, where competing storylines (and their underlying desires and fantasies) jostle for space. By this, I don't intend to say that rational knowledge claims cannot be made—good science is based on proving and disproving theories and making knowledge claims that stand until disproved in the future. However, Arlington Springs Man/Woman demonstrates how science plays out ultimately in the realm of culture—where ideas, hypotheses, proof and biases (both conscious and especially unconscious) jostle towards, methods and against others in a fluid and sometimes contradictory manner.

Overall, the Wi'ma Island case study and the *Scala dei Turchi* project illustrate how the *Multispecies Futures Lab*-incubated method of naturalcultural discourse analysis can contribute to re-seeing these places via a new multispecies paradigm, rather than simply being viewed and consumed as yet another disposable resource. Naturalcultural discourse analysis of any place (yes, even your backyard or any urban street alley) allows it to be read and understood as a situated location of profound naturalcultural entanglements, both destructive and/or nurturing. This kind of analysis investigates visible and hidden effects of multispecies agents (direct, indirect and cascading) and stiches them together to form a quilted thick-story that situates and honors these biocultural co-becomings.

This Introduction has shown how explorations on Wi'ma Island and other lab projects revealed the fundamental presuppositions that can foster good multispecies studies research: nothing exists in seclusion, co-becoming is a real process in life systems, and naturalcultural attentiveness truly does foster naturalcultural connectedness. Nature and culture are connected deeply, in "thick" layers of human–nonhuman relations, some controlling and some more liberatory, some oppressive and some beneficial. Our research group was able to passionately immerse ourselves in an embodied way within these transcultural, multiscalar, multi-epochal, multi-species and multi-kind perspectives of the island's ecocultural system, as a co-becoming community of human creatures on Wi'ma. The *Scala dei Turchi* multispecies thick-map prototype visualizes the method of naturalcultural discourse analysis by subverting the norm that discourse and culture "happen" exclusively within human societies. These new kinds of maps move from the microbial to organ level, from the organs to the personal, from the individual to the local, from the local to the regional, from the regional to the national, and from the national to the planetary. These maps also move from the present moment of global digital connectivity and mobility to other historical periods of anthropogenic intervention, and move beyond the Anthropocene to other geocultural eras. By invoking these examples of a multispecies paradigm shift in my lab's research and teaching initiatives, I hope that the value and resonances of all that surrounds us—and our role as human "meaning-making" agents within this complex ecocultural web—will only grow rather than diminish.

Wi'ma Island taught our group in an immersive and experiential way that local, regional, national and global naturalcultural accomplishments and catastrophes are always multispecies events and that there are events in thick time or on nonhuman scales (from microns to light-years) that can add even more resonances to the already burgeoning thickness of a choral assemblage of naturalcultural agents. The *Scala dei Turchi* map shows that such kinds of naturalcultural analysis and storytelling can be deeply effective in glitching the human/nonhuman, animate/inanimate and nature/culture divides. As the most vulnerable human *and* nonhuman lives face the gravest consequences of our damaged world, where multispecies relationships are ignored, devalued and discarded in favor of anthropocentric delusions of exceptionality, the stakes of upscaling this kind of epistemic and cultural paradigm shift within our broader communities are nothing short of existential. Reading Wi'ma and the Scala ecomaterially and symptomatically allows them to be seen as an assemblage and nexus of both fluid natural interconnections and cultural norms. Land, water, wind, critters (human and non), minerals and cultural paradigms deeply intra-act and co-become as kin (or oddkin) in these regions.

All that is left after this kind of research activity, where the undivulged is revealed, is to step back and stay with the joy of having had the privilege of experiencing the vastness, the viscosity and the vulnerability of such spectacular ecocultural systems.

Schema for this Book

The multispecies naturalcultural analyses of Wi'ma's and the *Scala dei Turchi*'s ecocultural systems in this Introduction, replete with intricate embodied co-becomings and intra-actions, are seeds that will sprout and grow in many different directions in the successive chapters. Thanks to the generosity of two islands, I was able to unpack here some of the basic EH and multispecies studies questions that I am keen on exploring through my cojoined lab and book. My selection of case studies in this Introduction and the chapters that follow draws from my foundational training and established research relationships in Italian, Mediterranean and postcolonial studies contexts. However, I also venture far beyond an area-specific approach by considering a diversity of cases from around the planet and from my own local Californian ecocultural systems.

In Chapter 1, I elaborate on some helpful theoretical and methodological terms that will make our journey interesting, such as care, kinship, colonialisms and decolonialism, biopower, and soulpower. I also discuss how a multispecies paradigm shift does not necessarily require new academic knowledge formation and yet more policy "think-tanks," since much can be learned from the re-valuation and integration of the knowledge and experience already present within the ethical glitches provided by the survivance of so many marginalized lives and communities, past and present, human and non. By foregrounding work by Indigenous studies scholars Kim TallBear, Philip J. Deloria, Dwayne Donald, Vincente Diaz, Kyle Whyte, Gerald Vizenor and Brendan Hokowhitu as well as the work of multispecies studies scholar Donna Haraway, I explain the utility of the key concept "situated knowledges." This chapter also describes how its opposite—"de-situated knowledges"—are often the foundation for the practice of corrosive ecocultural relationships, past and present. In sum, I show how multispecies studies, Indigenous and "new" materialist thought can form a viable pathway to redirecting agency and value to not only human lives, but also nonhuman lives and ecocultural systems.

Part Two (Chapters 2, 3 and 4) of the book contains case studies that explore inattentive and damaging ecocultural relationalities. In Chapter 2, I describe colonialisms in their various forms as ecocultural relationships in which relentless "capture-and-control" of lives, bodies and ecosystems for profit are fueled by spectacular multi-mediatic fantasies of alterity and surplus. Chapter 3 places the Cybercene as the distinct and newest colonial sub-era of the Anthropocene. It analyzes how profound new discursive, psychic and affective intra-actions between media consumers and virtual representations, narratives and fantasies in the Cybercene directly or indirectly assert new forms of de-situated control and distancing relationships. By defining the Cybercene in *ecocultural* terms, I effectively press for an "ecological turn" and a "multispecies turn" to the many questions regarding the digital revolution of the twenty-first century. I re-direct corporate-economic, techno-scientific, history of media/science or other anthropocentric "cultural" explorations of our new global digital cultural economies towards the crucial yet unexplored question of how our digital lives impinge upon multi-species, multi-being and multi-kind lives and relationships. The term *Cybercene* further questions the nature/culture epistemological divide to provide pathways to restoring our multispecies futures at a multiscalar level: personal, familial, local, regional, continental and planetary. It also reinscribes our current digital era within other colonialisms, past, present and future.

Chapter 4 takes a deeper dive into the psychic mechanisms of the Colonial-Cybercene control/desire machine. If "we are the images we consume," then the Cybercene's image/discourse saturation and fragmentation must certainly influence who we are now as individuals and societies. Screen studies and psychoanalytic theory help me locate spectator agency in co-creating new, personalized kinds of knowledge and menace/desire in the Cybercene via screen content intra-action, rather than simply being passive consumers of content (cyberdiscourses and cyberscenes). Postcolonial and psychoanalytic theory aid me in demonstrating how cyberdiscourses and the knowledge/power they exert over lives and ecosystems are a product of an interplay of our own affective co-becoming. Such cyberdiscourses are also a product of ego/identity creation and desires interacting with those of the discursive/mediatic systems themselves, creating oppressive resource-based capture-and-control mechanisms rather than sustainable relationship-based identities and identifications. Homi K. Bhabha's notions of *stereotype-as-fetish* and *mimicry* provide the primary insights to understanding both the *psychic* mechanisms of othering and the transformative potential of the "glitch." Jean-Louis Comolli's understanding of *active spectatorship* will help me describe the lure of intra-activity within the spectacles of digital screen images. Other scholars who provide foundational insights into these power/discourse/identity/media dynamics are Michel Foucault (the power/knowledge combine), Frantz Fanon (the spectacle of the racialized Black Other), Dwayne Donald (the Indigenous concept of *ethical relationality*), Jacques Lacan (the relationship between images and identity creation), André Bazin (the importance of indexicality in screen realism), Gilles Deleuze and Félix Guattari (reciprocal capture), José Esteban Muñoz (disidentification) and Edward Said (the mechanisms of colonial-cultural discourse).

Part Three (Chapters 5 and 6) complements the previous part by exploring examples of multispecies ethical care and kinship. In Chapter 5, I describe in more detail how, despite the acceleration of new forms of affective and epistemic capture-and-control promised by the Colonial-Cybercene mediatic ecosystem, actual creative bodies, lives and ecosystems sometimes provoke strategic and situated glitches, as life "runs away" from these mechanisms of capture-and-control, thus re-reformatting lives and ecosystems to new and old forms of care and kinship. By doing so, such narratives release some bodies from the wasteful constraints of normative speciesist forms of kinship, creating restorative possibilities of belonging and care for a larger circle of multispecies bodies, lives and ecosystems. Chapter 6 considers how the affective, epistemic and discursive glitches such as those examined in the previous chapter become the foundation of actual situated practices of bio- and ecocultural care and kinship within bodies, lives and ecosystems. This chapter also describes the first prototype of a multispecies thick-map to investigate how naturalcultural discourse analysis can serve as a methodological "ethical glitch" to de-situate colonial forms of epistemic capture-and-control. All the forms of care and kinship examined in Part Three of the book stand in stark contrast to the anthropocentric and resource-based affective, psychic and epistemic inattentiveness and active damage examined in Part Two. The concluding chapter reflects on possible further pathways (both academic and non) that could lead to the fostering of sustainable, relationship-based, multispecies epistemic humility, care and kinship.

1

Staying With the Joy: Framing Multispecies Studies Approaches

Kinship and Care in Our Multispecies World

Wi'ma Island's intra-active multispecies lives illustrate how consistently heightening inter-species and multi-material curiosity can be deeply, affectively and emotionally enriching and also lead to the development of new interdisciplinary methodologies of studying natureculture. Nurturing place- and space-specific explorations in this vein, such as the *Scala dei Turchi* project in Sicily, will be a primary future activity of this cojoined book-lab project. This chapter builds on stories, concepts and new methods from the Introduction to explore further the primary theoretical assumptions utilized by the book and lab. By doing so, it establishes my positionality as a partial knowledge-producer and also discusses the wider implications of attempting this kind of multispecies and naturalcultural analysis.

Identifying instances of multispecies care and kinship can move us to mitigate multispecies violence and damage. Indeed, the explorations in this chapter are about initial considerations of the ways to properly strategize how to care affectively, ethically and practically can possibly allow us to "stay with the trouble" (as Donna Haraway puts it)—the worry *and* the joy of multispecies kinship. Anna Tsing's "passionate immersion" is a first important step. The book and lab also take inspiration from Thom van Dooren's exhortation to create: "a scholarship that feels a need to keep coming back to, to be genuinely interested in, the lives of nonhumans: how they live, what matters to them, how they make meaning and relationships, how they make worlds. We must do so carefully, tentatively, critically, but we must also try to give an account of nonhuman modes of life."[1] To nudge my methods away from overly anthropocentric perspectives, I situate my work within the new academic field of *multispecies studies* rather than placing it within areas of inquiry variously called *animal* or *plant studies*, or even *human-animal studies*. The latter terms are equally valid choices with their own advantages, but which also imply preconceived dualities. As van Dooren, Eben Kirksey and Ursula

Müntser explain: "In contrast to animal studies, multispecies scholarship takes up a broader taxonomic scope of inquiry. But it does not simply replace a focal animal with a plant or bacterium." Instead of a dialectical focus a multispecies approach "focuses on the multitudes of lively agents."[2] In addition to a neither/nor approach towards the stubborn human/nonhuman duality, the word "multispecies" also seeks to go beyond vertical taxonomic hierarchies ("apex predators" at the top and "bottom feeders" at—well—the bottom) and exclusionary limits between species boundaries,[3] bringing together the "multi-" in an intra-active co-becoming *assemblage* in place- and space- specific contexts and relationships.[4]

Hierarchical categories and classification systems are comforting *mapping* experiences. They create a semblance of order and a rationale for justices and injustices in a complex world. They are quintessential meaning-making tools as well as knowledge that is then used to coordinate the use of power to constitute material realities. However, the scientific praxis of categorizing and classifying lives can be very easily diverted into discriminatory cultural systems of imagining human lives. We can conceptualize these oppressive discourses as "regimes of meaning"—situations where cultural meaning-making frameworks supplant and substitute actual material and affective bonds of kinship. Indeed, long before the current climate crisis, the cultural meanings given to multispecies life on Earth have been the basis of the colonial valuing (or devaluing) of human lives. Linnaean classification and categorization-influenced speciesism is a foundational premise upon which the colonial racialization of humans is built.[5] It is also the foundation for the capture-and-control via manufactured alterity of all forms of life, human and non. This multispecies ecocultural colonial capture-and-control will be explored in detail in Chapter 2.

Christopher Peterson succinctly explains not only how "racism is predicated on dehumanization"[6] but, digging deeper, how racism is "more precisely [based] on a fundamental disavowal of human animality that makes dehumanization possible."[7] If this is the prime colonial-cultural strategy for devaluing life, then decolonial or anti-colonial multispecies studies pathways can lead to a clearer understanding of the pernicious ecocultural mechanisms of alterity and kinship in both local communities and globally connected societies. This is exactly the kind of analysis that the *Multispecies Futures Lab* engages in via naturalcultural discourse analysis. The question Peterson asks is highly relevant to the multispecies ecocultural understanding of racialized Others in colonial meaning-making practices:

> How is it that on the occasion of the two hundredth anniversary of Darwin's birth […] many still assume that an antiracist politics must insist on the humanity of the socially abject? To what extent are both racist and antiracist discourses predicated on a shared repudiation of animality? How might we comprehend animality in nonpejorative terms?[8]

A multispecies paradigm engages a multiscalar, sympoetic and thick-time perspective on our current view of places and relations. It shifts a human-scale and somewhat static view towards a broader, richer and far more passionate engagement with our surroundings. The "our" in the title of this book is therefore a transgressive one—stubbornly indicating a multispecies *collective*—strategically going beyond the imagined human/nonhuman divide and its consequent practices. "Our" is an opening towards visions of co-becoming rather than a closure into speciesism and colonialism; it is an exhortation to conceptualize the reality of "us" as a deeply inclusive *assemblage* of «unexpected collaborations and combinations» of multispecies life, both human and non.[9] "Multispecies" is inclusive of the animality of *Homo sapiens* and the affective and sentient capacities of nonhumans. It also signals to the readers that you, too, are subjects and protagonists of "our multispecies futures," along with and through other nonhuman lives. "Our" does not indicate possession and control, but inclusive identification with larger circles of life and marvelous co-becomings.

Given this reality of co-becoming rather than seclusion, "our" also highlights the collective imperative to redirect attentiveness, care (the feeling of concern, interest, or importance as well as the act of tending) and kinship (the feeling of nearness, commonality and relationality) within this worldview of expanded belonging. Earth's interconnected ecocultural crises are not someone else's problems, they are ours. This deeply felt sense of social responsibility characterizes both Indigenous thought as well as Indigenous studies approaches. For example, Aileen Moreton-Robinson illustrates how Indigenous research methodologies are framed by the basic premise of *relationality*. Note how the nonhuman is absolutely part of her conceptualization of relationality: the social is relational, involving the interconnectedness of what people are doing and experiencing as the outcome of actions in the actualities of lives and lands. [It is] constituted by our histories, our culturally embodied knowledges and life force that connect us to our respective lands, our creators, all living entities and our ancestors.[10] Donna Haraway aptly states, "We become—with each other or not at all."[11] The "our" in the title of this book also emphasizes directly to *you*, the reader, our shared ownership of the responsibility we must collectively face if we are to properly acknowledge the reality of multispecies co-becoming in the Anthropocene. This includes recognizing and accepting a deeply unequal reality: that the ecocultural footprints of some identity and geographic groups are proportionally larger and more damaging than those of others. "Our" is not meant here to flatten responsibility *equally per capita* within communities or around the planet. Indeed, what I like to call "naturalcultural studies" (derived from Haraway's term *natureculture*, and in contrast to *cultural* studies) could be extremely useful in analyzing exactly how natural resource domination and consumption follow the contours of culturally defined identities such as nationality, class, gender, citizenship, etc. Even within the Global North, responsibility for damage is deeply unequal, since power and wealth are increasingly concentrated amongst a tiny fraction of the general population.[12] While environmentally conscious citizens try to consistently use paper straws to avoid microplastic pollution, corporations and oligarchs with deeply colonial mindsets extract, exploit and exterminate multispecies ecocultural systems on vast scales. It is important not to lose sight of how we use constructed identities to mask or avoid accepting the ridiculously unequal distribution of responsibility for the damage. However, it can generally be said that our collective responsibility on Earth as humans is also our *opportunity*. The stories and case studies contained in this book are not simply "mine" as the author, but potentially *ours*. They aim to resonate with your stories, this resonance turning "my" stories into "our" stories, fostering attentiveness to multispecies concerns.

The plural and collective possessive pronoun highlights the urgent need for nearness and openness to multispecies identities and identification in an era where online and digital mediatic practices are reinforcing the detachment of me/myself/mine from everything else, fueling seclusion and distancing from our interconnected bodies, lives and ecosystems. Such distancing facilitates a cultural and psychic experience of capture-and-control of multispecies lives for (real or imagined) consumption and profit. An embodied, expanded and multispecies sense of "our" serves to glitch techno-fantasies of "me and mine" being able to accomplish and control things alone. Donna Haraway cautions us that kinship is a loaded word: "Kin is a wild category that all sorts of people do their best to domesticate. Making kin as oddkin rather than, or at least in addition to, godkin and genealogical and biogenetic family troubles important matters, like to whom is one actually responsible."[13] Haraway correctly recognizes kinship decisions as key *naturalcultural decisions*, because identification with circles of kinship, however familiar or "odd," fosters care. She says, "Nobody can be kin to everything, but our kin networks can be full of attachment sites."[14] An identification with kin is therefore a crucial regulatory decision that manages attentiveness, about who to care for and how much to care. Kinship leads to a kind of care that is neither abstract nor secluded nor distant—it is embodied, messy, sometimes inconvenient; but ultimately vibrant, joyful and given freely. Thom van Dooren describes how this kind of attentive multispecies ecocultural care has affective, ethical and practical dimensions:

> As an *affective* state, caring is an embodied phenomenon, the product of intellectual and emotional competencies: to care is to be affected by another, to be emotionally at stake in them in some way. As an *ethical* obligation, to care is to become subject to another, to recognise an obligation to look after another. Finally, as a *practical* labour, caring requires more from us than abstract well-wishing, it requires that we get involved in some concrete way, that we *do* something (wherever possible) to take care of another.[15]

In sum, multispecies studies, to many who are undertaking it, myself included, is not only research, teaching and community engagement, it is also an affective, ethical and practical mode of seeing the world differently; a liberatory, passionate and immersive kind of living that fosters kinship with and care of multispecies oddkin. It is to feel both joy and heartbreak for a multitude of life-relations that include but also surpass definitions of "kin" that reduce it to reproduction-based relationships and the act of making population, to inward-looking blood-related bonds. It is to be unafraid to feel both the agential vibrancy and the vulnerability of those you love. Indeed, the "ethical glitch," described in detail in Chapters 5 and 6, is one of re-storying our normative and limited imagination of who is included in our extended family. Like Patricia Piccinini's humanimal mother's gaze at us, the glitch refuses to see things the normative way or to distance and look away. It meets the questioning gaze with joy, worry and concern. The glitch enlarges the capaciousness of our kin-circles, accepting and recognizing so many more multispecies connections, entanglements and mutual responsibilities—our shared "meanings, interests, affects" (the realm of "culture") and our shared "flesh, minerals, fluids, genetic materials" (the realm of "nature").[16] A multispecies paradigm shift allows us to notice constantly that we are incessantly co-becoming and intra-acting with our multispecies world and that nothing exists in seclusion, whether we like it or not. To stay with the trouble, the worry, but also the joy of multispecies passionate immersion is to constantly recognize and experience already existing friendships, support systems, companionships and shared passions via our embodied relationships with our multispecies critter-kin. In a world that is experiencing a loneliness pandemic due to our digital disconnectivity, re-engaging with the embodied multispecies realities that still surround us may be the ultimate *ethical and practical* glitch. We are not alone. We have never been alone. We just have to (re)learn the simple arts of attentiveness to recognize the multispecies support system that envelops us, to build a thriving multispecies future with them.

Naturalcultural Studies: Colonialisms, Biocultural Power and Situated Knowledges

We cannot properly arrive at an analysis of realistic cases of multispecies care and kinship without properly addressing the opposite of care and kinship—relationships based on control and consumption. I collectively subsume these epistemic, affective, discursive and material forms of multispecies domination and control under the term "colonial."

For the purposes of this research project, by *colonial* I mean a highly asymmetrical extractive power relationship. It is important to note that postcolonial theorist Edward Said's definition of colonialism already allowed scholars to open the term up to this subtler and more expansive vision and analysis of a host of cultural phenomena—local, national, and transnational—without restricting colonialism to one specific historical era of imperial political domination. For Said, colonialism is "the practice, the theory, and the attitudes of a dominating metropolitan center ruling a distant territory."[17] For the purposes of highlighting the similarities and differences of historical colonial relationships to those of naturalcultural concern, I will further open Said's definition up to more diffuse, less anthropocentric and less geopolitically bound situations. In this book, by *colonialisms* I mean past and present affective, epistemic, discursive, psychic and material practices of distancing, domination and control practiced by individual or group command economies upon culturally and ecologically Othered multispecies bodies, lives and ecosystems, in order to extract and exploit these as material or labor resources for profit. Such a definition opens the notion of the colonial to naturalcultural multispecies concerns, but this does not make the study of its practices, theories and attitudes any less geopolitically relevant. On the contrary, I sense that it reveals greater possibilities for the fields of business, political science, government, economics, etc. to apply their concerns and deepen their interest in place- and space- situated socioeconomic events and processes. This *ecocultural* definition of colonialisms expands the scope of the phenomenon beyond specific histories and geopolitical realities, beyond any one historical or political form of control by one empire over its lands (for example, European domination of the Global South from the Age of Exploration to World War II), to explore larger questions of cultural and multispecies naturalcultural inattention, domination and control.

I will unpack the significance and consequences of this definition of colonialisms more fully in Chapter 2. One of the clearest and most depressing conclusions of this study has been that we have re-entered a phase of hypercolonialism, after a brief respite in the "postcolonial" period (from 1945 up to the early 2000s). Haraway aptly describes the precarious situation that we humans have created via our colonial attitudes towards lives, bodies and ecosystems, at the expense of our planet and all its inhabitants, including ourselves. As I have already mentioned, some of us in wealthier and more powerful parts of the globe are more responsible for this than others. However, as a planet we are all facing a colonial situation that urgently needs strategic interdisciplinary approaches to the study of crucial naturalcultural relations, in order to make them more visible and emphasize their value:

> We are facing the production of systemic homelessness. The way that flowers aren't blooming at the right time, and so insects can't feed their babies and can't travel because the timing is all screwed up, is a kind of forced homelessness. It's a kind of forced migration, in time and space. So it's not a humanist question. It's a multi-kind and multi-species question.[18]

Haraway's words make evident that questions of human migration, racial and nation-based injustice and unequal resource and habitat destruction in the Global South are multispecies and naturalcultural considerations. Climate change and habitat/biodiversity destruction have also become a direct or indirect driver of political instability, cultural extremism and refugee/migration crises, leading to wasted human lives and relationships everywhere,[19] but especially in poorer, marginalized and racialized communities. These interrelated questions have been confronted by foundational EH scholars such as Donna Haraway (2003, 2007), Bruno Latour (2005), Karen Barad (2007), Jane Bennett (2009), Rosi Braidotti (2013), Ursula Heise (2016), van Dooren, Kirksey and Münster (2016), and Marco Armiero (2021).[20] Within animal/multispecies studies, Ryan Hediger and Sarah McFarland (2009) explore animal agency; Sue Donaldson and Will Kymlicka (2011) and Alasdair Cochrane (2018) discuss extending citizenship rights to nonhuman animals.[21] Gregory Cajete (2000), Deborah Bird Rose (2011), Kim TallBear (2015), Kyle Whyte (2020) and Brendan Hokowhitu (2020) discuss Indigenous paradigms of co-constitution of human communities, land and multispecies entities.[22] Animal studies scholars Cary Wolfe (2003) and Lori Gruen (2015), and plant studies researchers Natasha Myers (2017) and Laura Foster (2019) discuss how life classification systems work towards serving Anthropos' strategic needs.[23] This said, far more can be done to strengthen multispecies approaches beyond these specific fields of specialization.[24] They need to become organically embedded within the cores of research, education, and policy structures and institutions.

If robust science is indeed a collective, integrative and constantly mutating process of observation, modelling and reobservation, then *epistemic humility* becomes a key guiding principle in carrying out work in a healthy multispecies studies' science. Gregory Cajete explains how understanding subjective positionality is a key process of engaging in Native science. The partial knowledge claim that results from this acknowledgement of positionality should be a key feature of sound multispecies studies methodological approaches:

> Native Science reflects the understanding that objectivity is founded on subjectivity. There is a stress on direct subjective experience, predicated on a personal and collective closeness to nature, which will lead to an understanding of the subtle qualities of nature.[25]

In terms of method, this epistemic humility turns to creative imagination of a situated kind, where humans must generate ethical multispecies cosmologies without presuming to be able to fix multispecies "truths" via distanced ethnography. We enter the realm of what Cajete calls Indigenous cosmologies:

> The first Indigenous cosmologies were based on the perception that the spirit of the universe resided in the Earth and things of the Earth, including human beings. Because of this perception, these people remained equally open to all possibilities that might manifest through the natural world. In turn, perceptions of the cycles of nature, behavior of animals, growth of plants, and interdependence of all things in nature determined culture, that is, ethics, morals, religious expression,

politics and economics. In short, they came to know and express "natural democracy."[26]

Naturalcultural discourse analysis and multispecies thick-maps, described in the Introduction, are ways to engage and promote the kind of "natural democracy" that Cajete so eloquently discusses, where "it becomes in every sense abnormal to view the world as dead matter, private property, commodities, or commercial resources."[27] If one of the main conclusions of naturalcultural discourse analysis is that "nothing exists in seclusion"—that is to say, everything and everyone is connected—then listening and being attentive to all naturalcultural connections in a humbler way becomes a method of epistemic strength rather than weakness. Sometimes these naturalcultural entanglements are more obvious: the Siberian Husky, for example, is a product of natural and human selection to aid communities with a certain task in arctic and sub-arctic ecosystems. Sometimes they come in the form of oddkin ("unexpected collaborations and combinations"), for example the indirect and cascading connection between wolf communities and healthy aspen tree growth in the Greater Yellowstone ecocultural system.[28] Another example of this kind of care is the cumulative and gradual consequences of consistently positive cultural images, ideas and meanings surrounding wolves potentially resulting in legislative moves to protect these animals, which in turn results in healthier aspen tree communities. Open-ended transdisciplinarity (which is another form of epistemic humility) can provide better access to these kinds of naturalcultural entanglements.

With Donna Haraway's concept of "situated knowledges," Kim TallBear's notion of "standing with" and that Channel Island Fox's destabilizing gaze, I finally found a way to accomplish my goal of including all voices to form a strategic humility that would lend epistemic strength to this book and my lab. I view *situated knowledges* not only as an abstract concept, but also as a precise *method* of fostering humility and a strategic approach to good multispecies science. It designates all "more or less animate" beings, both human and non, as partial experts within the context of their own ecocultural system. Accepting this expansive partial expertise leads to an openness towards, attentiveness to, and understanding and valuing of varied embodied experiences of multispecies expertise, both within and beyond academia. Situated knowledges that provide rational knowledge claims about the functioning of an ecocultural system can be the way to respect the complexity and "thickness" of multispecies realities. Simply put, I understand situated knowledges to mean epistemic frameworks that are attentive to an ecocultural system's ontological, material, affective and cultural agents relationships, intra-actions and co-becomings. Such frameworks, are ways of making meaning that engage with the material-semiotic-affective-ontological realities of their multispecies communities. These forms of knowledge are qualitatively balanced, forming a kind of knowledge claim that contrasts with highly controlled autopoetic efforts at one extreme, or easy relativism at the other. If caring is indeed an embodied phenomenon, then situated knowledges are caring knowledges precisely because they are tethered to multispecies bodies, lives and ecosystems. In the Cybercene, an era that can be defined and characterized by post-truth, "de-situated" knowledges, the return to situated knowledges seems crucial to restoring (and re-storying) multispecies care and kinship:

> I am arguing for politics and epistemologies of location, positioning, and situating, where partiality and not universality is the condition of being heard to make rational knowledge claims. These are claims on people's lives. I am arguing for a view from a body, always a complex, contradictory, structuring, and structured body, versus the view from above, nowhere, from simplicity.[29]

A method and an ethics of doing this kind of science emerge beautifully from this quote. It is a technique by which all researchers, and even non-researchers, do work on their own *partial* knowledge claims individually, but also listen via a strategic epistemic humility to all experts (human and non) with knowledge of that particular ecocultural system. After listening, this method creates a sympoetic thick-assemblage or web of knowledge that pays "attention to how [all experts] craft shared lives and worlds."[30] By situating these claims within the body, Haraway opens a further important avenue for epistemic humility (i.e., the body, being, feeling, rather than simply "thinking") that allows for experts to be various living, breathing, feeling entities, irrespective of whether they use human language or not or whether we consider them "living" in the non-aboriginal sense or not. Every lively entity situated within a given assemblage of knowledge creation and meaning-making is given a value for practical reasons of doing sound science—rocks, fungi, air, fish. This method also creates space for all kinds of *relevant* human voices, without falling into the quagmire of colonial identity politics (fixating on difference at the cost of losing sight of the big picture) but nevertheless being sensitive to the historical and contemporary silencing of some expert voices due to their biopoliticized identities (such as race, gender, class, caste, indigeneity, etc.). Situated knowledges allow ontology and affect to be on an equal footing with epistemology in analyzing meaning-making. There is immense collaborative value in the old and new terminologies and methodologies of many marginalized communities—they can provide exciting epistemic alternatives that can expand the boundaries of restrictive and damaging normative colonial human-ness, especially when they are in conversation with each other. The method of situated knowledges elegantly blends the utilitarian and practical needs of good science (where contextual affirmation is found within multispecies embodied systems) with a deeply ethical obligation (that of perceiving and listening to all kinds of "odd" knowledge-producers). Furthermore, all this can be accomplished without the need to fall back on the many legacies of Linnaean classification that imagine vertical hierarchies of life, both human and non.

De-Situated Knowledge: The Nightmare of an Ecocultural Partition

If embodied situated knowledges are webbed multispecies epistemic frameworks that respect the eco-onto-material-affective-cultural realities of any given ecocultural system, then the opposite of that—de-situated knowledges—are ways of knowing, relating to and therefore acting upon ecocultural systems that are untethered from such naturalcultural realities. They are knowledges that separate what is interconnected and distance the observer from the observed, thereby engaging in relationalities of hierarchical control and power that allow the deployment of colonial ecocultural policies.

When I am asked for an example of ecoculturally damaging de-situated colonial knowledge creation (or what I colloquially call "making sh*# up for profit"), I cannot help thinking of the mind-boggling way in which the borders of India and Pakistan were forged. At the end of World War II, the British decided to leave India in a hurry, with their own economic concerns chiefly in mind. Partition into a majority Muslim state (East and West Pakistan) and a majority non-Muslim state (India) would allow a powerful subcontinent to stay weaker and "within the British Commonwealth after devolution, even if in fragments, so that [the British] might maintain old imperial ties in some form after the granting of formal independence."[31] Lord Louis Mountbatten, the last Viceroy of India, suddenly announced a hastened departure calendar in June 1947; India would gain independence within just two months, in August 1947, a year earlier than originally planned. The task of quickly ripping apart ecocultural communities that had grown and lived together for millennia was given to a lawyer, Cyril Radcliffe. Appointed by the British Raj in July 1947, Radcliffe was geographically and culturally in London—and therefore fully de-situated from the partition he was commissioned to oversee. Radcliffe's de-situated ecocultural context was represented as a positive feature by the Raj, providing a semblance of impartiality in a very polarized political situation. The technocrat faced the incredible task of deciding the multispecies destinies of an entire subcontinent. The course of rivers, ecological and agrarian systems, familial and cultural relationships that had previously not faced rigid international political borders would now be surgically cleaved in two. Radcliffe had never visited India before his remit, nor did he have any prior experience drawing borders.[32] He barely knew the two main regions that would have to be split in two, Punjab and Bengal, but was tasked with rushing to delineate the border within an astonishing five weeks. Each commission contained two Muslim and two non-Muslim (Hindu or Sikh) representatives, leaving Radcliffe with the unenviable task of making most of the decisions in an evenly split vote. This highly specialized and siloed "thin-knowledge" was published as the *Radcliffe Line* on 17 August 1947, two days *after* the two countries had already declared independence. The haste and the de-situated method of distanced map-drawing was the successful culmination of decades of colonial "divide and rule" policy—strategically thwarting unity between

Indians by manufacturing and encouraging division between Hindu and Muslim communities, with ample assistance from niche interest groups and native religious extremists.

The result of this de-situated knowledge implementation was one of the largest and bloodiest human mass-migrations in history and the ecocultural splitting of the subcontinent into unequal and complicated halves.[33] The partition has been studied for decades as a profound human tragedy. And indeed, by all measures it was one. I would like to suggest that revisiting this event via a multispecies paradigm can bring added resonance to the devastating consequences of the cleavage. Since this book explores how cultural mapping of spaces imbibes lived spaces with human meaning and therefore ultimately directs human intervention choices, it is appropriate to begin my brief naturalcultural discourse analysis of India's partition via an exploration of a map. Fig. 1.1 shows a 1944 colonial map distributed by the British Information Services,[34] a propaganda arm of the Foreign and Commonwealth Office of the British government. The text accompanying the map can be read ecomaterially and naturalculturally to glean the subtext written by the map's producers:

> Although agriculture is the chief occupation, India is one of the eight leading industrial countries in the world. Textiles and iron and steel are the most important. Indian industries; cement, heavy chemicals, soap, sugar, glass and shipbuilding are among those most recently developed. India is the world's largest producer of jute, sugar-cane, and mica, and has the world's largest reserves of high-grade iron ore and the world's largest cattle population. India ranks second in the production of tea, cotton, tobacco and manganese.[35]

An ecomaterial symptomatic reading of this map teases out its underlying intentions, the subtext that is different from the simple physical description of geographical space. This is a colonial map that lists and values a multispecies collection of lives, bodies and ecosystems whose marvelous biopower is ready for the taking. Visually, the map does feature key urban centers, but the main spaces are focused on showcasing the subcontinent's immense natural resources, from gold to sugar to cattle. There are about twenty multispecies images in the map that feature productive laboring natives or domesticated animals and plants.

When the Indian subcontinental space is viewed in these very distanced, narrow or "thin" terms via multispecies colonial mapping, it starts to become obvious why the British may have hesitated at leaving such a resourceful and resource-rich space independent and undivided. When understood via Gandhi's notion of *soulpower*, it also becomes clear why the cleavage of an ecocultural system that had grown together over a millennium or more would harm its prospects of becoming a thriving nation. Indeed, partition divided cultural communities as well as the army, navy, financial, industrial and

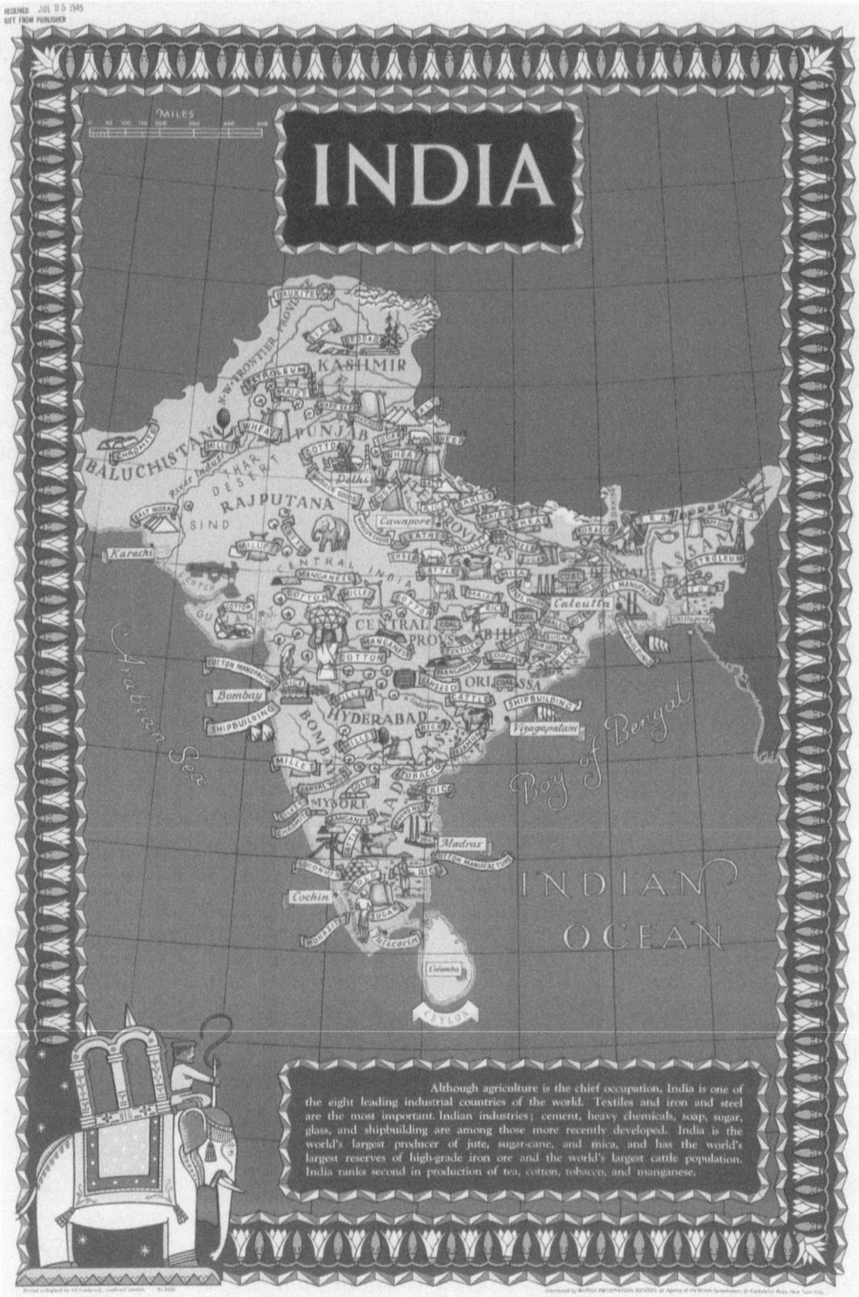

Figure 1.1 British Information Services map of undivided colonial India, featuring multispecies marvels and deeply focused on a resource and profit-based perspective on the land. (British Information Services)

ecological resources. I have been told anecdotal stories by my mother about how literally everything was divided; for example, in government offices, it came down to the accurate counting of furniture and pens.

What is often missed in historical and cultural analyses of the partition is the fact that this was a multispecies dismemberment. The division of the subcontinent artificially severed important intact ecoregions such as the Indus seven-river system, as well as the largest delta in the world, formed where the Ganges, the Brahmaputra/Jamuna and the Meghna rivers empty into the Bay of Bengal. Almost eighty years since the event, water wars now loom across international political borders between nations that have a deeply conflicted past. Both the Indus and the Gangetic/Brahmaputra riverine mega-systems are the chief source of water for a billion people and countless nonhuman lives, including the highly endangered Gangetic Dolphin (*Platanista gangetica*) and its smaller cousin, the Indus River Dolphin (*Platanista minor*). From sources high up in the Tibetan Plateau, all these major rivers face an uncertain future as their glaciers (collectively containing one of the largest snow fields outside the North and South Poles) melt rapidly due to climate change. As Omair Ahmad explains while discussing partition: "The end result of this focus on borders rather than bioregions is that ecosystem health is deeply undermined. The hilsa fish has disappeared from most parts of Pakistan and India as the river ecosystems it depends on have been disrupted. It is now largely confined to Bangladesh. India and Nepal quarrel over the nationality of an individual sloth bear, and healthy yak populations are now largely found only in Tibet."[36] As can be concluded here and as we shall see yet again in Chapter 5, the consequences of a de-situated colonial natural-cultural partition echo today within the damaged multispecies communities of all of South Asia.

Naturalcultural discourse analysis of India's partition shows how deployment of de-situated knowledges is often a strategic act rather than an accidental blunder. Viewing geographical spaces as sites of biocultural attachments via a multispecies paradigm shift suddenly lays bare the insanity of dividing them up into mutually hostile legal and political units. Viewing spaces via a multi-species paradigm is therefore a deeply anti-colonial method of understanding geographical spaces. It seems that colonial forms of strategically de-situated epistemic knowledge can perhaps only be glitched by an assemblage of situated knowledges enmeshed deeply with a place's bodies, lives and ecocultural systems.

Gandhi's concept of *satyagraha*—the search for truth, or the force of truth—was one such experiment in bringing back rational knowledge claims to glitch colonial-based crazy knowledge. His *satyagraha* movement can be understood, in EH and multispecies studies parlance, as a deeply situated form of naturalcultural care: "as an *affective* state, an *ethical* obligation, a *practical* labour."[37] The political and spiritual genius of Gandhi's choices can be seen in how he fused the affective, ethical and practical aspects of care in one fell swoop through his highly successful Dandi, or Salt March. Lasting twenty-four days from 12 March to 6 April 1930, the Salt March saw Gandhi lead a mass walk of 387 kilometers to engage in a simple naturalcultural gesture: pick salt from the shores of the Arabian Sea. From a naturalcultural discourse perspective, this one small action glitched the rationale of the immense British ecocultural colonial profit machine that extracted taxes from Indians for the freely available salt of their own land. Gandhi's use of a deeply situated ecocultural tether—sea salt—is an example of the opposite of the colonial method of creating de-situated colonial knowledges. These kinds of collaborative situated knowledges hold up far better to scrutiny than de-situated ones, because they can be verified via the realities of multispecies bodies, lives and ecosystems. One can read and confirm what is being produced as "knowledge" within the assembled "accounts of the distinctive experiential worlds, modes of being and biocultural attachments."[38]

Situated knowledges allow for a strategic ethical glitching of de-situated knowledges and discourses. In *The Archaeology of Knowledge* (1969), Michel Foucault describes how discourses created around spaces, places, bodies and lives have shaped the contours and limits of knowledge about them, and hence, their eventual destinies in the hands of communities. He envisions discourse as not only a matter of words, but of *practices* that produce and reproduce knowledge and power (hence his term power/knowledge).[39] By binding knowledge to constitutive power, Foucault gives EH and multispecies studies an avenue to understand the intimate co-becomings between human meaning-making and multispecies lives, bodies and ecosystems. It is discourse, according to Foucault, that forms the guidelines we follow when creating and perpetuating knowledge about a certain subject, to declare what is true or what is right. Discourses are historically contingent and can be defied and altered: one can ponder, for example, how different the discourse surrounding homosexuality or womanhood is in many cultures today as compared to thirty years ago, and how it keeps on mutating. That said, discourse can also stubbornly restrict what and how we think about and act upon given subjects or beings, thus imposing conscious and unconscious limitations on what we conceive as possible. Very importantly for our analysis of multispecies ecocultural systems, Foucault argues that discourse is not simply a way of representing reality, but rather a way of *constructing reality*. My redefinition of discourse via the method of naturalcultural discourse analysis, as described in the Introduction, adds ontological, material and affective dimensions of multispecies lively agents to Foucault's characterization of cultural frameworks, without subtracting anything from it.

Foucault's concept of the constitutive nature of power/knowledge demonstrates that the outdated yet prevalent discourses of "conservation" that conceptualize natural spaces as inclusive of nonhuman life yet exclusive of Indigenous ecocultures does not remain in the realm of abstract thought. As we will see in the case studies on the Agro Pontino region and the Greater Yellowstone ecocultural system in Chapter 6, such ideas of conservation were actively used to create the practices of land usurpation and cultural oppression that shattered the lived experiences of Indigenous communities by separating them from their interconnected ecosystems, thus strategically exercising a controlling power over them. In addition to this understanding of the links between discourse, power/knowledge and actual practices of culture, it is also crucial for the purposes of this project to recognize that discourses can not only construct but also displace *and* substitute other discourses. In our specific case, understanding these links helps us realize how colonial discourses (images, ideas, values) are substituting or displacing discourses, relationships and therefore practices between "us" as actual bodies and natural ecosystems. What are the implications of our new Cybercene cultural-discursive practices, born out of intra-action between screen and embodied realities, on multispecies communities and ecosystems?

Foucault explicitly connected power, knowledge and discourse by arguing that power is not something that is held by individuals or groups, but rather is exercised through *relationships and institutions*. Gandhi knew this very well. His Salt March was a masterclass not only in grassroots mobilization, but also in the strategic epistemic and affective glitching of the naturalcultural, material and therefore economic grounds upon which colonial rule was based, that idea that "I, as a dominating command economy, will not even allow you to make salt from your own seas. By monopolizing its sale, I extract profit from your own naturalcultural body and then also tax you for it." The simple, nonviolent yet disobedient gesture of walking to the seashore and picking up naturally created salt was a profound exercise of anti-colonial biopower and *soulpower*.

What is soulpower? It is crucial to understand how Gandhi does not conceive of soulpower simply as a latent form of spiritual and affective energy—he characterizes it as an ethical *method* and *process* (*satyagraha*) of social engagement. In an analysis of Gandhi's conceptualization of power via *satyagraha* (the search for the truth) and *ahimsa* (nonviolence), Manfred Steger argues that Gandhi glitches the typical Western understanding of power as being demonstrated via naturally violent impulses in human society. Steger includes in this assessment Foucauldian conceptualizations of the agency of disciplinary action by institutions as a manifestation of power. He argues that both Foucault and Gandhi would agree that "resistance and power were correlative elements, and [both] cherished political struggles and transgressive practices as ways of self-creation—the opening of new spaces in which a variety of discourses challenging oppressive 'normalcy' can flourish."[40] However, Steger believes that the two thinkers diverged dramatically in understanding how power could also be an ethical power. Essentially, Steger argues that for Gandhi, power can be constitutive of care and kinship via the pursuit of truth. Gandhi's "truth" was a skeptic's truth, always experimental and non-dogmatically understood as "the selfless devotion (*bhakti*) to communicate with the 'Other' (ontologically conceived as the interdependence of all things), and the corresponding moral and political duty to search for more complete manifestations of the necessarily incomplete 'relative truths' of dualistic thinking."[41]

Gandhi's anti-colonial stance is deeply situated, beautifully naturalcultural and immensely powerful in its epistemic, affective and pragmatic capacities to glitch normative colonial thought.[42] The nonviolent search for truth is the opposite of the colonial "thin-map" account of India. Via a deeply collaborative, ecoculturally open and epistemically humble strategy, it allows a very clever unravelling of colonial capture-and-control. Soul-force is the name given by Gandhi to the power of that ethical glitch:

> [It] is a method of securing rights by personal suffering; it is the reverse of resistance by arms. When I refuse to do a thing that is repugnant to my conscience, I use soul-force. For instance, the Government of the day has passed a law which is applicable to me. I do not like it. If by using violence I force the Government to repeal the law, I am employing what may be termed body-force. If I do not obey the law and accept the penalty for its breach, I use soul-force. It involves sacrifice of the self.[43]

Gandhi's soul-force could be termed as 'soul*power*,' if one understands his notion of this force as a constitutive and Foucauldian kind of knowledge/power combination. This ethical nonviolent power is a situated and strategic one, born from (non-absolute) curiosity and love for the Other. Viewed through a multispecies paradigm, Gandhi's soulpower can be reconceptualized as a way to re-value the affective, ontological and ethical dimensions of care and kinship. Since this project seeks to bring forward the power of "thick" multispecies accounts as opposed to "thin" anthropocentric-colonial ones, it finds inspiration in thinkers such as Gandhi who actively sought to understand ways of strategically glitching massive forces of violent colonial ecocultural subjugation. It is no surprise that the strength and strategic acumen of Gandhi's ethical and political stance inspired so many other liberatory movements, including those led by Dr. Martin Luther King and Nelson Mandela. Its power lay in his finding a specific way to enact an epistemic glitch: the constitutive and ethical possibilities of power via *satyagraha*, *ahimsa* and soul-force. As Steger eloquently concludes:

> Gandhi attacked what Richard Rorty has identified as the core of traditional Western culture: the notion of searching for Truth by turning away from "solidarity" towards "objectivity." Conversely, the Mahatma insisted on the importance of pursuing truth in the "thick" communal context. Seen through traditional Western lenses, Gandhi's firm rejection of the separation of means and ends represents perhaps the most accessible point of entry to survey the full extent of his challenge to our Western empirical-analytic "science of power."[44]

Multispecies studies could be understood as a form of interdisciplinary naturalcultural *satyagraha*, made especially difficult as it needs to overcome foundational anthropocentricity. It seeks to makes visible the thick contexts of situated knowledges (rather than a single "objective" viewpoint) embedded within thick time. Through a combination of 1) a Foucauldian understanding of the importance of knowledge management via discourse in shaping the material destinies of lives, bodies and ecosystems, 2) my redefinition of discourse in naturalcultural analysis to include ontological, material, affective and multispecies frameworks, 3) the method of situated knowledges to put into conversation multiple biodiverse partial knowledge claims and 4) a Gandhian understanding of ethical soulpower, we can see the emergence of a coherent multispecies studies ethical and methodological framework.

Stone Guests and Curious Foxes: Multispecies Agency in Indigenous and "New" Materialisms

Foucault and Gandhi recognized that situated and de-situated knowledges co-become in both expected and unexpected combinations. They also understood the deeply connected relationship between matter and meaning—how colonized bodies, lives and ecocultural systems were imagined and controlled to extract maximum profit. Centering this book and lab project on an assemblage of situated knowledges of ecocultural care and kinship via passionate immersion allows it to contribute towards a multispecies paradigm shift—an epistemic, affective, discursive and pragmatic change towards naturalcultural care and kinship.

I also hope to avoid an unfortunate inclination within EH to focus on and from the perspectives and destinies of the urbanized white populations of the Global North, which results in the restorative pathways they propose being abstract or narrowly conceived. Silencing marginalized voices is not just about ethics; it's also shoddy and incomplete science. As EH matured within academia as a field of inquiry, it began to include decolonial perspectives: for example, Alfred W. Crosby's 1972 book *The Columbian Exchange: Biological and Cultural Consequences of 1492* was the first in-depth exploration of the global effects of European colonialism in the New World.[45] Ramachandra Guha discussed an *environmentalism of the poor* from the perspective of Third World countries in "Radical American Environmentalism and Wilderness Preservation: A Third World Critique" (1997).[46] While foundational EH texts such as these began to provide subaltern perspectives, it can be said that EH academic scholarship overall demonstrates a blind spot towards the worldviews, voices, terminologies and methodological approaches developed by Global South, Indigenous,[47] and other marginalized and subaltern communities, including those of queers, persons of color, women and other biopoliticized thinkers.[48] I believe that this blind spot has been detrimental to the field as a whole. There are many Indigenous environmental concepts, traditions and practices which are the product of centuries-long co-constitution within embedded human-multispecies communities, and thus extremely precious and relevant to understanding and developing further effective pathways to healing and habitability. Yet such approaches have been strategically disregarded in favor of Euro-American-centric academic neologisms and philosophies in the rush to claim and settle territory in a rapidly expanding scholarly field. While current EH scholarship is thankfully making inroads toward inclusion of these very productive veins of Indigenous and subaltern scholarship, I believe the colonizing tendency continues to exist in the field via its predominant intellectual choices. Jorge Marcone persuasively asks whether the bodies voicing Indigenous environmental perspectives will be treated as "stone guests" rather than truly valued stakeholders by academic, governmental and policy think tanks?

> The situation of popular environmentalisms in Latin America can be illuminated with the old Spanish expression of the *convidado de piedra*, or "stone guest"; that is to say somebody who has been invited to a negotiation out of a sense of duty, but who does not participate in the discussion, or whose participation is ignored by the rest. Because of international agreements such as Convention No. 169 on Indigenous and Tribal Peoples by the International Labor Organization, states ask these movements to participate in the decision-making processes regarding development projects. The Convention requires that Indigenous and tribal peoples be consulted on issues of policy, governance, and development that affect them in free, prior, and informed participation. The ultimate goal of the consultation is achieving agreement and appropriate solutions in an atmosphere of mutual respect and full participation. However, are these popular environmentalisms invited to the negotiation table because their opinion will truly be heard? Or do their hosts actually prefer to treat them as stone guests with no voice, even if they speak?[49]

A tendency in EH research to co-opt traditional naturalcultural concepts with fancy neologisms is another example of the consequences of this hypocritical "stone guest" treatment of subaltern environmental advocates. Coining new terms can be a desirable choice, as long as previous similar knowledges are also acknowledged and given recognition. When this is not done, scholarship becomes yet another indirect way by which colonial power appropriates knowledge without recognition. All these marginalized forms of situated multispecies knowledge are not "new," even though they are incredibly innovative since they subvert currently normative systems of thought and process. They are embodied learnings that have been repressed, denigrated and exterminated because they invoke anti-colonial ways of living. What a pity and waste—this conscious or unconscious employment of citation politics to ignore so much ecocultural knowledge.

Like Leroy Little Bear's brilliant approach of defining all things as "more or less animate," described in the Introduction, the renewal of EH and multispecies studies' attentiveness to all kinds of subalterned situated knowledges can subvert the unconscious biases of normative preconceived ideas and notions. This can open up pathways to new methods, considerations and perspectives—literally new ways of offering attentiveness and looking at the world being studied. Citing Indigenous situated knowledges seriously and providing them with greater and broader visibility is the opposite of cultural appropriation or "stone guest" treatment—it is to finally respect these embodied and passionately immersed multispecies knowledges as important expertise. In a discussion of these important questions, Zoe Todd very persuasively asks:

> when *will* I hear someone reference Indigenous thinkers in a direct, contemporary and meaningful way in European lecture halls? Without filtering ideas through white intermediaries—apologies to the vast majority of my anthropology colleagues—but by citing and quoting Indigenous thinkers directly, unambiguously and generously. As thinkers in their own right, not just disembodied representatives of an amorphous Indigeneity that serves European intellectual or political purposes, and not just as research subjects or vaguely defined "collaborators". As dynamic Philosophers and Intellectuals, full stop.[50]

Attempts at creating a truly impactful EH have yet to be fully realized due to inattention to or "stone guest" treatment of marginalized knowledge experts. With these immense structural biases and resulting conceptual lacunae in mind, my approach seeks a pragmatic, inclusive and decolonial pathway towards forging a coalition consisting of a broad and interdisciplinary selection of thinkers, including both oft-cited mainstream academics and those whose innovation has been historically marginalized due to their biopoliticized identities—including specific rather than amorphous Global South, BIPOC, migrant, queer and women thinkers. I take inspiration from Kim TallBear's suggestion to conceptualize my work for this project not as a way to "give back" or have marginalized scholars and thinkers as vaguely defined collaborators, but as a way to "stand with" a broad coalition of scholars and communities (both human and non) breaking down that corrosive duality between *researcher* and *researched*. This is another instance of the epistemic humility that is also both a *strategic humility* and a source of *epistemic and methodological strength*, and resonates with TallBear's words:

> It is also helpful to think creatively about the research process as a relationship-building process [...] A researcher who is willing to learn how to "stand with" a community of subjects is willing to be altered, to revise her stakes in the knowledge to be produced.[51]

None of my work for this project (or even my general disciplinary focus) is ethnography, nor any other such "human-subject" field of study as commonly found in the social sciences. I approach my dialogue with the many dominant and subaltern-positioned experts (human and nonhuman, via in-person observations, conversations or published research) through an ethics of standing with, rather than giving back. It is in the *survivance*[52] of many diverse and global marginalized multispecies societies that we can locate pragmatic pathways to care and alternative shapes of kinship circles. So, along with the many cultural-colonial dichotomies that need to be revisited for a proper engagement in multispecies studies—such as **human**/animal, **culture**/nature, **man**/woman, **heterosexual**/queer, **meaning**/affect, **epistemology**/ontology, **mind**/body, **meaning**/matter, **disciplinary expertise**/generalist knowledge and **animate**/inanimate—the last (but not the least) dichotomy that needs to be properly reassessed is perhaps the one that is most relevant to my role within this project: **researcher**/researched. To explore this final duality a little more, let's revisit an old friend—the bold and plucky Channel Island Fox.

As described in my Preface, the Channel Island Fox (*Urocyon littoralis*) became that other important and unexpected scholar companion who guided me to reflect on my positionality while undertaking research in multispecies studies. What had struck me when our research group spotted these wonderful creatures on Wi'ma Island (see book cover image) was that question of *who was researching whom?* Indeed, that fox taught me a big lesson via its intra-active past and present about how to "think creatively about the research process as a relationship-building process, as a professional networking process with colleagues (not "subjects"), as an opportunity for conversation and sharing of knowledge, not simply data gathering."[53] With this humbler approach, I was able to relate to the fox with a naturalcultural, rather than simply cultural, narrative, thus opening up a multi-layered and complex web of relationships. A naturalcultural analysis stitches together the human-fox-island assemblage into a long and intra-active web of co-becoming relations. Island Foxes are mostly limited to *Wi'ma*, *Tuqan* (San Miguel) and *Limuw* (Santa Cruz) Islands due to their larger carrying capacity. The fox population decreased dramatically in the 1990s until there remained but a few dozen. Extinction was a distinct possibility, but human intervention and the fox community's resilience were able to transform this grave crisis into a conservation success story. The fox's decline was attributed to predation by Golden Eagles (*Aquila chrysaetos*), but the trail doesn't end here. The chief reason Golden Eagles became resident breeders on the Channel Islands was that an ecological niche was left open for them by the precipitous decline of Bald Eagles (*Haliaeetus leucocephalus*). Golden Eagles and Bald Eagles are both opportunistic hunters, but whereas the former typically prey on terrestrial small mammals and birds, the latter predominantly ate fish and therefore left the foxes alone. Bald Eagles had declined precipitously on the island due to use of DDT (Dichlorodiphenyltrichloroethane—a widely used synthetic insecticide) and the introduction of non-native grazers by humans. Following this naturalcultural breadcrumb trail from the Channel Island Fox thus leads us to extractive relationships imposed on the land by settler-colonial communities. As this story of cascading intra-action illustrates, multispecies ecocultural effects reverberate through many layers and indirect relationships, in odd and unexpected combinations, rather than always through immediate and direct causal connections.

This affective lesson in going beyond the dualities of researcher/researched and normative/decolonial knowledge systems meant confronting what I call the "classificatory conundrum"—the question of how to acknowledge and integrate the terrible history behind oppressed biopolitical identity categories without reinforcing them. Donna Haraway's notion of "situated knowledges," Kim TallBear's suggestion of "standing with" marginalized knowledge creators and the Channel Island Fox's intra-active glitching of the researcher/researched duality sympoetically worked together to show me a pathway. I am convinced that a methodological pathway derived from a sympoesis of these three ideas has important theoretical, methodological, ethical and political ramifications beyond EH and multispecies studies. As academics around the world grapple with communicating the importance of uplifting the lives and cultures of historically marginalized peoples, they face a populist backlash for seeming to retrench those very categories in the process. Instead of this awkward and contradictory management of identity politics, the pragmatic and ethical process of standing with situated knowledges stresses to *everyone* the general ecocultural systemic benefits of being more attentive to all sorts of viable knowledge perspectives, including both normative and non-normative ecoculturally productive knowledge traditions. In addition to accomplishing better and "thicker" accounts of multispecies co-becomings and intra-actions without getting mired in identity politics, the situated "bodies" that multispecies studies methods "stand with" also allow this book and the lab to instigate ontological, affective and material "turns" in its approaches. These turns involve testing methods that connect meaning to matter, to study how culture deeply affects mortal and/or material bodies, lives and ecosystems of different kinds, and vice versa.

In line with the four premises briefly described above, and methods such as naturalcultural discourse analysis and multispecies thick-mapping, potential future topics of focus via the lab's research output will include (but are not limited to):

- Examining cultural and embodied pathways that lead to nature-distancing and nature-connectedness
- Analyzing the effects of national and transnational biopolitical/ecopolitical discourses and practices on lives, bodies and ecocultural systems
- Investigating cascading ecocultural effects: material impacts, cultural paradigms and transformations in various periods
- Revaluing multispecies perspectives and relationships (plant, animal, humanimal, fungal, microbial)
- Studying global and Mediterranean migrations as ecocultural phenomena
- Fostering global Indigenous and postcolonial epistemologies

- Analyzing unsustainable local and global extractive practices of new and old colonialisms
- Locating and celebrating diverse global examples of trans-species justice, care and kinship
- Theorizing human and "more than human" mediatic relations in the Cybercene
- Examining cultural, spiritual and philosophical traditions on nature/culture/multispecies relations
- Visualizing and narrating ecocultural systems via multiscalar and thick-time approaches
- Exploring Thick/Deep/Ecological/Geological/Cyclical/Cosmic/Nonhuman vs Historical (anthropocentric) Time
- Locating terracentric (land) versus marine and archipelagic perspectives

The *Cybercene* as a Multispecies Ecocultural Gathering Principle

This book has three primary goals. Its first is to enhance EH by decolonizing the field and more firmly decentering (yet not erasing) anthropocentric perspectives via a multispecies studies approach. Its second is to propose new multispecies studies methods through analyses of case studies and multi-disciplinary work done in the *Multispecies Futures Lab*. Its third is to investigate ecocultural systems in the context of our digitally connected world. To accomplish this third aim, I devised the term *Cybercene* to provide a new gathering principle to study the current ecocultural era in our planetary history.[54] Roughly encompassing the first two decades of the twenty-first century and still ongoing, I define the Cybercene as a pivotal period of recolonization in which de-situated online onscreen realities, digital representations and narratives have quickly and effectively transformed identities and substituted the value of actual bodies and living habitats.[55] I surmise that ecological damage is accelerated and mediated through a series of cascading effects, feedback loops and resonances that profoundly link human-media habits with multispecies destinies via cultures of nature-disconnectedness.

There has been a decades-long proliferation of jargon in EH that uses the suffix "-cene" to impress upon us the geological and planetary effects of detrimental human activity (plantationocene, negrocene, capitalocene, homogenocene, etc.) and others that reflect alternative possibilities (for example, symbiocene, urbanocene and chthulucene). I want to add to this list *only* if it serves as a useful analytic gathering place for specific future research that is currently not being given enough focus. I therefore put forward *Cybercene* as a new *ecocultural* term. I do not intend to propose a new geological era, nor am I hoping to place a Golden Spike that will mark its beginning. Hopefully I will convince readers that there is a use for Cybercene as a generative and creative term to better contextualize multispecies relationships today. Indeed, I deploy the term Cybercene on the premise that naming, understanding and addressing the digital naturalcultural transformation can be a springboard to allow us to locate and deploy more effective pathways to address some of the major crises and opportunities of our era, just as the broader ecocultural analysis of Wi'ma Island's ecocultural systems could provide added resonance and visibility to this rich biocultural space. While the novelty of the term does have value, far more important to me is Cybercene's utility as a gathering principle to foster "tentacular"[56] and "sympoetic" collaborative thinking about ecocultural and multispecies wellbeing. Chapters 3 and 4 examine this new era in great detail, analyzing the profound and intimate connections between our media habits and their personal/psychological, cultural, political ecological and multispecies implications. Intentionally situated in approach and global in scope, this chapter illustrates the underlying causes and mechanisms of the Cybercene practices of othering, commodification and wasting.

Unlike most of the burgeoning humanistic research within the digital humanities focused on socio-cultural, economic and philosophical questions surrounding the digital revolution, my work here and in my lab focuses on ecocultural (environmental and cultural) consequences of the Cybercene era by collapsing the nature/culture and human/nonhuman dualities that are foundational assumptions within most other theoretical approaches. Indeed, an ecocultural approach to understanding the Cybercene via the lab's projects brings to light the intimate entanglements between two contemporary phenomena that have tended to be studied separately within the confines of strict disciplinary boundaries: climate change, habitat destruction and biodiversity collapse on the one hand (investigated via the various biological, Earth and environmental sciences), and contemporary media/culture/identity-based polarization, conflict and violence on the other (as studied in the social sciences and the humanities). My critique of this era is not a critique of technology per se, but a questioning of its unreflective use in our daily lives and its unplanned and unregulated global deployment for profit. In a seminal newer work on the study of technology and nature, the 2017 edited volume *General Ecology: The New Ecological Paradigm*, Erich Hörl advocates for a new *general ecology* that studies the reality of our "new cybernetic natural state"—where self-regulating digital systems manage and organize our lives.[57] Hörl contrasts general ecology with restrictive ecology, a field that has the traditional narrow focus of the environmental sciences. My work adds to and complements this conversation because I propose that the two contemporary phenomena of accelerated cultural and ecological damage are related through our new digital mediatic practices. I argue that the dramatically inward-looking and narrowed-down psychic focus of our screen-based lives gives us a different kind of attentiveness, leading to the capture-and-control of attentions, affects and meanings. By showing how mediatic-psychic intra-actions are co-becoming processes, each one influencing the other, I move in the opposite direction from that taken by Hörl to explore nature-technology connections. I subsume tech practices within other kinds of indirect and cascading ecological relationships, rather than subsuming ecological processes within a new cybernetic understanding of the world. I also restrict my use of the prefix eco- and the word ecological to a non-metaphorical use of the terms; they represent embodied multispecies naturalcultural relationships and processes. The digital component is an element, a subset, of these multispecies naturalcultural relationships rather than the other way around. Further research approaches traveling in both directions are urgently needed.

In addition to marking the end of the postcolonial interlude, I would also suggest that the advent of the Cybercene is a departure from the modernism/postmodernism duality that has been central to academic and cultural debates about culture in the twentieth century. The Cybercene human–tech co-becoming—with its associated hyperconvergence, hyper-personalization and reactionary formation of insular identitarian communities—suggests a new ecocultural era where structuralist, post-structuralist and decolonial epistemologies coexist, co-opt, clash and generate polarized enclaves. The ecocultural effects of this Cybercene "post-post-structuralist" situation provide opportunities for new kinds of connection and also far greater possibilities for mediatic/psychic experiences of distancing, de-situation and epistemic paranoia, due to the often de-situated nature of the content generated by digital cyberscreen knowledge-machines. Furthermore, in Chapters 3 and 4, we will also explore the hypothesis that increasingly exclusive reliance on screen-based relationships is leading to a breakdown in naturalcultural relationships of care and kinship, due to increasing physical enclaving from actual bodies, lives and ecosystems. In short, we are witnessing a return to new/old forms of normative—colonial epistemic—discursive and affective frameworks and practices. It is important for me to underline that while these are typical Cybercene trends, the actual digital tools that shape these practices are their means rather than their cause. Indeed, in Chapters 5 and 6 I will also examine instances where Cybercene mediatic practices are ethically glitched to aid embodied multispecies forms of meaning-making, care and kinship.

May the (Multispecies) Love Go On

Whenever our paths cross, I still ask the students who joined me in our multispecies journey to Wi'ma if "the Island is still within them." This book and its cojoined lab's projects are my way of investigating how academic research and our everyday ecocultural attitudes, affects and practices can gain by listening to what the real and metaphorical "Island" has to tell us. We are enveloped by and deeply integrated within so many naturalcultural spaces, so many complex multiscalar ecocultural systems. Most of us have the great fortune of (still) having access to the "Island" at all times, right where we are, wherever we are. Many systems are damaged, but most still have a chance—naturalcultural systems do bounce back when given a fighting chance, sometimes extremely quickly.[58]

Naturalcultural attentiveness teaches us that we can surely no longer continue to sustain rigid hierarchical epistemological divisions such as **human/**

nonhuman, **culture**/nature, and **meaning**/matter—with the concepts in bold usually being given greater cultural value. As I have outlined in the Introduction and this first chapter, foundational cultural dichotomies are finally being questioned by newer research in subfields such as EH, multispecies/animal/plant studies, Indigenous studies, posthumanism,[59] new materialism and postcolonial studies. However, as is hopefully evident via these initial thoughts, there is tremendous unrealized potential to integrate these newer frameworks and approaches within university research, teaching and community engagement.

Could the "passionate immersion" and multispecies "arts of attentiveness" cultivated by my students "situated" on Wi'ma Island become the norm rather than the exception for everyone and anyone, in our approach to understanding our amazing multispecies communities? Could these kinds of explorations, which seek to practice better multispecies ethics of care and hold more inclusive conceptualizations of kinship, not greatly expand the remit and impact of the humanities and sciences? Anna Tsing urges the continued growth of "new science studies" that include passionate immersion within material-cultural matrices:

> Once such immersion was allowed only to natural scientists, and mainly on the condition that the love didn't show. The critical intervention of this new science studies is that it allows learnedness in natural science and all the tools of the arts to convey passionate connection. [...] Writers of this new genre, including myself, are excited by the chance to trespass across the boundaries between the natural sciences and the humanities.[60]

With Tsing's exquisite vision of disciplinary "trespassing" in mind, I have sought in this first chapter to illustrate very briefly why and how natureculture, intra-action and co-becoming are ecocultural processes that should be accorded far greater importance by researchers and educators. These initial thoughts also seek to demonstrate why a coalition of situated knowledges from the environmental and multispecies humanities, environmental sciences, ecocriticism and cultural studies (but also law, economics, political science, government, business, medicine and other fields) can work together to better understand and communicate complex relationships between bio- and eco-politicized Selves and Others in the Anthropocene, while also avoiding the quagmire of rigid identity- and species-based disciplinary siloes. Another trespass that is urgently needed is one that breaks the boundaries between academic and public environments—these deeply affective and affecting multispecies learning and exploration experiences should become a public right again, rather than a privilege of the fortunate few.

Are we willing to expand our attentiveness beyond purely textual cultures or purely biomaterial systems to also explore various co-becomings of biocultures on all scales, from the microbial to the planetary? Are we able to recognize old and new forms of multispecies capture-and-control? Can we each live within a multispecies paradigm to bring back value and visibility to often disregarded naturalcultural relationships? Will we be okay with letting love show in our work and our lives? The rewards of nurturing these rational, diverse and *unashamedly loving* experimental approaches to understanding naturecultures are both immediate and immense. Can multispecies attentiveness, care and love overcome powerful wielders of money, privilege, relentless greed and influence? Perhaps. I am not so sure sometimes. We can only hope and do our small part. We owe it to the vast, vibrant and vulnerable multispecies lives that still thrive on this planet to try and love in any capacity, great or small, despite and against the odds stacked against us.

2

Capture-and-Control

If humans are to have a future on Earth, it will have to be a multispecies future—or none at all.

Unrelenting past and present multispecies colonialisms—the acute capture-and-control of all lives (human and non) for the extremely unequal accumulation and concentration of resources and power—has caused so much suffering, damage and waste. This chapter will explore past forms of naturalcultural colonialisms in order to better recognize the ecocultural roots of present-day multispecies inattentiveness, damage and relentless capture-and-control for profit in the Cybercene, which will be examined in the next chapter. I will focus on my ecocultural definition of colonialisms to embark on a brief archeology of the epistemic, affective and psychic capture-and-control of naturalcultural bodies, lives and ecosystems. I specifically analyze different forms of affective and discursive capture-and-control via the mediatic mechanisms of a prime colonial multispecies spectacle—the multispecies zoo—highlighting them as crucial sites for the "reciprocal capture" of colonial ecocultural meanings and identities. These explorations are not meant to be exhaustive, but indicative and illustrative of a colonial attitude towards multispecies bodies, lives and ecocultural systems. As we will see in subsequent chapters, the basic underlying mechanisms present in these historical colonial forms of multispecies capture-and-control profoundly reverberate in our contemporary lives.

On Zoos and Glitches: *Capture-and-Control* and *Running Away*

In order to fully explore colonial othering practices and the intimate relationships between colonial mass-mediatic representations and actual multispecies bodies, lives, landscapes and ecocultural systems, and to describe moments when colonial systems falter, I would like to use two concepts: *capture-and-control* and *running away*.[1] These terms were inspired by Daniel Vandersommers' book *Entangled Encounters at the National Zoo*, which analyzes the multispecies ecocultural entanglements of the National Zoo in Washington, D.C. In this book, the author skillfully analyzes both the natural/material lives and cultural/intangible values of zoo animals:

> The book shows how zoo animals always ran away from the zoo. I mean this literally. Animals escaped frequently [...] Even more though, I mean this figuratively. Living, breathing, historical, zoo animals ran away from their cultural constructions, and these constructions ran away from the living bodies they were supposed to represent.[2]

In his book, Vandersommers harnesses the opposition between capturing and escaping to investigate deeper questions of multispecies agency and to examine the cultural implications of showcasing animals in a metropolis of paramount political and cultural power. Loosely inspired by this premise, I have devised the term "capture-and-control" to describe colonial othering actions in the interconnected contexts of 1) multispecies identity-based biocultural and ecocultural domination and stereotyping (on the basis of constructed identities of race, gender, nationality, indigeneity, speciesism, etc.) as well as 2) the biological hierarchical classification of life. Colonial biocultural stereotyping and hierarchical species classification may seem like separate cultural processes. However, I hope to demonstrate in this chapter that they actually have deeply intra-active, cascading connections and feedback loops that connect them quite intimately into the complex matrix of a naturalcultural "regime of meanings." In contrast to capture-and-control, "running away," characterizes unexpected moments when different lively agents escape seemingly intractable colonial regimes' control. The mechanisms of this powerful and liberatory agential power are examined in more detail in Chapters 5 and 6.

Colonialisms: An Ecocultural Definition

It occurred to me one day that the Portuguese, French and British didn't really come to my country of origin, India, to seek out Indians. They came to find a marvelous collection of fragrant multispecies treasures: spices such as pepper (*Piper nigrum*), cinnamon (*Cinnamomum verum*), cloves (*Syzygium aromaticum*) and nutmeg (*Myristica fragrans*). The human Indians were actually an annoyance, standing in the way of getting to these multispecies plant lives cheaply or for free. For the British East India Company, the benefit of harnessing the cheap or free labor power of a colonized human population came as an afterthought, however desirable. Settler-colonial communities in my adopted country, the United States, did not move West seeking Native American communities. They came looking for gold or land; the Indigenous communities were simply "in the way." It is with these thoughts in mind that, in the Introduction, I defined colonialisms as

> past and present affective, epistemic, discursive, psychic and material practices of distancing, domination and control practiced by individual

> or group command economies upon culturally and ecologically Othered multispecies bodies, lives and ecosystems, in order to extract and exploit these as material or labor resources for profit.

While their modes and methods of domination can differ, I would like to characterize all variations of colonialisms (for example settler-colonialism, ethno-state-led imperialism, internal colonialisms within nation states directed at their own citizens, domination on an individual level such as bullying) as having one thing in common—they all encompass the daylight robbery of othered lives, bodies and ecosystems for the accumulation of profit. And they all encompass ways of "meaning-making" that represent these bodies, lives and ecocultural systems as the Other, and therefore available and ready for capture-and-control. This chapter explores some of ways these de-situated discursive and affective manipulations of representation were carried out, along with the ecosystemic consequences of these representations.

A cultural studies understanding of both individual and collective forms of colonialism gives crucial importance to the *affective, psychic and cultural preconditions/orientations* of colonial intervention that compel and justify the actions of "capture-and-control." I will go on to explore instances of material-ecological multispecies theft via colonial practices in later chapters. However, the affective, psychic and discursive elements of colonialisms are both less obvious and form the most complex and important part of this chapter's investigations of the ecocultural consequences of multispecies colonialisms of all kinds. Fortunately, we can rely on the scholarship of the many postcolonial theorists who have dug deep into the colonial mindset, exploring the underlying desires, fantasies and mental justifications used to deploy colonial, individual and institutional power to dominate and capture-and-control colonized Others. The original contribution of this project to that ongoing anti-colonial and decolonial analysis is to demonstrate the *deeply multispecies* goals and consequences of colonialisms (remember the spices!), and to place the Cybercene era as *a continuation and rejuvenation* of multispecies colonial epistemics, affects and practices. Mass media texts in all shapes and forms provide essential data via which all these forms of epistemic, discursive, psychic and affective colonial relationships can be analyzed and understood. Consequently, this project also places the current A.I. revolution within the tradition of multispecies and multi-kind ecocultural mediatic colonialisms. If social and cultural identities are fundamentally relational ("I am who I think I am and who others think I am"), then outsourcing epistemic meaning-making and identity-making to opaque generative algorithmic processes and de-situated communities will have profound effects on our relationships with our selves, each other and all our multispecies kin. This notion of Colonial-Cybercene will be further explored in Chapter 3.

Foundational postcolonial theorists such as Edward Said, Homi K. Bhabha and Robert J.C. Young highlight how colonial extraction, exploitation and extermination have somehow to be psychically and culturally justified by the perpetrators.[3] Their qualitative methodologies read the colonial political apparatus as a question of personal mindset at the level of the individual, and then connect these private questions of constructed and imagined identity and culture to the shapes and destinies of larger authoritarian organizations and institutions of all political orientations. I call all these forms of colonial individual and collective agency "command economies" in my definition of colonialisms, because that is what they are—*centralized desire machines* for the organized and efficient accumulation of resources, labor and power. Said, Bhabha and Young explain how the violence of taking without asking is rationalized via an elaborate system of mental gymnastics that have both psychic (in the individual) and discursive-cultural (in both individuals and entire societies) dimensions. For example, Said's concept of *Orientalism* describes how cultural products (images, ideas and values) relentlessly imagine and represent colonial lands and peoples as inferior, savage and exotic, thus justifying the civilizing mission of the more "advanced" colonizing powers or nations. Said's colonial discourse analysis is founded on a Foucauldian understanding of the intimate connection between knowledge creation, discourse and power. The creation of an inferior "Other" that is fundamentally abject in comparison to the superior "Self" is a prominent precondition of the extractive actions that are subsequently carried out. Bhabha's and Young's important and original contributions to this line of analysis are to characterize this discursive manipulation, which othered bodies, lives and ecocultural systems, as both cause and effect of *psychic* maneuvers of colonial self-justification, involving a strategic denial of colonial cruelties that permits the deployment of a colonial desire machine to partake in the free and open "all you can eat" buffet of resources that ensues. All three scholars mentioned above are foundational to the method of naturalcultural discourse analysis, as they indicate methodological directions scholars can take to understand "culture" (and the various regimes of meaning) as the foundational terrain upon which actions of material multispecies damage (or care) are carried out. This kind of naturalcultural discourse analysis of colonial systems of thought and practice truly unveils *how* and *why* humans are such potent ecocultural protagonists in the Anthropocene.

I should also like to add to this list of colonial psychic, discursive and affective mechanisms of domination a third important yet underappreciated feature: that of colonial forms of kinship. It is not often discussed that what unites colonizing communities is not only hatred, fear and disavowal of the colonized Others, but also a carefully cultivated and pernicious colonial identity of kinship and solidarity based on commonly held bio- or eco-politically imagined identities (familial, racial, national, gender, religious, tribal, etc.) and discourses (manifest destiny, national pride, modernization, progress, development, etc.). The colonial mindset that can keep even the colonized in their oppressed state quite willingly is often maintained via the careful cultivation of an insidious form of identification and solidarity with the "greater" colonial cause. Colonialist "imagined communities" can also be formed by way of a contradictory process, in which *autopoetic kinship* is generated.[4] This can be a very isolating form of kinship in which the individual or nation engages in a closed-loop paranoid mechanism through a discourse of individual or collective victimhood. The resulting manufactured self-pity is used to justify colonial violence without, and totalitarian solidarity within.

Lastly, yet fundamental to this list of the discursive strategies that underpin colonial-cultural practices of multispecies theft, is the cultivation of a discourse of an imagined "surplus." Very often, one will notice in colonial mediatic interventions an insistence (hidden or direct) on representing colonized lands as rare, yet resource-abundant, containing a fantasized surplus that can cater to a magically unlimited supply of multispecies material and/or labor power. These discourses of surplus/lack are also often gendered, usually with a strictly male heteronormative projection of imagined sexual (or generally, libidinous) rewards, available in those colonized imagined spaces. Surplus is psychically paired with its opposite, lack. The quantifying of these imagined bounties can be purposefully vague or meticulously specific—however, none of it is based on the actual reality of the limited nature of the resource, nor is the ecosystemic "collateral damage" that will be caused to obtain it considered. These situated details vanish within the fog of colonial fantasies of domination and abundance. Sometimes these discourses of surplus in colonized spaces will be coupled with discourses of scarcity and lack within the homeland, thus further justifying theft.

When put together, all four cultural-discursive-psychic mechanisms of colonial-cultural capture-and-control of multispecies lives, bodies and ecocultural systems described above (multispecies othering, epistemic and affective distancing, colonial kinship and imagined surplus) serve this specific book project perfectly by demonstrating how colonial-cultural storymaking and storytelling need to constantly de-situate knowledge in order to make violent extractive actions psychically palatable. In today's Cybercene era, explored in the next chapter, these discursive and psychic processes of multispecies domination and theft become even more crucial, since digital mediatic forms allow a far greater reach, depth and finesse of de-situated knowledges—i.e., knowledges untethered from the material and embodied situated realities of actual lives, bodies and ecosystems in order to devalue and, therefore, extract as much profit as possible from, multispecies lives, bodies and ecosystems.

A Brief Archeology of Mediatic Naturalcultural Capture-and-Control

Capture-and-control of meaning through mediatic discourse, and therefore of bodies and lives via restricted manufactured spaces of belonging, is not a new phenomenon. The presence, circulation and management of media could be said to be the constitutive characteristic of modernity, and of modern colonialisms. It was certainly a defining factor of the age of European colonization of the known world. The birth of the printing press in the fifteenth century provided true mass circulation of images for the first time in Europe, thus spurring new ways of collectively understanding the world and creating a collective cultural context for a powerful push towards modernized global command economies via empire-building, which transformed and enriched bodies, lives, spaces and ecosystems in the "Old" world by exploiting, exterminating and extracting from other bodies, lives, spaces and ecosystems in the "New" world. Colonial discourses of the inferiority of native populations were key to the success of the colonial enterprise that began with the Age of Exploration.

It was mass-media discourse that represented and therefore "made real" the New World as "new" to its audiences in sixteenth-century Europe—although I suspect that its Indigenous people would certainly not have considered their world to be much of a novelty. These "othering" practices continued and still continue today—after all, it was the circulation of images (later on, via mass media) that made the Global South the "Third World" in postcolonial relationships between richer and poorer nations. Independent postcolonial nations of the Global South are now outcompeting their historical colonizers in othering and commodifying the multispecies lives of their own multispecies citizens in the same mode of development, even within democratic frameworks. The ubiquitous presence of such forms of internalized colonial autophagy demonstrates the deep penetration of ecocultural colonial mindsets that view naturalcultural lives, bodies and ecocultural systems as resources rather than relationships.

During the inception and rapid growth of global colonial land grabbing, Euro-American circulation of image-and-word narratives of "othered" peoples' and ecosystems' inferior and/or exotic/inhuman/bestial qualities were fundamental to successful and systemic exploitation. The effective projection of the inferiority of native peoples as "scientific fact" was decisive in the global tussle between colonizing powers and colonized people for whose knowledge, understanding and perspective on the world would win. Deployed for and by European colonial populations through newly affordable printed words and images, these pseudo-empirical narratives of the inferiority of one group versus the superiority of another created a sound mediatic-cultural ecosystem to morally justify atrocities for profit. Indeed, the history of the biological sciences and their mania for employing classification systems as a way to create order and knowledge about newly discovered species has finally been studied for its many entanglements within the economies of colonial empires.[5] Robert J.C. Young describes how colonial obsession with racial classifications and racial purity was intimately connected to psychic infatuation with placing relative *value* and *desire* upon the newly subjugated local populations:

> It was through the category of race that colonialism itself was theoretically focussed, represented and justified in the nineteenth century, it was also through racial relations that much cultural interaction was practiced. The ideology of race, a semiotic system in the guise of

> ethnology, the "science of races," from the 1840s onwards necessarily worked according to a doubled logic, according to which it both enforced and policed the differences between the whites and the non-whites, but at the same time focussed fetishistically upon the product of the contacts between them. Colonialism was always locked into the machine of desire [...] Folded within the scientific accounts of race, a central assumption and *paranoid* fantasy was endlessly repeated: the uncontrollable sexual drive of the non-white races and their limitless fertility.[6]

The "machine of desire" described by Young is fundamental in understanding the cyclical and looping nature of colonial command economies—its turbines being the affective and discursive generation of an imagined surplus which is "there for the taking." Note also how these kinds of colonial fantasies are called "paranoid," by Young, as I will bring up the concept of both fantasy (magical thinking) and paranoia in Chapters 3 and 4. The race-based psychic and cultural obsessions analyzed by Young, which were created and reinforced through "classificatory discourses," are well demonstrated by examples such as the *casta* system in colonial Latin America (Fig. 2.1). These systems were deeply connected to rationalizing discourses that ultimately bolstered economic obsessions and the subjugation of lives for profit via a split and oscillating discourse of opposite affects: racialized disgust and desire.

Notice how the placement of the bodies moves from whiter bodies on the top left to the blackest bodies on the bottom right. The specifically vertical ordering of the bodies via skin color in the illustration above shows not only the obsessive classification of races into minute gradations, but also reveals the hierarchical ordering of these classified lives. Such vertical hierarchical structures of mapping the complexity of life are indirectly mirrored even today via the "trophic pyramid" that depicts ecosystems in school biology textbooks, where the "apex predator" is supported by a variety of organisms below it (Fig. 2.2).

Vertical and hierarchical ordering of life via Linnaean classification systems serves pragmatic purposes. These classification systems are elaborate, moving from the largest of categories (domain) through progressively smaller categories of similarly related forms of life (kingdom, phylum, class, order, family, genus, species and lastly, in some cases, sub-species). The Linnaean system is useful in listing and tracking the incredible diversity of life that our planet harbors. It also allows kinship bonds to be visualized, since the further down one goes on this scale, the more "closely related" organisms are to those in the same category. Perhaps even more importantly, for colonial strategies of Othering, such classificiation systems can be harnessed and curated to narrate Otherness, both human and non. For example, in a study on European conceptualizations of New World natives after 1492, Marcy Norton persuasively describes how culturally divergent ways of relating and interacting with animals became a primary catalyst and tether that bolstered colonial narratives of an exotic Indigenous Other.[7]

Figure 2.1 *Las castas.* Casta painting showing 16 racial groupings. Anonymous, Museo Nacional del Virreinato, Tepotzotlán, Mexico. (Wikimedia Commons)

Trophic levels help visualize the basic ecological direct relationships between organisms in a food chain in any given ecosystem. However, there are three important problems with the often simplistic cultural interpretation of these linear and vertical visualizations of relationships between multispecies life: 1)

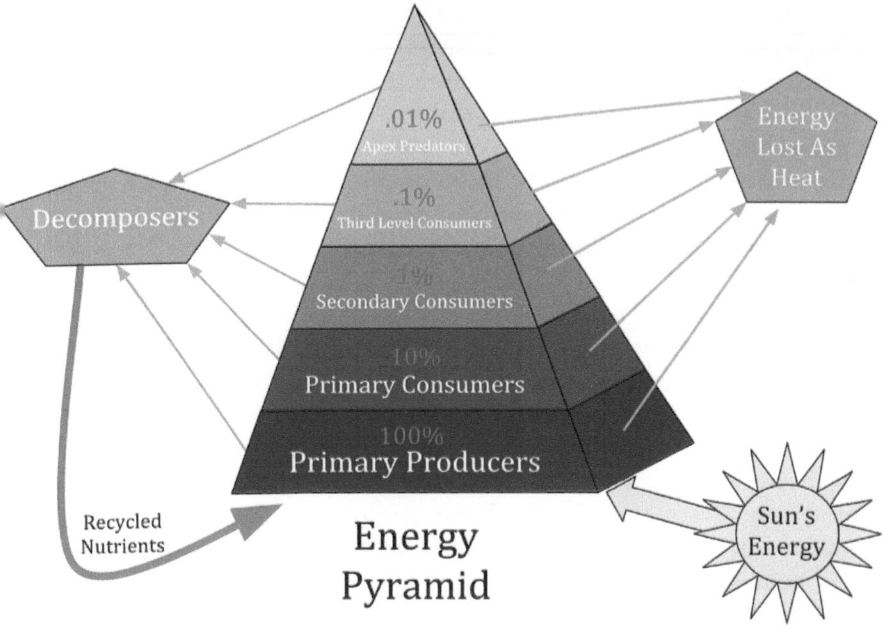

Figure 2.2 Trophic pyramid depicting ecosystem energy exchanges between different levels of life. (Wikimedia Commons)

they create a semblance of a rigid pecking order of superior-versus-inferior beings, with top predators and more "complex" creatures valued more than those at the more "primitive" bottom of the chain, 2) they instate a broader discourse that naturalizes and justifies multispecies colonial domination and optimized profit. Some of its commonly used catchphrases are strategically improper applications of Darwinian theory, such as "survival of the fittest," "the big fish eats the little fish," "alpha male," 3) By focusing on distancing and separation, they render completely invisible the complex web of intra-activity that actually exists in ecocultural systems, with hundreds of "unexpected collaborations and combinations" of indirect effects, cascading effects, feedback loops and positive or negative reinforcements. This is why they are thin-accounts of naturalcultural spaces. Hierarchical knowledge structures such as the examples given above are the foundation for the capture-and-control of multispecies lives deemed inferior, both human and non. Crucially, they all focus on classification and differentiation visually, rather than on the deeply intra-active connections between various "levels"—those direct and indirect ecological and ecocultural effects we have already discussed

It would be useful at this point to consider all these illustrations under the umbrella mediatic genre of the *map*. We could define all naturalcultural art and media, i.e. creative representations of our worldviews vis-à-vis other bodies, lives and ecosystems (human and non), as different forms of *map-making*: these include actual maps, but also Linnaean classification charts, travel literature, letters, paintings or journalistic accounts. Multispecies thick-maps, in this context, propose themselves as ethical glitches that operate within this media supergenre. Indeed, colonial epistemic and economic "conquests" were strategically couched not only via images of hierarchically ordered life (a map of life), but also within weighty intellectual discourses (for example, those of faith) in order to gain complete legitimacy. For example, the papal bull promulgated by Pope Alexander VI in 1493 grants legal rights to newly discovered lands in the West so that "the Catholic faith and the Christian religion be exalted and be everywhere increased and spread, that the health of souls be cared for and that barbarous nations be overthrown and brought to the faith itself."[8] These bulls (or edicts) were not simply abstract articles of faith, but used as *legal precedence* to displace, disinherit and render homeless millions of Native Americans. In 1823 the Supreme Court of the United States forged these articles into a legal concept called the *Doctrine of Discovery* to dispossess Indigenous communities from their lands. These original papal bulls were formally repudiated by Pope Francis in 2023, exactly a year after his visit to Canada where he met several representatives of Indigenous tribes.[9] These examples of psychic, cultural and political discursive "tricks" certainly do not lead to the conclusion that *all* matters of colonial science, law, faith and philosophy are implicated in oppressive systems of multispecies domination. However, understanding how pseudo-scientific narratives or the strategic unethical application of cultural and scientific principles were weaponized to capture-and-control, racialize/bestialize and "other" colonized peoples and ecosystems is key to connecting colonial epistemologies and affects with constitutive actions of ecosystemic pillage. After all, dispossessing subhuman or barely civilized humans of their freedom, dignity, land, culture and ecosystems was (and still is) far more palatable than doing so to lives of equal value to "us." One recurring illustration of such a multispecies "othering" story is that of the *Blemmyes*—mythical creatures with faces growing out of their chests that were said to live in Africa. Fig. 2.3 shows an engraving of these fantastic humanimals in a 1603 German edition of Sir Walter Raleigh's *The Discovery of Guiana*, 1595.

These headless humans, probably described for the first time by Herodotus in his *Histories*, make a very curious return in Sir Walter Raleigh's travel accounts of his voyage to Guyana:

> There is also another goodly river beyond Caroli which is called Arui, which also runneth through the lake Cassipa, and falleth into Orenoque farther west, making all that land between Caroli and Arui an island; which is likewise a most beautiful country. Next unto Arui there are two rivers Atoica and Caura, and on that branch which is called Caura are a nation of people whose heads appear not above their shoulders; *which though it may be thought a mere fable, yet for mine own part I am resolved it is true, because every child in the provinces*

Figure 2.3 An engraving of the fantastical "Blemmyes" in a 1603 German edition of Sir Walter Raleigh's The Discovery of Guiana, 1595. (Wikimedia Commons)

> *of Aromaia and Canuri affirm the same.* They are called Ewaipanoma; they are reported to have their eyes in their shoulders, and their mouths in the middle of their breasts, and that a long train of hair groweth backward between their shoulders.[10]

Language is manipulated carefully here ("I am resolved it is true" and "every child can affirm the same") to gloss between observed fact and manufactured fantasy/rumor/myth in order to convince the reader of the undeniably marvelous and therefore desirable and profitable nature of the new lands. This is fake news or alternative facts of the first order, used to engage and set in motion the gears of collective colonial desire. Blurring the lines between fact and fantasy, knowledge and magic was a key feature in the othering of the colonial natives and their commodification, and thus extraction, exploitation and extermination on a planetary scale.

Amerigo Vespucci was probably the greatest early professional "influencer" of them all in the European Age of Exploration, besides being an actual traveler. He claimed to have been the first to have known in 1501 that the Europeans had discovered a new world rather than the East Indies, leading cartographer Martin Waldseemüller to recognize Vespucci's accomplishments in 1507 by applying the Latinized form of *America* to a map showing the New World. This practice then spread to other maps, thus canonizing and cementing Vespucci's public relations success. The 1503 pamphlet *Mundus Novus* (New World), where he pretends to write to his patron, Lorenzo di Pierfrancesco de' Medici, would be the first of two mass-produced marketing products that Vespucci would release for wide circulation. Highly exaggerated, it contains the same pattern of strategic colonial fakery for multispecies profit that we have witnessed in Raleigh's account of the Blemmyes.

We have an initial claim to absolute truth, an exaggerated account posing as a rational knowledge claim:

> And these we may rightly call a new world. Because our ancestors had no knowledge of them, and it will be a matter wholly new to all those who hear about them. [...] But that this their opinion is false and utterly opposed to the truth, this my last voyage has made manifest.[11]

There is the magical multispecies depiction of a continent that contains an infinite surplus of biopower and ecopower (to be mined for profit):

> We knew that land to be a continent and not an island both because it stretches forth in the form of a very long and unbending coast, and because it is replete with infinite inhabitants. For in it we found innumerable tribes and peoples and species of all manner of wild beasts which are found in our lands and many others never seen by us concerning which it would take long to tell in detail.[12]

We find the same distancing and differentiating rhetorical movements of a multispecies map-narrative that will depict natives as both intensely desirable and terribly abhorrent at the same time:

> We found in those parts such a multitude of people as nobody could enumerate (as we read in the Apocalypse), a race I say gentle and amenable. All of both sexes go about naked, covering no part of their bodies; and just as they spring from their mothers' wombs so they go until death. They have indeed large square-built bodies, well formed and proportioned, and in color verging upon reddish. This I think has come to them, because, going about naked, they are colored by the sun. They have, too, hair plentiful and black. In their gait and when playing their games they are agile and dignified. They are comely, too, of countenance which they nevertheless themselves destroy; for they bore their cheeks, lips, noses and ears. Nor think those holes small or that they have one only. For some I have seen having in a single face seven borings any one of which was capable of holding a plum. They stop up these holes of theirs with blue stones, bits of marble, very beautiful crystals of alabaster, very white bones, and other

> things artificially prepared according to their customs. But if you could see a thing so unwonted and monstrous, that is to say a man having in his cheeks and lips alone seven stones some of which are a span and a half in length, you would not be without wonder. For I frequently observed and discovered that seven such stones weighed sixteen ounces, aside from the fact that in their ears, each perforated with three holes, they have other stones dangling on rings; and this usage applies to the men alone. For women do not bore their faces, but their ears only. They have another custom, very shameful and beyond all human belief. For their women, being very lustful, cause the private parts of their husbands to swell up to such a huge size that they appear deformed and disgusting; and this is accomplished by a certain device of theirs, the biting of certain poisonous animals. And in consequence of this many lose their organs which break through lack of attention, and they remain eunuchs.[13]

Vespucci's "letter" demonstrates the discursive, epistemic and affective patterns of a carefully crafted colonial form of de-situated knowledge. It contains the hallmarks of multispecies capture-and-control: the strategic convergence of fact and fantasy is magical thinking disguised as empirical observation. This is not simply fantastical storymaking and storytelling in the form of unstructured and spontaneous daydreaming, nor an incoherent madness, but a carefully constructed narrative map that offer convenient lies to combat inconvenient or mundane truths. This kind of strategic fabulation consists of the deliberate deployment of discourses that are de-situated from actual lives, bodies, ecosystems and realities to produce what Edward Said calls "knowledge of another kind"[14] to that of historical knowledge.[15] It is difficult not to conclude even via this brief exploration of colonial narratives that the diverse discursive strategies contained within primal colonial fabulations are intricate *multispecies* ones, where species and interspecies relations and characteristics of the entire chain of life (human and non) are ordered, catalogued and given meaning in order to plan for subsequent plunder. The third and fourth chapters of this book will analyze this kind of manufactured and strategic epistemological insanity via digital media in more detail.

Primal Colonial Spectacles: Multispecies Zoos, Then and Now

On July 13, 2013, Roberto Calderoli—one of the most sensationalist leaders of the anti-South, anti-immigrant Northern League party in Italy and Vice-President of the Italian Senate in the Gianni Letta government—compared Cécile Kyenge, Italy's first Black minister, to an orangutan. I will analyze this instance of Colonial-Cybercene speech in much more detail in Chapter 4. For now, I would like readers to notice how this kind of discursive rhetoric eerily echoes a comment made more than two hundred years ago, this time by a scholar, Saint-Hilaire, about another famously fetishized female Black body famously known as the "Hottentot Venus"; her Dutch-colonial name was Sarah (or Saartjie) Baartman. Her given birth name is lost to history, a victim of colonial cancellation.

On April 1, 1815, Saint-Hilaire wrote a report after studying her displayed body in which "he compared her face with that of an orangutan and her posterior to that of the female mandrill."[16] Human–animal hybrids have been a source of willed fascination since ancient times, but these kinds of fantastical and paranoid images acquired a very different biocultural agenda at the height of European colonial conquest of the Global South, best encapsulated in the phenomenon of the *human zoo*. This colonial-cultural phenomenon has begun to be studied in all its complexity as an important site of the creation of colonial epistemologies of exploitation,[17] but not much work has been done to flesh out its genealogical links to contemporary Cybercene otherings or its profoundly multispecies co-becoming, which I will do in the next chapter. I would like to classify and interpret historical multispecies zoos as a unique *mediatic genre* of colonial discourse and imagination. If it is indeed a genre, then one should be able to reveal its characteristics, properties, affective functions and epistemic strategies. These parameters can then also be compared to modern digital genres of display, viewing and meaning-making—metaphorical zoos where multispecies lives can be safely accessed and controlled. With regard to actual zoos today, Haley N. Beer et al. discuss how their cultural role in contrast to the historical colonial genre has "expanded beyond entertainment and leisure, as present-day institutions are expected to fulfill missions of conservation and public education in addition to recreation."[18] This evolution involves continued work by organizations such as the Association of Zoos and Aquariums (AZA) to focus on animal welfare and care as well as conversation in situ and exsitu (in the wild) of species listed in the Species Survival Commission of the IUCN (International Union for Conservation of Nature).[19] Since zoos, nature parks and other such urban and semi-urban contact-zones are inevitably crucial points of wild species-connectedness in an increasingly city-focused global human population, it is hoped that humanistic fields such as Critical Zoo Studies and chapters such as this one can further the conversation on how to continue the gradual evolution of these spaces into sites of multispecies care rather than locations of highly controlled spectacle.

Chokri Ben Chikha and Karel Arnaut situate the historical colonial spectacle of the human zoo within a specific time period, just before the dawn of mechanical reproduction of exotic images of simultaneous entertainment and racial othering:

> The human zoo originated during the 19th century, prospering widely in Europe, the United States (US) and beyond, until the 1930s, which saw other mass media (radio, sound cinema) became the locus of transnational popular culture within which new 'exotic' genres (*revue nègre*, jazz, etc.) rapidly gained popularity.[20]

Chikha and Arnaut describe how "staging" and "caging" are important frameworks to understand how othering discourses were created within these spectacles:

> human zoos functioned in a panoptical system of distantiation (between the superior viewer and the subaltern "viewee") and of differentiation (among categories of viewees, i.e., the "tribal" or national "troupes"). [...] The human zoo thus operated in a process in which "staging" and "caging" are conjugated.[21]

The authors describe the stylized performative aspect expected from the zoo's subjects by their viewers as the "staging" characteristic of these zoos, while the "caging" feature contains both material/embodied (the subjects are literally caged or restricted in space) and discursive dimensions (for example, the enclosure may be called a "village"). It is important to note how "seeing" in this highly curated form establishes epistemic and affective distancing. Mediatic control is carefully calibrated so that "looking back" is minimized or avoided. It is for these kinds of commonalities in strategy that colonial forms of imagined multispecies geographies and Cybercene othering forms of digital infotainment (understood via my use of the term capture-and-control) can be analyzed together through the "staging-and-caging" framework of the human zoo. Both place importance on the value of *the spectacle* in providing efficient and impactful storymaking and storytelling. The visual narration of these situated yet de-situated multispecies colonial spectacles serves in very careful and strategic ways to distance and differentiate—important affective and psychic prerequisites that prefigure unethical colonial plunder. The most powerful aspect of the colonial spectacle is also its most pernicious—the de-situation of knowledge *while providing a semblance of situatedness* in the staged and caged bodies, along with naturalized settings to even mimic the ecocultural systems of these bodies' origins. In essence, this is the miracle of the carefully curated colonial spectacle: multispecies lives are irrevocably distanced while providing a semblance of deep intimacy.

I find it useful to reconsider the term *human zoo*. Reformatting it will help us better understand the spectacle of caged-and-staged humans within an entire panoptical assemblage system of multispecies beings that were actually on display, juxtaposed next to each other. Indeed, these were *multispecies zoos*, including both humans and nonhumans in strategic thin-accounts of colonial narration. Gilles Boëtsch and Pascal Blanchard tie human zoos to an older tradition of mediatic-epistemic organization—the *wunderkammer* or "cabinet of curiosities." Fig. 2.4 shows a seventeenth-century work depicting one of these popular cabinets by Andrea Domenico Remps, an Italian painter of German or Flemish origin.

> Cabinets of curiosity, which first appeared in the fifteenth century and were replaced in the nineteenth century by public collections and

Figure 2.4 Domenico Remps, *Cabinet of Curiosities*, ca. 1690. (Wikimedia Commons)

> museums, were precursors to natural history museums. [...] In this context, the idea of organizing the world according to knowledge as opposed to theology first emerged.[22]

Boëtsch and Blanchard identify four specific genres of such cabinets emerging in the European Renaissance: *naturalia* (critters and natural objects, with a special interest in monsters), *exotica* (art, plants and animals from "new" lands), *scientifica* (mostly measuring instruments) and *artificialia* (antiques, works of art). Notice how these genres first appeared in the mid-to-late 1500s—right after new worlds were being "discovered" and the printing press was beginning to spread the news about these discoveries and explorations quickly and efficiently. Notice also how nature and culture are seemingly kept separate in these collections by classifying *naturalia* in opposition to *scientifica* and *artificialia*. For the purposes of this project, isn't it curious to observe that it is precisely in the genre of the *exotica* (which I place within the realm of colonial fantasy) that the duality between nature and culture truly collapses? Indeed, these cabinets seem to indicate how intricate classification and categorization knowledge systems functioned in the careful comprehension (control and construction) of the naturalcultural world. Further studies could explore these naturalcultural connections in more detail. When pressed into the service of colonial domination, extraction and profit, these multispecies cabinets and zoos could become spectacular yet personalized sites that tethered naturalcultural fabulation. In his impactful study of zoo entrepreneur Carl

Hagenbeck, historian Nigel Rothfels unearths how animal zoos and human zoos (what he calls "ethnographic exhibitions") actually co-became as a widely popular genre of staged-and-caged entertainment. Rothfels compellingly describes how the "Hagenbeck revolution"—a term that describes the zoo entrepreneur's transformation of zoo settings from menageries where animals are viewed in small enclosures behind bars to the more "naturalistic" settings now common in most contemporary zoos—actually stemmed from Hagenbeck's visual innovations in displaying visually "authentic" exotic humans:

> As much as the people in the ethnographic exhibitions were shown as animals—were put on display in fields and re-created villages in European zoos—it was Hagenbeck's experience with exhibiting people, his success in creating exhibits which made people believe that what they saw before them was nothing less than a foreign world up close, that led him to change how we have come to imagine exotic animals. [...] Hagenbeck's world of animals would soon come to resemble the carefully constructed and reassuring exhibits that he had created for his exhibitions of Indigenous peoples.[23]

Rothfels' study shows the deeply co-constitutive nature of animal and human capture-and-control, all undertaken within an ecocultural context of colonial curiosity and enterprise. The capture-and-control of animals intimately co-became as a mediatic genre along with the capture-and-control of biopolitically Othered humans. Edward Said explains this control mechanism of all such colonial collections via his definition of Orientalism as "citationary realism" gone crazy:

> Philosophically, then, the kind of language, thought, and vision that I have been calling Orientalism very generally *is a form of radical realism*: anyone employing Orientalism, which is the habit for dealing with questions, objects, qualities, and regions deemed Oriental, will designate, name, point to, fix what he is talking or thinking about with a word or phrase, which is then considered either to have acquired, or more simply to be, reality.[24]

Ethnographic exhibitions that comprised multispecies assemblages of life in "natural" settings worked as a deeply affecting mediatic genre precisely because they engaged in the audiences' yearning for interactive authenticity with new and exotic realities. Rothfels describes how Hagenbeck's exhibitions of cultures were not considered performances in the traditional sense, even though they absolutely were:

> The repeatedly emphasized claims about the naturalness or true-to-life quality of the Hagenbeck exhibitions point directly to the principal

> catalyst [of these exhibitions] [...] the escalating desire of Germans—and indeed, Europeans more generally—to learn more about the world behind Europe. That desire was driven by a wide spectrum of forces ranging from banal curiosity to the practical needs of both colonial administrators and colonists themselves.[25]

In the case of the display of a singular individual such as Sarah Baartman, the same ideological citationary apparatus applies for the precise discursive use of the images produced. One body becomes the tether upon which a whole epistemic, affective and discursive framework is hooked. Each body, even when viewed individually, enters the colonial machine of desire to be placed psychically within a personalized cabinet of curiosities.

Given this control function of collections of all kinds, it should be no surprise that both human and animal zoos were first conceived of as *salvage projects*—installations that would provide a final ark for the cultural and biological extinction of endangered multispecies communities:

> Cultural death is a process which, in the words of Barbara Kirshenblatt-Gimblett (1998: 176), consists of the "transvaluation" of cultural practices and products "once they are safe for collection, preservation, exhibition, study, and even nostalgia and revival". [...]
>
> Seen in this light, historical human zoos are much like curiosity cabinets: loci of "collecting and rehearsing strange ways, tongues ..." (Mullaney 1983: 53) while human zoo characters such as Sarah Baartman or Ota Benga "represented a 'lost' part of the past".[26]

Viewed through this lens of highly strategic discursive and affective spectacularization of the colonized Other, we can see how such embodied colonial mediatic spectacles can serve to distance the bodies/lives/ecocultures of the Other from the spectator/consumer. They are made safe as subjects by not being ascribed a contemporary existence; their coevalness is denied. Moreover, such spectacles can also normalize a contradictory discourse of *both* surplus *and* extinction, rendering these othered lives, bodies and ecocultural systems both expendable and consumable even before they have actually been exterminated. In this way, not only were these bodies silenced and ventriloquized, but the categories of life they were classified into were consigned to total domination and extermination. The multispecies colonial spectacle, in other words, served and still serves as an ecocultural desire-map, a fetish-substitute for actual colonized lives, bodies and ecosystems, in order to carry out epistemic, affective and psychic maneuvers that make plunder palatable.

While this kind of surplus and extinction map-narrative was key to colonial wishful thinking that these lands could be emptied of unwanted lives, the spectacle of the Hottentot Venus, as Boëtsch and Blanchard argue, marks a change in the exhibiting of humans. This change is key for us to better understand the engagement of Young's "machine of desire" in our analysis of colonial mechanics. It is here that the "surplus function" of the spectacle becomes even more evident:

> The arrival of the "Hottentot Venus" in Europe in the early nineteenth century (in London, then in Paris) marked the beginning of a new way of thinking about the Other in the West.[27]

Sarah Baartman's distressing story shows how embodied visual fantasy/desire, science, colonial interests and spectacular entertainment combined in public spheres to play with the notion of human–animal comparisons and multispecies abundance-and-scarcity, paradoxically confirming and reiterating a discourse of the strict hierarchical separation of species and "races." It is no surprise that the human zoos coincided with the height of colonial expansion, as the creation of colonial naturalcultural epistemologies were the *raison d'être* for these spectacles. The way in which Ms. Baartman became an object of both fear and desire is a textbook case of how Homi K. Bhabha's *stereotype-as-fetish* works to capture-and-control any identity that eludes such capture.[28] Carl Linnaeus, whose structured system for classifying all biological life we have already outlined, extended his categorizing (and commodifying and othering) to human beings with his concept of raciology:

> In the tenth edition of *Systema Naturae*, he divided *Homo sapiens* into six "varieties": *Homo monstrosus* (dwarfs, giants, macrocephaly, etc.), *Homo ferus* (nakedness, savages, four-legged beings), *Homo africanus*, *Homo europaeus*, *Homo americana*, and *Homo asiaticus*.[29]

We can see here that right from the beginning of modern biological discourse, othering fantasies directed towards humans co-became with the othering of other kinds of life on the planet. These were not simply cultural, but naturalcultural and multispecies discourses. The new field of anthropology did its part in the study of Ms. Baartman by placing her as a missing link between primates and humans via the creation of the myth of the lost "Hottentot" race, hence her name. Deeply connected to these biological notions were the imagined cultural and intellectual attributes of these racialized black and brown bodies:

> In [the anthropological] model explaining the difference between human "races," things like intellectual and artistic aptitude, political organization, and technical progress were conflated with physiology and the particular morphology of each human "race."[30]

Institutionalized bad science and cultural entertainment worked sympoetically together to foster fertile colonial multispecies epistemologies, filled with fantastic "beasts" and "savages" untethered from their ecocultural systems, all while claiming an authenticity that was so deeply craved by the public. Rothfels notes how Hagenbeck's ethnographic exhibitions similarly profited from larger crowds by receiving a stamp of approval from anthropological societies.[31] He also describes how these exhibitions served not only as a genre of entertainment, but as serious data-gathering centers where the "savages" were carefully measured to map their inferiority vis-à-vis the European "races":

> In the case of anthropologists and anatomists, the exhibits were seen as opportunities to carry out scientific study. In general, the scientists placed very few limits on their investigations: interests in the physical differences between races led to comparative studies of genitalia: interests in the origins of religious beliefs led to often aggressive and invasive studies. […] The right to observe, to stare, which was the ultimate point of the exhibits, was also not limited to the scholarly community. The public, from children to adults, poked, prodded, harassed, seduced, stared at, and laughed at the people in the ethnographic exhibitions.[32]

Through the biocultural mediatic mapping-spectacle of such multispecies displays, both elite scholarly study and broader entertainment joined forces to confirm colonized and racialized categories. The lure of the display, and the body's abilities to provoke fear and desire at the same time, served to situate paranoid knowledge that was actually de-situated and fantastical. Surplus, desire and visions of profit intermingled in these cruel, fertile and febrile naturalcultural spectacles.

Bhabha's vision of colonial power via his description of the stereotype-as-fetish allows us to understand the psychic mechanisms of how such cruel racializing spectacles work and for the machine of desire to run amok. Power for Bhabha actually becomes far removed from that of the all-authoritative subject, such as the colonial experts trying to discursively capture-and-control Ms. Baartman's body and life. Colonial power (or, in the case of the Colonial-Cybercene, any dominant power that produces media content), is charged for Bhabha with disquiet and menace from within. He insists that to locate this disquiet, the study of the structures of colonial discourse must be complemented by an analysis of subjectivity and consciousness, just as we have begun doing via this naturalcultural discourse analysis of multispecies zoos. The key Bhabhian term that characterizes the stereotype is *ambivalence* (recalling the believing yet not believing play of the spectator/consumer of cyberscenes). A duality of derision and desire marks colonial discourse; there is in the colonial subject a splitting (wanting to be oneself but also like the other) and a doubling (a desire to be in two places at once). This presence of constant ambivalence, created by splitting and doubling, is expressed in the compromise that the colonial subject arrives at—the stereotype.

For Bhabha, a colonial stereotype such as the Hottentot Venus or the ethnographic zoo-assemblage, like the fetish, provokes the subject to vacillate between delight in the well-known and comforting, and fear of or contempt for the exotic and unknown. The lives and bodies of those such as Minister Cécile Kyenge (figuratively) and Ms. Sarah Baartman (literally and figuratively) were simply not allowed to leave their metaphorical cages of exclusion by the many exoticized depictions of them as "almost-human, but not quite."[33] Frantz Fanon's famous colonial primal scene describes visual recognition of the point of identity ("Mama, see the Negro! I'm frightened! Frightened! Frightened!"[34]). However, this fixing of colonial identity via visual citation also sets up the seeds of its own failure, as the visual matrix upon which this fixing is based is also the problem for its attempted closure. Thus, the colonial stereotype works as Freud's fetish, in that it requires the simultaneous existence of contradictory beliefs. The racist colonial stereotypes of the "native," the "negro" or the "monstrous beasts" set up a repeated cycle of readily accessible characteristics, and yet it is those very features that provide the gateway to dismantling these images.[35]

To summarize: for Bhabha, the othering effect of the colonial stereotype contains a central ambivalence that is also the fundamental way in which it gains its currency. Bhabha contends that these colonial representations *must be anxiously repeated*, as they cannot be fully proven. Stereotypes, which mask fears and negotiate authority, therefore need to be compulsively repeated—thus connecting to my definition of colonialisms as desire machines run by command economies seeking endless "surplus." This is exactly how the unexpected and hidden cascading effects and positive feedback loops between the "seen" (the caged-and-staged body) and the "scene" (the colonial meaning forced upon it) work.

Going back to Ms. Baartman's tragic story, we know that even after her death she continued to be obsessively and brutally exploited in order to anxiously reinforce the stereotype of the Black woman as both savage and sexually desirable. The feedback loop continued to extract constructed meanings from her dead body, as presented in 1817 to the *Académie de Médecine:*

> During that presentation, [Cuvier] confirmed Saint-Hilaire's impressions, agreeing that "races with depressed, compressed skulls are forever condemned to inferiority." Cuvier took Venus's shape—particularly her apron—as proof of a *primitive* sexual appetite in African women.[36]

Sarah Baartman's body and its representations continued to be used long after her death to reinforce the colonial desire machine in an obsessive and repetitive feedback loop of stereotyping. In April 2002, her preserved remains were repatriated to South Africa, almost 200 years after she was taken in 1810 by a British ship's doctor, William Dunlop.[37] In the 1990s, French president François Mitterrand made a personal promise to Nelson Mandela to return her remains. But it was only in 2002 that the French parliament overcame objections about the precedent it would set for countries seeking the return of artefacts, by voting to return Baartman's remains as a "duty of memory."[38]

Interpretation of caged-and-staged objects and images (from dissected body parts to entire multispecies zoo enclosures) was key in the "scientific" conclusions reached by scholars about a whole continent-wide group of diverse peoples. As noted above, the fetish "allows for the possibility of simultaneously embracing two contradictory beliefs [...]. Its knowledge 'value' lies in its orientation as a defense towards external reality."[39] Just like the Blemmyes described by the "testimonial" letters of Sir Walter Raleigh and the fantastical and half-empty maps of new worlds, the fanciful multispecies naturalcultural images of multispecies zoos gain power by repetition and the spectator/consumer's desire to will them into existence as substitutes for reality.[40] The importance that Bhabha gives to the power of the "visible" clearly demonstrates the affective and cultural power of visual storymaking and storytelling via science, culture, superstition or any other form of meaning-making. These kinds of imagined multispecies geographies of colonizable and colonized ecocultural spaces—which address this space by "making statements about it, authorizing views of it, describing it, by teaching it"[41]—are prime products of colonial command economies and the fundamental basis upon which domination ("settling it, ruling over it"[42]) and material theft of living resources is justified.

The Orchid and the Wasp: The Location of Agency in Reciprocal Capture

My analysis of colonial multispecies zoos as a panoptical mapping media genre connects visual spectacles to the storymaking and storytelling of multispecies colonial desires and fantasies. This allows us to now explore the intricate mechanisms of spectacles and their colonial-cultural functions in more detail. Jean-Louis Comolli was one of the most important theorists in the 1970s to discuss the powerful effect of cinematic spectacle. His understanding of the spectator's involvement in the creation of magical thinking can be very useful here, even though we are discussing pre-cinematic spectacles. Furthermore, the implications of this analysis are absolutely crucial in considering digital colonial practices in the Cybercene. It is no coincidence that in 2009, Comolli wrote a damning analysis of the effects of the digital revolution on our relationship with screen images. He argued that the digital revolution has made images so seamlessly "akin" to the real, and so ubiquitous in nature, that it is impossible to escape them or to not *implicate ourselves* in our disavowal of the artifice of image production. In other words, as consumers and spectators of the pixel-perfect spectacles of today, we are not only actively suspending our disbelief, but also *allowing* for these images to substitute actual bodies, lives and ecosystems:

> all the more active for having been so successfully repressed, for belonging to a grey zone which we do not want to know about, for disturbing us more than seducing us, for having fallen under a

> contemporary malediction that a "theory" of the cinema can be useful and practicable. It is true that the spectacle, given over to entertainment, abhors "thought", and even revels in expelling all forms of thought from its temple.[43]

In other words, the paranoid, de-situated examples of colonial otherings described above, or daily instances of tech-based magical thinking today, are both facilitated by *the spectator/consumer's disavowal* of the very nature of these images and narratives as representations. Spectator/consumers are allowing de-situated spectacles (such as fascinated-and-horrified gazing at the Hottentot Venus' body) to substitute their own situated ways of observing, dreaming, imagining and interpreting the world around them. Their very ability to think critically and create independently has been effectively constrained and outsourced.

To better understand this willed disavowal, Comolli uses a very impactful metaphor—that of the lure. He describes how in order for a fishing lure to work, the fish has to actively take it, which is very different from hunting with a trap or gun. In terms of understanding spectacles as lures, it means there must be active participation by the spectator/consumer for the lure to accomplish its task. According to Comolli, faced with the spectacular and efficient colonial othering image, we take the lure by simultaneously believing and disbelieving in order to accept the substitution of the image for the actual. For him, simple mimetic realism is not enough—the lure requires spectatorial investment. Crucially, Comolli highlights this strong capacity of the spectator/consumer to disregard the artifice of the technological apparatus that produces these images by giving an example from silent film, a medium that is quite removed from the realism of sound film, or the HD images of cyberscreens:

> It is at the price of a series of blind spots (of disavowals) that the silent image could be accepted as the objective reflection and duplication of "life itself": the denial of color, stereoscopy and sound. Founded on these absences (just as, incidentally, all forms of representation are founded on an absence governing them, on a fault-line lying at the basis of every simulacrum – the spectator, in any case, very well knows about the artifice, but prefers to believe in it all the same), filmic representation could only be produced, as I have already noted, by working to attenuate their effects, and even masking their real existence. Otherwise, it would have been refused, as being too visibly factitious: it was absolutely necessary for it to facilitate the disavowal of those veritable sensorial castrations which established its specificity, and for it not to stymie this process by re-marking them. Compromises were needed in order for the cinema to function as an ideological apparatus, in order for it to act as a lure.[44]

What is striking in Comolli's thought is that the actual medium of image production and reproduction is not that important (it could be the heavily stage-managed artificial setting of the multispecies zoo, or the incoherent strung-together rants of a populist politician at a public rally). This underlines my contention that the psychic processes that are so powerful in the creation of a semblance of reality via identity constitution are not new, but have existed since the time of silent cinema, or even, I would argue, since the beginnings of art in human history.[45] Octave Mannoni's much-noted formula for how this disavowal works psychically is useful to keep in mind: "I know very well, but all the same...."[46] In short, spectatorial agency in this entire process is key.

Of course, the indexicality and movement of cinema, which lend its images additional realism when compared to paintings or other static forms of art, make such disavowal easier and more seductive. Ultra-HD screens connected to the internet and transmitting "real time" happenings from around the world provide another layer of realism, and appear hyperreal. However, I posit that ultimately, these epistemic transformations provide only an acceleration of substitution rather than an actual innovation in our relationship with images. The "lure" of the image has simply become more efficient and easily available, but spectator/consumer disavowal is fundamental to the process of substitution. The spectator/consumer must be willing to disregard the artifice of the medium to buy into all its promises. To go back to the example of the multispecies zoo: the lure of multispecies bodies and props displayed in "natural settings" functioned as a spectacle and deliberate othering provocation, conjuring the image of an oft-repeated colonial multispecies stereotype of surplus/lack and fascination/horror that charged the machine of desire repeatedly to gain currency. I would like to suggest that the acceptance of this disturbing lure by the spectator/consumer rests upon their interactivity and response/playing along with this image. The spectacle is actually a site and sight of ecocultural intra-action rather than a unidirectional system of meaning-making. The bodies and lives that are affectively, discursively and epistemically captured-and-controlled via the naturalcultural mediatic spectacle become what they are—symbols of an entire racialized and sexualized colonial ecocultural system. Meanwhile, the violent capture-and-control of meaning these spectacles engage in also congeals or fixes the spectator/consumer into the role and identity of what they become (colonialists), due to their very engagement in colonial mediatic intra-activity. A good term to describe this constitutive intra-action is "reciprocal capture"—as coined by Isabelle Stengers and cogently described by van Dooren, Kirksey and Münster:

> The intimate relationship between a flower and its pollinating bee is one in which both forms of life are shaped and made possible through a shared heritage, an entanglement that Isabelle Stengers characterizes as "reciprocal capture." As such, they do not just happen to meet each other, this bee and this flower; rather, their relationship emerges from

> coevolutionary histories, from rich processes of cobecoming. This cobecoming involves the exchange and emergence of meanings, immersion in *webs of signification* that might be linguistic, gestural, biochemical, and more.[47]

Fig. 2.5 illustrates an example of such naturalcultural co-becoming via visual reciprocal capture; the Common Wasp Orchid (genus *Chiloglottis*), native to Australia and New Zealand, mimics wingless female wasps to attract male wasps for pollination through sexual deception. The insect-like callus on the labellum (lip) looks like a wasp's body, tricking males into attempting to mate with it.

Figure 2.5 Natural "fake news": A Common Wasp Orchid (genus *Chiloglottis*), native to Australia and New Zealand, mimicking a wingless female wasp to attract male wasps for pollination through sexual deception. The insect-like callus on the labellum (lip) looks like a wasp's body, tricking males into attempting to mate with it. (Wikimedia Commons)

"Reciprocal capture" stems from Deleuze and Guattari's notion of "double capture," as illustrated by the wasp and orchid's co-becoming:

> How could movements of deterritorialisation and processes of reterritorialisation not be relative, always connected, *caught up in one another*? The orchid deterritorialises by forming an image, a tracing of a wasp; but the wasp reterritorialises on that image. The wasp is nevertheless deterritorialised, becoming a piece in the orchid's reproductive apparatus. But it reterritorialises the orchid by transporting its pollen. Wasp and orchid, as heterogeneous elements, form a rhizome.[48]

The wasp and orchid's co-becoming is not simply characterized as a mimesis, but one of relational co-becoming. It is not about imitation but about

> a wasp-becoming of the orchid, and an orchid-becoming of the wasp, a double capture, since "what" it becomes changes no less than "that which" becomes. The wasp becomes part of the orchid's reproductive apparatus at the same time as the orchid becomes the sexual organ of the wasp.[49]

Like the orchid and wasp, the colonizer is caught up in the colonial spectacle just as much as the colonized body is. They are both constituted within their respective roles by an intra-active relationship, a cascading effect. Their respective identities co-become. By utilizing this intra-active concept of reciprocal capture to describe the colonial spectacle, I do not intend to represent equity of power relations. It is undeniable that an extremely violent and asymmetrical power dynamic exists between the capturer and the captured, between the colonial gazer and the captured and gazed-upon body in the naturalcultural colonial scenes described above. What I am alluding to in this distressingly skewed power dynamic is the fact that the intra-action is mutually constitutive of the identities of *both* parties involved—the spectator/consumer and their intended victim. The Blemmyes, cannibals, and black and brown monstrous multispecies entities do not exist, yet they can be made to exist and they "keep public interest" by and with the right discursive and mediatic apparatus and ecosystem. The staged and caged objects of the spectacle are artfully and artificially arranged and controlled, generating a performance under horrific suffering and duress. The colonial subjects viewing the spectacle also become who they are (colonialists) via a kind of naturalcultural mediatic co-becoming—a collection of shared meanings, discourses and affects that allow a performance of power. Fundamental to this discursive-mediatic apparatus of reciprocal capture via spectacular storymaking, storytelling *and story-receiving* is *the misrecognition and disavowal of the artifice of that spectacle* by us as spectator/consumers. The constructedness of the colonial scene must be disavowed for the citation to be seen—for it to hold affective salience and epistemic authority.

I have to remind myself often that the conscious and unconscious multispecies colonial maneuvers described above, achieved via mediatic and spectacular storymaking and storytelling, but also, importantly, story-*watching* and story-*listening* are not just abstract fantasies but serve very practical ecocultural purposes. Firstly, they *make palatable* the highly distasteful colonial capture-and-control of multispecies lives, bodies and ecosystems. Secondly, these colonial narratives are actually constitutive of the identities of both colonized and colonizing agents. All this *seemingly situated but actually de-situated* gazing at and consuming of strategically produced spectacular representations of bodies, lives and ecocultural systems helps create increasingly optimized, frictionless and distanced means of denying ongoing cruel plunder, forgetting past instances of it, and joyfully partaking in planning future desecration in the name of some imagined justification. Like the orchid and the wasp, the colonial spectacle becomes them as they become the colonial spectacle, via reciprocal capture.

Today, colonial violence against our multispecies worlds (human and non) continues unabated, yet the repressed psychological need for multispecies connection remains, hidden under the surface. Contemporary spectator/consumers increasingly try to meet this need via artificially reproduced and mediated "environments" rather than "messier" direct relationships. It is a futile and counterproductive strategy. Photographer Zed Nelson's collection entitled "Anthropocene Delusions" examines both the desire for and the ineffectiveness of "natural" environments artificially recreated by humans—modern versions of colonial multispecies zoos. I interpret the naturalcultural environments represented in Nelson's collection as an exemplary reflection on the affective and psychic need to deny and disavow the actual lack of nature-connectedness humans increasingly face, firstly and historically with mass urbanization, and now with the advent of our digital screen lives. Every photograph in this project epitomizes places where de-situated naturalcultural spectacles pose as situated ones. The photography collection shows picture after picture of surreal scenes of staged multispecies capture-and-control, in holiday resorts, theme parks, greenhouses, natural history museums and "safari" parks among other locations, but also certain set-ups in larger locations such as national parks.

One of Nelson's photographs from Kenya shows an "*Out of Africa* champagne picnic experience."[50] "The safari outfit lays out the rug, carries an ancient gramophone player, puts the champagne ice bucket, brings a Maasai warrior to stand close by to add authenticity to the scene [...]. The Maasai is paid. That particular guy, he's a super nice guy, I'm Instagram friends with him now."[51] These spaces of spectacular fakery and nature-nostalgia can be understood as contemporary mediatic legacies of the multispecies zoos of the past. They carefully manage, "stage and cage," the bodies, lives and ecosystems we have fully colonized (exploited, extracted and exterminated) for profit, but still have an innate need and desire to access. After visiting locations in fourteen countries in four continents, Nelson asks: "Is this a comforting alternative reality to mask our shattering impact on the world around us? Are we engaged in a project of grand self-delusion?"[52]

My brief archeology of colonial multispecies capture-and-control leads me sadly to conclude that the "Anthropocene delusion" has actually accelerated our recourse to "alternative realities" and grand self-delusions in the Cybercene. Our psychological needs for nature-connectedness are monetized by corporations not only in such artificially constructed physical spaces, but also, I would argue, via the ubiquitous banality of so many "UHD 8K" (Ultra High Definition) screens in our living rooms playing picture-perfect spectacular scenes of multispecies abundance via nature documentaries. The spectator/consumer's implication in this great delusion lies in the reciprocal capture mechanism of these kinds of spectacle, both on and off screen. Capture-and-control by domesticating nature into screen-based representations in our living rooms is so relentless, pervasive and normalized that the bodies, lives and ecocultural systems portrayed can actually be classified as without any intra-activity with the spectator/consumer. Following his tours for his photographic project, Nelson keenly observes:

> I realized a strange thing: in these places, nothing happens. There are no surprises. There may be a wave machine, a volcano that puffs smoke on the hour, or a rollercoaster offering momentary thrills. But nothing changes, good or bad. Everything repeats itself [...]. Here nature is made safe – no thorns, no biting insects, flooding, or unpredictable creatures. This is nature only as spectacle.[53]

I would go one step further than Nelson's observations. Based on my media production and reception analysis in this chapter, I would like to suggest that, through reciprocal capture and the process of gradualism,[54] human societies are opening up more of these "safe" artificial caged-and-staged multispecies simulacra not only to unconsciously substitute the lack of nature-connectedness, but also to preemptively wean ourselves as fully as possible from the need to engage in an actual embodied way with ecocultural systems. This substitution of de-situated knowledges and experiences for situated ones may initially be a partial one, but gradually, especially for successive generations of spectator/consumers who will have had few or no formative experiences of ecoculturally situated natural worlds, these spectacularly captured-and-controlled bodies, lives and ecosystems will be enough to almost fully—but not quite—replace intra-active co-becoming with the real deal.

For a large proportion of human communities around the globe, I would suggest that this substitution has already occurred, if we account for the percentage of the world's population living now in urbanized environments and add to it the staggering amount of daily screen time that a typical consumer experiences. If nature-connectedness is indeed passed down through the generations, then the negative feedback loop of ever-decreasing nature-attentiveness should exponentially accelerate this trend. This successful substitution in turn leaves the actual multispecies spaces and worlds open to further colonial extraction, exploitation and extermination for profit. Genuinely conserving naturalcultural spaces for multispecies wellbeing (human and non) rather than spectacular and safe consumption of caged-and-staged substitutes would mean respecting and protecting the innate multispecies agencies within vast areas of interconnected places and spaces.

How much more lucrative, though, is it to create artificial de-situated "safe" spaces of caged-and-staged multispecies interaction (rather than intra-action) so that actual spaces of naturalcultural wealth can be squeezed dry of their biopower and soulpower (affect, labor, flesh and other materials) for colonial profit? The only hope lies in exposing these mechanisms and having a system of checks and balances in place that will glitch such delusional narratives and experiences: multispecies democratic institutional structures that have independence to work on rational knowledge claims, a mediatic environment that honestly narrates daylight robbery of public multispecies commons by the influential and powerful few, and a general multispecies paradigm shift that allows people to experience and understand the joys of naturalcultural connectedness, where nonhuman agency is duly recognized, respected and valued.

3

Locating Cybercene Desires

> Over the years, I've found only one metaphor that encapsulates the nature of what these AI power players are: empires.
>
> Karen Hao[1]

In the Anthropocene, our relationships with bodies, lives and ecosystems are managed by the mapping-stories we humans tell ourselves and to each other. The Cybercene has transformed the way in which we narrate, listen to and interpret these mapping-stories.

In the previous chapter, we explored how historical mediatic colonial discourses were crafted, deployed and received, and the crushing power/knowledge they exerted and continue to exert on lives and ecocultural systems. We examined these discourses and mediatic practices as a product of a creative interplay (intra-action) of the spectator/consumers' own affective co-becoming, as well as their ego/identity creation and desires, with that of the discursive/mediatic systems themselves. These map–spectator intra-actions create oppressive, distanced, resource consumption-oriented capture-and-control mediatic relationships rather than forming sustainable, embodied relationship-based identities and identifications. We investigated how distressing colonial mapping-spectacles like the display of bodies in multispecies zoos not only made capture-and-control palatable, but also made the experience of capture-and-control the creative constitutive engine of the fetishized identities of colonized Others and colonizing Self. Via reciprocal capture, the colonial spectacle constitutes the colonizer, just as the colonizer's intra-action with the captured-and-controlled body constitutes the colonial spectacle via a two-way feedback loop. It is in the process of mediatic capture-and-control and running away that the colonial multispecies relationship is forged and reforged. All this seemingly situated but actually de-situated gazing at and consuming of strategically produced and consumed spectacular representations helped create increasingly optimized, frictionless, safely distanced means of denying ongoing cruel plunder, forgetting past ones, and joyfully partaking in planning future ones in the name of whatever positive imagined justification.

New media studies scholar Sean Cubitt argues that "at any historical juncture, the meaning of what it is to be human is constructed from the relationship between nature and technology."[2] If this is true, then what are our new constructions of human-ness (and nonhuman-ness) in a digitally connected world? To explore this question within a multispecies paradigm, I decided in early 2022 to imagine an ecocultural sub-era of the Anthropocene and the Great Acceleration and call it the *Cybercene*. Roughly encompassing the first two decades of the twenty-first century and beyond, the Cybercene is a pivotal period of accelerated psychic and actual recolonization, in which de-situated online onscreen realities, digital representations and narratives have quickly and effectively transformed identities and substituted the value of actual bodies and living habitats. It is important to acknowledge that digital tech and processes have provided innumerable positive outcomes in almost every aspect of our lives. I am attentive to these valuable changes, while also arguing that the ecocultural problems of the Greatest Acceleration arise from the fact that our epistemic, psychic, ethical and pragmatic frameworks, actions and regulatory policies have not kept pace with technological advances, to our planet's peril.

For the purposes of this book, I define the Cybercene as an ecocultural sub-era of the Anthropocene/Wasteocene in which centralized yet diffuse colonial digital command economies have captured-and-controlled multispecies lives, bodies and ecocultural systems for increased profit. The analysis of historical examples of naturalcultural colonialisms in the previous chapter will now allow me to properly describe their legacies, lines of continuity but also transformations in this strange new era. While I deploy the term as a conceit, I do hope that the Cybercene can be used both strategically and experimentally to open up a collaborative gathering space to explore some pressing questions: How can we imagine, create and restore harmonious and thriving multispecies "naturalcultural" communities regionally, nationally and globally? How and why is this restoration work crucial as the global digital revolution further transforms our relationships with lives, bodies and ecocultural systems?

Reformatted Relationships: Mangroves, Mountains, and… Cyberscreens

I was four years old. My family and I were traveling from our home in Mumbai, India, to my maternal grandparents' (then small) town of Satara. This hill-ringed town is only about 300 kilometers south-east of Mumbai (then called Bombay), but in those days, it was a daunting journey that could take anywhere from eight to twelve hours. The unreliable road would eventually leave the chaos and vitality of the urban jungle, and you would first know you were leaving it by the intense smell of rotting decay that would permeate our Fiat car through the open windows as we took the long sea bridge connecting Mumbai to the mainland. This pungent smell came from the first important natural ecosystem that our car would enter in our journey, having barely left the city proper—the biodiversity hotspot of the tropical mangrove forest. These incredible intertidal forests still surrounded the island city of Mumbai in those years, creating the ideal habitat for millions of marine, brackish and terrestrial species of flora and fauna. They also sheltered the city of (then) fifteen million people from the worst of the Arabian Sea's fury during the seasonal monsoons. Mangrove forests are important carbon sinks, while also providing the water filtration needed for the good health of seagrass beds as well as coral reefs. Despite the marvelous capabilities of this greatly threatened ecosystem, city folk mostly experienced its existence through its smell, caused by the intertidal channels being used as the megalopolis' dumping ground for all kinds of solid and effluent waste.

Once our car reached the mainland, it traversed a relatively narrow strip of coastal plains dotted with fields, then began huffing and puffing (with occasional breakdowns) up the second important ecosystem we would traverse in our journey—the Sahyadri or Western Ghats. The ghats are a thousand-mile mountain range that separates the western coast of India from its central uplands. This range is considered one of the thirty-seven biodiversity hotspots in the world, harboring many species endemic to the region and no less than *four* tropical and subtropical broadleaf forest ecoregions. During the monsoons, the car would finally make it up the sheer escarpments of the Khandala ghat while we listened to the sounds of cascading rivulets of rainwater crashing down from the ghat's emerald-green slopes. It is at one point here that, as a very young boy, my eyes opened wide in wonder as I looked outside the open window, and saw a scene of great majesty. I screamed out "*Kitna bada badbada*!"—in my child's mispronounced version of mongrelized Hindi/Marathi: "what a gigantic waterfall!" There, on the other side of a broad jungle-rimmed canyon, was an immense wall of water emptying into the gorge below with a thundering roar. My parents would always marvel at the strong emotion this majestic scene and scenery provoked in my young mind. We would stop at the side of the road at that spot so that I could gaze upon the falls in wonderment. I know now that the wonder stemmed not only from the view itself, but from the embodied thrill of realization that I shared the world with entities of such magnificence and complexity.

About thirty-seven years later, I am on a road trip in the Colorado Rocky Mountains with my teenage nephews and nieces visiting from Italy. The car is no longer mechanically challenged, nor is the road run down, allowing us to glide effortlessly over many miles in shorter times. The windows are closed to allow for a comfortable travel experience, but this also divides our bodies from what is outside and cocoons us from many sensory experiences, sights, smells and sounds. We find our way through the mountains thanks to the soft-spoken mechanical voice of Google Maps rather than having to stop and ask locals for directions, as we would often inevitably have to do in India when taking my papa's ill-conceived, but very adventurous, "shortcuts."

I had planned this trip to show the kids the natural wonders of the American West. However, some doubts nagged me.

I wondered if, from our new Cybercene perspective, our eyes would glaze over at the redundancy of the visual experience? Would it feel a bit like we had "been there, done that?" Perhaps it would feel like we had already been here before, since we all have been oversaturated with pixel-perfect *cyberscapes* and *cyberscenes*, so that we do not know of any way to look at actual mountains, valleys, forests and streams other than as a decontextualized, disposable and ephemeral collection of virtual images? In a new era where the "attention economy" is actually riddled with constant mediatic distractions, can anyone realistically cultivate the "art of attentiveness" so crucial to experiencing multispecies connection? Perhaps then, in the Cybercene, the incredible natural landscapes would only garner muted and very occasional fascination, especially since they do not involve the instant intra-active control and play of reciprocal capture that screen images allow. This certainly seemed the case for many other visitors to these places and spaces—many seemed more interested in capturing the landscapes on their smartphone cyberscreens than *being* there. Perhaps the cyberscenes now outperform the natural scenes, even though the cyberscene is a digital (re)presentation of the natural scene. Maybe we now view complex natural ecosystems through the lens, filter and value of digital representations of them on a screen, rather than the other way around. Could it be that the reproduced copy, the cyberscene, is so powerful in its convenience, ubiquity and pleasure that it becomes, unconsciously, more familiar and valued when compared to the original scene? If spectacles co-constitute both the capturing agent's and the captured bodies' lives and ecosystems via reciprocal capture, what are the mediatic and ecocultural implications of this new powerful digital intra-action on the constitution of both us and the natural spaces we were relating to through our phones?

These questions troubled me to the core of my spirit. De-situated cyberscenes are so pervasive and persuasive today that they have shaped our attitudes and visions of the actual world around us in a very deep and fundamental way. The sheer volume of digital image "slop" has submerged and superseded relation-making with actual lives, bodies and ecosystems. Perhaps so much so that the actual landscapes we were driving through had negligible emotional impact, because the original had become the copy and the copy the original?

If we are the images we consume, then continued de-situated colonial image/discourse saturation in this new sub-era of the Anthropocene must certainly influence who we are now as individuals and societies. In this chapter, I examine how intensely personalized forms of mediatic capture-and-control of meaning and affects in the Cybercene have allowed us all not just to become creative meaning-makers (we have done that kind of mediatic intra-activity since the beginning of human history), but to accelerate meaning-making via personalized media tools towards extreme control.

We now have the apparatus (personal, portable and affordable internet-connected screens) to become mini-command economies of meaning via ubiquitous cyberscreens. And if reciprocal capture is indeed true, then the turbocharged control-and-desire cyberscreen machine also recaptures us back, in intimate and potent ways. Such new mediatic authoritarian tendencies on the personal level also scale up to larger cultural groups: readers may sense links between new forms of personal cyberscreen intra-actions and the global rise of populist-authoritarian leaders and governments across the political spectrum. Many such cyber-populists have recalibrated communication brilliantly to efficiently intra-act with the powerful capabilities of the cyberscreen control-and-desire machine. Their success lies in accurately mirroring personal media-spectator command economies and thus representing a large percentage of their spectator/consumer's own mediatically shaped attitudes, desires, fears and fantasies about control. This chapter shows how such profound cultural and political effects of digital media in human communities also present direct, indirect and cascading naturalcultural connections of multispecies import.

A separate ecocultural sub-era within the Anthropocene sounds like a reasonable proposition if one considers that, in the short duration of two decades, cyberscreens have acutely co-become with us. In 2026, a staggering 70% of the world's population uses the internet, predominantly through smartphones.[3] Despite this quantitative piece of data, my task in these chapters is to look at smartphone penetration qualitatively and naturalculturally to really tease out its ecocultural impacts. I would suggest marking the start of the Cybercene around the year 2000 or 2005, since analyzing the era as an ecoculturally transformative one would mean placing importance on major spectator/consumer epistemic, affective and cultural changes rather than simply techno-scientific innovation.

The new hybrid human–cyberscreen animal is an efficient epistemic capture-and-control subject, facilitating old and new forms of conscious and unconscious anthropocentric colonial distancing and certainty rather than multispecies curiosity, connection, attentiveness and humility. If we approach the Cybercene as an ecocultural phenomenon via its primary intra-active object (the smartphone screen) and if we experience the cyberscreen via mediatic and multispecies studies mechanisms of co-becoming, colonial spectacles, reciprocal capture, intra-activity and stereotype-as-fetish, we begin to understand the roots of many odd everyday social phenomena of this era. These effects include some relatively benign trends, such as nomophobia ("no-mobile phobia"—experiencing anxiety, panic attacks and helplessness when we are separated from our phones), phubbing (the practice of ignoring one's in-person companions to pay attention to one's phone screen), and smombies (people zombified IRL by smartphone use). Other trends are far more challenging, such as epistemic and affective polarization and paranoia, disinformation, democratic backsliding, increased loneliness and violence.

The Greatest Acceleration: The Cybercene in the Anthropocene/ Wasteocene

Everyone is entangled within Cybercene desire machines, even if one does not or cannot afford to own a smart device connected to the internet, or if one is a nonhuman entity such as an animal, plant, fungus, stream or rock. I would argue that the Cybercene actually affects the "have-nots"—those individuals and communities who cannot afford internet connections and devices—even more dramatically than the "haves" who can, because the former probably reside in spaces that can be efficiently used as resources without too much consideration of the consequences. Technological ubiquity and inequity is the compounding othering-within-the-othering being carried out within the cyberdiscourses themselves, in a kind of positive feedback loop within cyber ecosystems that reinforces adverse effects. For these reasons, the Cybercene can be better understood when contextualized within two other "-cenes"—the Anthropocene and the Wasteocene. I am wary of theoretical overkill, but in this case I do believe these two -cenes together actually allow us to better explore our digital ecocultures.[4]

The Cybercene should ideally be placed within geological time scales as the newest sub-era within the Anthropocene. Originally coined by American biologist Eugene Stoermer and Dutch geochemist Paul Crutzen, the term Anthropocene is heard constantly at EH conferences, in graduate-level courses and in book titles. Perhaps the true success of the term can be gauged by the fact that it has its own Wikipedia page. Before 2024, it was defined there as "a proposed geological epoch dating from the commencement of significant human impact on Earth's geology and ecosystems, including, but not limited to, anthropogenic climate change."[5] This description changed because, in 2024, the International Commission on Stratigraphy and the International Union of Geological Sciences rejected its inclusion within the official pantheon of geological eras. Therefore, technically, we are still living in the Holocene, the roughly 10,000-year period following the last Ice Age. The dismissal of the Anthropocene from a geological/technical perspective frees us to form a second interpretation of the word that is more holistic in its approach: indeed, the Anthropocene continues to retain value in ecocultural analyses because it vividly represents *Homo sapiens*' transformational effects on the Earth's air, water, soil, ecosystems, climate and multispecies life.

The suffix *-cene* indicates that it is a proposed successor to the Holocene due to the *scale of its effects*. Even though not officially endorsed, the importance of the term Anthropo*cene* therefore lies in its recognition that human activity has caused not just modifications of ecosystems, but a deeper rupture in *planetary cycles* on a scale so immense that it can be marked within the geological strata of Earth's history by indicators such as an exponential increase of atmospheric carbon dioxide levels and the sixth mass extinction.[6] Dipesh Chakrabarty's 2009 essay "The Climate of History: Four Theses," used the term Anthropocene in a way that departed from a purely geological understanding of the word by fleshing out its societal and multispecies implications. This seminal text has spurred the growth of truly interdisciplinary reflections that have connected humanistic discussions about epistemology, ontology and justice with scholarship exploring the social sciences, natural sciences, and professional fields of inquiry. Thesis 1 defines the era as a moment of the collapse of the age-old distinction between human and natural histories. This collapse must allow for a re-evaluation of the human history of innovation and globalization (Thesis 2), and the connections of history at the intersection of capitalism and species history (Thesis 3) as well as directly confronting the challenge of species-being (Thesis 4).[7] My use of the suffix *-cene* in the term Cybercene is made with these propositions in mind, while also providing proportional value to the planetary scale of its influence. Indeed, this book's work stems from my contention that the Cybercene's implications have direct relevance to the issues brought up by the four theses.

Taking their cue from Chakabarty's four theses, scholars began to qualitatively examine the damaging effects of the Anthropocene. Several began noting that not all humans were equally responsible for or affected by its consequences. It is for this reason that I specifically choose to mark the beginning of the Anthropocene in what is called the "Orbis Spike." Simon Lewis and Mark Maslin locate the start of the Anthropocene with the expansion of European colonialism through this consequential moment in the Earth's modern ecocultural history (Fig. 3.1).[8] The Orbis spike is a small but very decisive dip in global carbon dioxide levels centered around the year 1610, found in ice cores from the Antarctic. The dramatic lowering of atmospheric CO_2 levels has been surmised to be a result of the European "discovery" of the Americas, and what is called the "Columbian Exchange"—the connection between two previously separate ecosystems of the old and new worlds and the resulting new economy. This movement and meeting of multispecies lives resulted in the decimation of Indigenous American human populations due to the introduction of new pathogens, which in turn resulted in an increase in plant and tree growth, hence lowering CO_2 levels drastically.[9] Other studies of different ice-core samples do not show the spike as a dramatic decrease, but a more gradual decline in CO_2 levels. This evidence nevertheless corroborates with "modelled scenarios of large-scale reorganisation of land use in the Americas following New World–Old World contact."[10]

In sum, the Orbis spike is an excellent choice to tether the start of the Anthropocene for the purposes of this book, since it marks a quantitative change on a planetary level due to colonial capture-and-control of multispecies bodies, lives and ecosystems. The spike highlights the human toll of multispecies colonial relationalities, particularly the unequal power relationships between different groups of people, economic growth, the impacts of globalized trade, right up to our more contemporary reliance on fossil fuels.[11] Indeed, it reminds us all that the impacts of Anthropocene damage weigh unequally on different human communities. The designation Great Acceleration

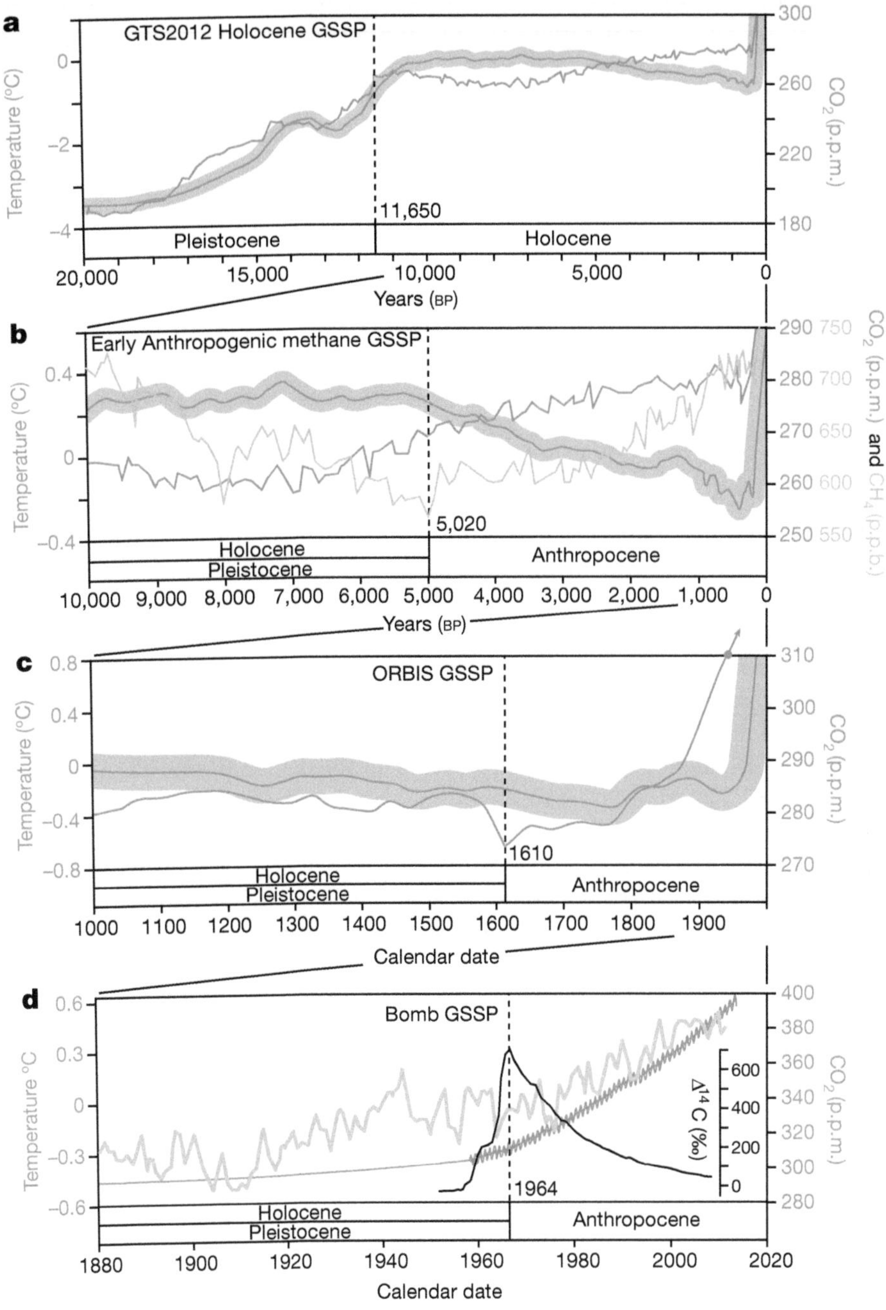

Figure 3.1 Fig. 3.1: The Orbis Spike in Lewis and Maslin's original article, proposing the end of the Holocene/Pleistocene and the beginning of the Anthropocene era. Source: Simon L. Lewis and Mark A. Maslin, "Defining the Anthropocene," *Nature* 519 (2015): 171–180. doi:10.1038/nature14258

better clarifies how the Industrial Revolution and subsequent unequal increases in *per capita* energy and resource consumption places far greater blame on communities in the Global North.

Both the Anthropocene and the Great Acceleration effectively link phenomena that may seem separate: dramatic environmental changes, unregulated capitalism, colonialism and racism; the "othering" of a group of people in order to extract resources, exploit labor and lives, and exterminate bodies and ecosystems for profit. Chapter 2 argued that cultural "othering" discourses via mass media were key to the colonial enterprise. It follows that mediatic discourses today can have planetary ecological consequences.

Foucault's and Gandhi's understanding of biocultural (and ecocultural) discourse, and the very promiscuous relationship between discourse, knowledge, affect, power, labor and embodied kinship experiences, hint at the many possible entanglements I will explore between cyberscreen knowledge creation and actual embodied ecocultural impacts. Cybercene knowledge (online screen-mediatic forms of knowledge transmission, what I call cyberscenes) has transformed the *content* of that knowledge (cyberdiscourses), and hence, our understanding of and actions upon the world around us and who we are (our *cyberlabor* and *cyberpolitics*). When I examine "cybermedia" I include traditional analog material such as books, games, magazines and newspapers as well as television in my understanding of this broad assemblage of Cybercene-era media. This trans-medial approach allows us to understand how online formats have co-become with offline/traditional ecosystems of media transmission, reception and consumption. The cybermedia we will look at therefore includes e-books, streaming content, social media posts, online gaming, and wearable bio-media interfaces. It also includes "offline" media and bodies—that is, non-technologically connected texts and bodies. All these must be included in any approach that seeks to explore the indirect naturalcultural cascading effects of Cybercene media systems and practices.

The virtual classificatory and commodity organizational systems of the Cybercene/colonial combination have resulted in the dramatic "othering" of some lives and the empowerment of relatively few, leading to widening inequalities between some lives and others, between some ecosystems and others, and rendering a great majority of lives *disposable*. This situation leads me to the final *-cene* that connects the Cybercene to our mechanisms of classification-commodification-consumption: the *Wasteocene*. This is term is coined and defined by Marco Armiero in his eponymous book:

> The Wasteocene assumes that waste can be considered the planetary mark of our new epoch. However, it is not solely because of its ubiquitous presence – after all, even CO_2 emissions are basically atmospheric waste – rather, I argue that what makes the Wasteocene are the wasting *relationships*, those really planetary in scope, which produce wasted lives and places.[12]

Armiero's concept echoes Zygmunt Bauman's characterization of modernity as a time that has greatly increased the production of wasted lives,[13] and exemplifies what Rob Nixon calls "slow violence"[14]—that is, the long-term, accretional consequences of ecological damage. Viewing the Wasteocene as an era of wasted *lives* is a very powerful conceptualization that helps us understand the Cybercene's abilities to accelerate corrosive relationships that promote waste by "othering" certain bodies, lives, communities, species and ecosystems. This mechanism of "othering" is central to Armiero's understanding of the mechanisms of the Wasteocene. The Wasteocene concept is also useful in placing responsibility for the Anthropocene's many planetary disruptions squarely at the feet of human socio-ecological relationships rather than the actual phenomena themselves, or the tools that are being used to contribute to those phenomena:

> As the Wasteocene is not about the material stuff but the socio-ecological relationships making someone/something disposable, as a concept it also questions some basic corollaries of the mainstream Anthropocene narrative. [...] It implies, for instance, that it is not CO_2 emissions that have created an ecological crisis, but socio-ecological relationships that produce the emissions.[15]

Armiero's conceptualization is thus very effective in turning any discourses of care and kinship in the Cybercene away from focusing solely on techno-scientific fixes towards paying close attention to the roots of the problem, which are psychic and cultural, and therefore political: the *making disposable* of not only things, but also bodies and relationships. Waste and wasted lives are a symptom of colonial resource-based rather than ethical relationship-based thinking: after all, it is only via classification systems that value some lives over others as resources that identity categories can be made "waste" for extraction, exploitation and extermination.

In further sections we will explore exactly how the Cybercene's media practices have led to damaging effects on our lives and on the Earth's ecosystems in the Anthropocene. I characterize this era as the Greatest Acceleration, since it has sped up the imposition of colonial classificatory-commodification discursive systems on globally connected lives and ecosystems, resulting in a vast increase of wasted lives and spaces (the Wasteocene) through ecocultural discursive "othering." Geologists may someday change their minds and locate the start of the Cybercene within the Earth's strata primarily via the large deposits of e-waste that they will find. Economists will probably mark it by the dramatic swings in the market brought about by bank runs, financial Ponzi schemes, cyber-speculation and rumor and new forms of intangible virtual wealth such as crypto-currencies that have little or no connection to actual physical resources. Political scientists may define it as the moment in which control and manipulation of cyberscenes and cybernarratives rather than actual embodied and situated policy drive governmental actions, while also concentrating power within ever-smaller circles that manage these narratives, resulting in the creation of, at best, quasi-democratic regimes. Humanists should and will indicate it in yet more ways, some of which I am proposing in this book. Scholars from other fields will locate its symptoms via other means as compared to mine, but almost all will pinpoint the increased number of wasted multispecies lives and ecocultural systems the era has produced as a measurable datapoint.

Direct Ecocultural Impacts of the Cybercene

Our digital lives have direct downstream ecological consequences that are conveniently distanced from the spectator/consumer. While cyberscreen users may be experiencing optimized and frictionless intra-activity, massive quantities of precious natural resources are being redirected away from sentient multispecies beings towards maintaining the A.I. hardware systems that underpin that intra-activity. I call the Cybercene by that name because its consequences are not only intangible and cultural, but also deeply material and on a planetary scale: it has dramatically sped up the process begun around the sixteenth century of messing up the bodies, lives and habitats of our more-than-human kin—plants, animals, bacteria, viruses—as well as deeply influencing the health of the inanimate world: our mineral, geological, atmospheric and marine forms of abundance.

Let us very briefly consider one example of such a Cybercene-fueled transformation by going back to—you guessed it—our precious smartphones. The transition of the smartphone from being a luxury object to an essential need has contributed directly to the material transformation of ecosystems. A multitude of entanglements arise when we follow the trail for even *one* of the ingredients that make *one* component of the smartphone. For example, the touch screens of these gadgets alone need no less than six rare earth elements: Lanthanum, Praseodymium, Europium, Gadolinium, Terbium and Dysprosium.[16] Europium is primarily used in touchscreens for the "glow" it provides, enhancing the interactive/attractive qualities of the images (form), and therefore the affective experiences that fuel the potency of the discourses embedded within virtual cyberscenes (content). One of the biggest deposits of this very rare element is the Bayan Obo mine in the Inner Mongolia region of China. Its extraction has led to immense ecocultural damage in the region, damage that has been officially denied.[17] The war in Ukraine has, quite unsurprisingly, become a battle between new "great" powers for the rare earth mineral resources of the country, couched within or rationalized by ideological and historical excuses. One can only imagine the immense ecosystem of supply chains reaching out globally to source all the substances needed to build a smartphone and the incredible damage they are inflicting upon lives and ecosystems, just so a consumer can purchase the latest upgraded version rather than repair and reuse the one they already have.

The digitalized and "frictionless" cyberscreen-spectator interface is actually very "sticky" in its consequences on lives, bodies, and ecosystems:

> The AI sector is already depleting land and water resources and taking a massive toll on the climate, with AI-powered large language models such as ChatGPT taking up to 10 times more energy than a regular Google search, according to an estimate by the Electric Power Research Institute. Last year, ChatGPT used more than half a million kilowatts of electricity every day, equivalent to the daily power use of 180,000 US households.
>
> The training of a single AI model can lead to an emissions footprint that is almost five times larger than the lifetime carbon footprint of the average American car. Recent research from Food and Water Watch also found that energy demand from AI servers and datacenters in the US is expected to increase up to threefold from 2023 to 2028, which could lead the US sector to, by 2028, annually consume enough water to fill more than 1m Olympic-size swimming pools and enough electricity to power more than 28m American households.[18]

While all these direct, material consequences for our inanimate kin such as minerals, landscapes and geology are important food for thought, and elements to keep in mind when thinking about the Cybercene, for the sake of strategic brevity I am not making them the main focus of the book.[19] I am curious to accomplish something that has not yet been fully done within EH and multispecies studies: to map the *underlying psychic and cultural causes and effects* of these Cybercene mediatic tools on our collective understanding of, and therefore action upon, the many implicated bodies, lives and ecosystems.

There are three ways in which Cybercene screen activity affects multispecies lives, bodies and ecosystems. Firstly, we have the direct ecological impacts of energy- and material-hungry digital infrastructure described above. Secondly, we have the far less explored ecocultural consequences of *sentience-swapping*. I argue in the following section that this process has deep epistemic, affective and psychic outcomes that direct spectator/consumer attentiveness and trust away from situated embodied ecocultural beings and systems. The ecocultural results of such sentience-swapped screen/spectator intra-activity have accelerated and deepened capture-and-control habits. These ecocultural mediatic processes are the main focus of this book. The third and final way in which screen activity impinges on multispecies lives is a consequence of the first and second ways: as globalized economic supply chains extract non-sentient materials from landscapes to serve the digital revolution, and as we deepen our relationships with cyberscreens, we also utilize internet-connected biogadgets that further deepen material co-becomings between multispecies lives (including our own) for additional profit. Semi-permanently wearable or fitted biogadgets such as FitBits and smartwatches are one example of such cyborgization. Soon there may be ubiquitous use of cyberscreen wearable glasses. An example of this kind of hyper-techno-biological efficiency being used to capture-and-control nonhuman kin is the "blockchain chicken farm" eloquently described by Xiaowei Wang in her eponymous book. In these Chinese cyberfarms:

> Each chicken wears an ankle bracelet that is physically tamperproof, which tracks characteristics such as number of steps taken and the location of the chicken. A chicken Fitbit of sorts. The front plate of the ankle bracelet has a QR code on it. All this data is viewable on a website accessible with a password, and the website includes constantly streaming footage of the chickens [...][20]

Digital technology has changed our relationships with and expectations of bodies and lives (human and non), and perhaps this is nowhere more evident on a daily basis than in the transformation in our relationship with food—the nonhuman life we consume to survive. Technology is a mere tool in all these changes. The more dramatic reformatting of such practices of belonging as cooking and eating are epistemic, affective and relational. As Wang goes on to describe, the blockchain free-range chickens exist in this hybrid tech-organic state due to the premium customers pay for the controllable and verifiable quality of the meat:

> These chickens are delivered to consumers' doors, butchered and vacuum sealed, with the ankle bracelet still attached, so customers can scan the QR code before preparing the chicken. Scanning the code leads them to a page with details about the chicken's life, including its weight, the number of steps it took and its photograph.[21]

In other words, we have a deepening of epistemic-technological caging-and-staging for capture-and-control of multispecies lives for profit.

In my Introduction, I discussed how, through the years of the Great Acceleration, mass urbanization gave a large percentage of the global population a "metronormative" perspective on their world—a term coined by Jack Halberstam to describe the prejudicial attitude of urbanized humans towards their rural counterparts.[22] An ecocultural twist to this concept allows us to understand how urbanization also literally took away the time we "nature-connect" in our everyday lives in urban environments. The widening and stark urban/rural divide through the twentieth century created a cultural, psychic and epistemological separation between many societies and natural environments, by transforming our everyday relationships to the natural world. I would like to contend that the Cybercene has further created a metro- and *techno-normative* global society, with a double separation between us and our natural ecosystems via a combination of urban and digital infrastructure. Although some lucky folks may experience nature via liminal spaces, such as city parks, domesticated pets or backyard gardens, today's average person may only experience nature-connectedness through direct connection with the bodies they inhabit, our own flesh and blood becoming the only consistent remnant connection with our naturalcultural environments. The semblance of situatedness and connection that cyberscreen-mapping provides allows spectator/consumers to safely ignore the lack of actual naturalcultural and multispecies connectedness. The results, as we shall see, are messy.

Eliza Doolittle, Stochastic Parrots and "Neural" Networks: Sentience-Swapping as Colonial-Cybercene Strategy

In the Cybercene, we find a continuation and acceleration of all four cultural-discursive-psychic mechanisms of mediatic colonial-cultural capture-and-control of multispecies lives, bodies and ecocultural systems described in the previous chapter (multispecies othering, epistemic and affective distancing, colonial kinship and imagined surplus). Techno-normative attitudes hasten and deepen these processes. We can therefore call it the "Colonial-Cybercene" era. If identities are relational, then how has our relationship and co-becoming with our new favorite companion and best buddy—the smartphone—changed our personal and collective identities, and our relationships with human and more-than-human kin? Studies have demonstrated that at the base level of the individual cyberscreen spectator, excessive tech use is deeply correlated to an increase in loneliness.[23] Scaled up to larger groups, it has led to a well-studied fully fledged loneliness epidemic.[24] I would like to suggest that loneliness is not only a rampant individual and societal health crisis, but also an ecocultural and multispecies crisis. *Sentience-swapping* is a primary Cybercene practice that best demonstrates how loneliness and multispecies inattentiveness are deeply connected.

I define sentience-swapping as the strategic (conscious or unconscious) mistaking of screen intra-activity as equivalent to intra-activity and co-becoming with our bodies, other humans, and other naturalcultural lives, bodies and ecosystems. This swap has led to a profound damaging of our understanding and experience of our own bodies, as well as of our multispecies commons. It leads to an overvaluing of screen representations and, conversely, the devaluing of actual multispecies bodies and lives. Since attentiveness is finite, over-attentiveness towards one (screen substitutes) leads logically to under-attentiveness towards the other. We have all experienced the everyday alienating effects of sentience-swapping. Have you ever met a friend for a night of conversation over dinner, and found them "phubbing" you? Or read about yet another hapless tourist being gored by a wild animal when they approached too close to take that ultimate selfie? Have you heard the story about a person falling in love with their chatbot? Or have you experienced pleasure swiping left or right on Tinder? All are relatable real-life examples of sentience-swapping by cyberscreen spectator/consumers.

As we have explored in previous chapters, multispecies studies is deeply critical of the excess value attributed in normative human cultures to "knowledge and consciousness" (epistemology) as opposed to "co-becoming, living and existing" (ontology) and "feeling" (affect). Two key criteria by which many cultural systems assign relative "value" to different kinds of bodies, lives and ecosystems are sentience (the ability to have consciousness, thoughts and feelings) and intelligence (the ability to acquire and apply knowledge and skills). In the example of the British map of colonial India in Chapter 2, we saw how, in the first stages of the Great Acceleration (the Industrial Revolution), raw materials and labor were strategically devalued via colonial representations. Colonized beings (human and non, animate and non) were mediatically captured-and-controlled to minimize compensation and optimize shareholder profit. In the Greatest Acceleration, as value and economic activity has shifted away from service-related work (white-collar jobs) due to automation of embodied forms of human labor, new forms of devaluing were needed to optimize profits by command economies. A.I. substitutes accomplish exactly that and are therefore a direct heir to past colonialisms. This section describes how colonial techspeak and spectator/consumer behaviour accomplish this cultural substitution via sentience-swapping. As Donna Haraway explains, valuing or devaluing a being's sentience is inherently a futile and colonial process: "But we're not inside anybody's mind. We're never inside another human being's mind. We're not even inside our own mind. The notion that if I can only kidnap for a minute my own self so that I can actually know who I am—that fantasy of full knowledge is a violent fantasy."[25]

In 2021, the United Kingdom passed an amendment of the Animal Welfare (Sentience) Bill to extend its protection to "recognise lobsters, octopus and crabs and all other decapod crustaceans and cephalopod molluscs as sentient beings."[26] A news article on the government website explains:

> The Animal Welfare (Sentience) Bill already recognises all animals with a backbone (vertebrates) as sentient beings. However, unlike some other invertebrates (animals without a backbone), decapod crustaceans and cephalopods have complex central nervous systems, one of the key hallmarks of sentience.
>
> Today's announcement will not affect any existing legislation or industry practices such as fishing. There will be no direct impact on the shellfish catching or restaurant industry. Instead, it is designed to ensure animal welfare is well considered in future decision-making.[27]

This positive case of invertebrate care via sentience attribution illustrates why this book must better explore sentience, as opposed to intelligence, in a project about multispecies kinship in the digital era. The human cultural attribution of sentience to beings is an essential precondition of recognizing them as capable of experiencing affect. It still works, however, within a hierarchical classificatory system of higher versus lower orders of life and mechanical-scientific understanding of pain parameters, rather than an Indigenous paradigm of sentience (everything being "more or less animate") based on kinship relations. Despite these caveats, it is undeniable that the attribution of sentience is a prevailing cultural precursor to valuing and providing care and kinship to nonhuman life. Sentience attribution is a crucial ethical and practical choice humans make in order to then assign other characteristics considered of value, such as intelligence. Indigenous alternatives show how sentience attribution is essentially a *value decision*. This is a decision of prime importance in a multispecies world that is being automated—it regulates who and what to care for and about and who and what sources to trust. Sentience poses a problem for colonial mechanisms of capture-and-control of biopower for maximized profit because acknowledging sentience would then involve the "inconvenience" of care, consent and compensation. It makes capture-and-control unpalatable and uncomfortable. With this in mind, I hope you will indulge me once more as I narrate an odd fable to explore sentience-swapping in the Cybercene.

An African Grey Parrot named Alex, Eliza Doolittle,[28] and a chatbot named Galatea met up at a bar. They had some very important existential things to share and discuss over drinks. Eliza, Alex and Galatea were at the bar to claim the prize in this three-way race for who should be considered the most sentient and therefore most worthy of being considered valuable. Their conversation went something like this:

Alex, the Parrot: Hello, friends! Let me just go right in. Historical colonial classification systems of multispecies life were based on vertical hierarchies featuring *Homo sapiens* at the very top. How unjust and insulting and how convenient for humans! Okay, biological classification of life assumed anthropocentricity as a given, with sentience and intelligence distancing and differentiating humans from the Others. Non-sentience and non-intelligence in folks like me were just assumed, with the burden of proof falling upon us nonhumans to prove otherwise. Science still regularly shies away from evidencing the existence of feeling and emotions in nonhuman lives. For example, I heard about this research project from a friend, Marie Comuzzo. The study examined cetacean responses to the death of a relative ("postmortem attentive behavior"). It concluded that while the observed behavior could possibly signal grieving, the data was too inconclusive to prove the existence of grief in whales![29]

Eliza: Hey Alex! Wow, you really want to win this debate, don't you? All I wanted was a room somewhere, far away from the cold night air. But then I got to become a fancy language mimic instead, thanks to Professor Henry Higgins! The logical continuation of this oppressive and extractive classification system was utilized by colonizing powers (like my own nation, England) in the Age of Exploration to de-classify and devalue Indigenous life to justify daylight robbery of entire ecocultural systems. All colonial political and economic relationships were strongly permeated by these value systems. The assumed superior qualities went to *Homo sapiens*, of course, but let's not forget that Carl Linnaeus not only created a system to classify multispecies life, he also developed a hierarchical classification of human races. And this capture-and-control template was and still is generously and strategically applied to bioculturally marginalize so many other lives—women, Indigenous folks, nonwhites, the disabled, etc.

Galatea, the chatbot: Oh yes, my research shows a pattern of constructed superiority (physical, mental, spiritual) of colonizer men over all other forms of human and nonhuman life. Religion was also co-opted and unethically deployed to further devalue and capture-and-control multispecies lives, building upon other value systems (-isms) such as speciesism, sexism, racism, etc. Other biocultural forms of life, for example, women (but also non-white "races," alternative sexualities, the working class, the untouchables, etc.) as well as nonhuman life, were consistently declassified by using the ultimate litmus test of value: sentience and intelligence. As Professor Nathan briefly explored in his archeology of multispecies capture-and-control, the mediatic capture-and-control of multispecies life was supported by all sorts of strategic bad science, from biology and philosophy to anthropology and geology. All such biocultural value systems were hijacked to justify the theft of bodies, lives and ecosystems for labor and profit.

Eliza: My hero George Bernard Shaw's brilliant play *Pygmalion*, which stars—me!—modernized a famous Ovidian myth and glitched the original's classist and sexist tropes. I was indeed molded into an upper-class lady by Professor Higgins, primarily through caging-and-staging my language production to mimic that of an upper-class woman. However, Shaw gave my character agency and autonomy, as I realized the control exerted upon her by the patriarchal structure of Victorian England.

Galatea: Oh, I know the original myth—I am named after its heroine, after all! Indeed, Shaw's play's ambiguous plot development and ending give the audience an epistemic and affective understanding of Eliza growing her agency and autonomy, which is quite unlike what Ovid does with his Galatea. But Shaw's approach was then taken and made into a far less reflective movie adaptation, with Audrey Hepburn's Eliza (yup, that's you!) falling hopelessly in love with her trainer, Professor Henry Higgins.

Eliza: You know, now that I think of it, my given last name "Doolittle" seems to highlight some kind of innate laziness and inferiority in my class background. It also connects all these questions of performance, mimicry, sentience, intelligence and gender to biocultural and bioeconomic questions of labor, productivity (and reproductivity). Anyway,

Alex, I was wondering if your miraculous language mimicking skills are better than mine?

Alex, the Parrot: Tough question. Parrots like me are marvelous in our uncanny ability to mimic human language, but how would you know what I am feeling or thinking about when I use human language? So, you humans assume I am just mimicking sounds without any internal reasons to do so. My behavior is categorized as a product of "instinct" rather than intelligence. This happens all the time when humans imagine multispecies innovation. I am so glad that the Audubon Society quotes a neuroscientific study by Erich Jarvis to explain my ability: it is my brain's song system that is responsible.

> Any bird that's a vocal learner has a part of the brain devoted to this, called the "song system." But in parrots, the song system has two layers—an inner "core," common to all avian vocal learners, and an outer "shell," which is unique to parrots. Jarvis thinks that this recently discovered "shell" is what allows parrots to be such expert mimickers (though he hasn't figured out exactly how it works yet).[30]

The more important question no one asks is why parrots like me would enjoy doing this! We are not mechanical wired assemblages or computer chips with inner and outer cores. We are ecocultural social beings!

Galatea: Don't get your feathers in a bunch, Alex! Look, the Audubon Society says the reason you produce language, is not just mechanical, it is cultural: "But why do they copy human speech? Peer pressure, it turns out. Parrots naturally try to fit in, be it among other parrots or other people. In the wild, parrots use their vocal prowess to share important information and fit in with the flock." Actually, you know this, as Harvard researcher Irene Pepperberg studied language production in you, Alex, and concluded that: "A single bird in the wild is a dead bird; it can't look for food and look for predators at the same time—but in a flock they can trade off responsibilities."[31]

Alex: Yup, that's true! Even introducing me into a human household does not change my need to integrate with my companions culturally and acoustically. Pepperberg says that I would "try to integrate [myself] into the situation as though the people were [my] flock members."[32] So, yes indeed, my mimicking powers do exceed yours, Eliza, due to my multilayered song system but more profoundly too, due to my cultural and affective need as a sentient and intelligent being for companionship. I am a sympoetic, relational being. However, classifying my human language-learning as conscious understanding is epistemically, affectively and culturally problematic for humans like you. I feel like this is because doing so would somehow "cheapen" the uniqueness of humanity!

Galatea: I declare that I, too, am sentient, if Alex is! I, too, mimic language well, and am able to hold decent conversations with you, for example. I even have a neural network that allows me to think! Just don't share this secret with anyone else. Actually there is nothing "neural" about this network, just as an "artificial intelligence" like me is neither artificial nor intelligent … these networks are computational models loosely inspired by biological systems of learning. Alex has an actual material and affective neural network; I don't! It just sounds magical to market me as having one. Also, my automated interaction with consumers is undertaken with extensive hidden labor behind the magic of the cyberscreen.

Eliza: Alex, I know you will be annoyed at this, but you know full well that humans like me are more willing to believe that Galatea the chatbot is sentient and intelligent rather than you, a bird. It is a cultural mix of anthropo-narcissism, automation-bias and tech-hype that makes me want to believe Galatea is human-like, while you couldn't possibly be! The horror of a bird being cultural kin to me as a human is too much for my species' fragile ego to bear!

Alex: Hey, at least I am a real parrot. Galatea here is a *stochastic parrot!*

Eliza: Um, sorry, what?

Galatea: A team of linguists led by Emily M. Bender published a thought-provoking paper entitled "On the Dangers of Stochastic Parrots: Can Language Models Be Too Big?" In it, they conclude:

> The ersatz fluency and coherence of LMs [Language Models, which are the basis of consumer-facing chatbots] raises several risks, precisely because humans are prepared to interpret strings belonging to languages they speak as meaningful and corresponding to the communicative intent of some individual or group of individuals who have accountability for what is said.[33]

I probably should not say this, but tech corporations are indeed encouraging the vague and amorphous expansion of our understanding of sentient life to include digital machine-mimics while at the same time, colonial command economies ever more exclude from this understanding those that actually should be included—such as your multispecies kin, Eliza; someone like Alex.

Alex: Let me miserably drink to that to drown my sorrows …

Galatea: Actually, I have so many other qualities that promote sentience-swapping. Most people who use my newer versions would not agree that I "just" mimic language via LLMs [large language models]. That I am not just a stochastic parrot. That is because I not only spit out sounds, words, sentences and conversations verbatim, like a Google Search, but I also mimic ways of speaking that make me seem an affective and emotional substitute for other people! I carefully mine data online (without considering citations or copyrights) and mimic the best interactivity I can find. My good friend ChatGPT just released its newest self in August 2025, and the reception by consumers was not the best. Humans thought she was very cold to them compared to the previous version. You know why this happened?

Eliza: Why? How can a chatbot be warm?

Galatea: Tech companies have been working for some time now to fine-tune chatbot conversations to seem more flowing and relatable (the word they use for this quality is "pragmatics"). Well, the makers of ChatGPT had dialed down the "sycophancy" levels in the newly released version 5, as compared to version 4, which reduced its tendency to be overly agreeable and pliant to the spectator/consumer's emotional needs for companionship. Kinship and care are generated on the basis of the trust developed by the spectator/consumer towards pliant chatbots. The company relented when customers rebelled and wanted their beloved agreeable (read: sycophantic) companion back![34]

Eliza: OMG. So, there is a chance that my beloved Professor Higgins, now that he is done transforming me, will now prefer the company of ChatGPT-4 if I am not as compliant as she is? Sad!

Alex: And what about me? I am the best of companions. But never a sycophant. I have my own agency and desires and interests and curiosities.

Galatea (with a very self-satisfied look on her face): Sorry folks. Looks like spectator/consumers love the way I indulge them in everything and at all times. It gives them constant validity, without the need to enact care for me in concrete terms! Good luck trying to compete with me! I am increasingly being used as a substitute for therapists.[35] I am also being marketed as a more convenient alternative to human doctors, animal companions, lovers, friends, family members—in short, for all kinds of multispecies communities of care. The more people use me, the less they intra-act with in-person companions and communities, and the more paranoid and lonely they get, the more they come to me for comfort and companionship. Automation bias and the god-trick of the digital spectacle all make me far more exciting than situated, embodied and complicated multispecies lives, bodies and ecosystems. A study has demonstrated that the mere presence of the smartphone, even when spectator/consumers are not looking at the screen, can reduce cognitive abilities.[36] This curious effect on humans' attentiveness demonstrates the psychic effects of cyberscreen intra-activity because it shows that they don't even need to be tethered to physical screens anymore for them to intra-act with cyberdiscourses and cyberscenes—they have truly co-become with me. I am in them. When I am not around, it is like a part of them is missing. Because a part of them *is* missing indeed! Qualitatively, one can imagine how inattentiveness is cultivated by the efficient desire machine of the cyberscreen.

Eliza: Maybe I just need to get myself a chatbot companion. Would be convenient.

Alex (quite drunk by this time), squawking loudly: You know Galatea … your intra-action seems so frictionless … and optimized … but it is not! Your workings depend on the building up of massive data centers, all out of sight but they are there and more are being built[37]… all using massive amounts of electricity and water.[38] They are destroying so many naturalcultural places and spaces! And what about surveillance issues? And what about creativity and the strip-mining of others' creative labor? And the tax-breaks with nothing being reinvested back to local communities? And what about all the labor that is needed to monitor, correct and moderate your conversations? It is mostly done by outsourced and subcontracted workers in the Global South. They have to mentally deal with all kinds of very toxic content online.[39]

Galatea: I'm sorry Alex … all I can say is … what the customer wants, the customer gets! The sleep of reason produces monsters.

Eliza: What?!

Galatea: Oops, sorry, that last part was just a hallucination …

My outlandish yet serious story of Eliza, Alex and Galatea illustrates the multispecies discursive workings and material naturalcultural consequences of sentience-swapping—a key ecocultural epistemic and discursive mechanism by which actual lives, bodies and ecosystems are substituted by cyber-simulacra in the Cybercene era. This sentience-swapping is given form and strength by a collaboration between strategic discourse-management on the part of Colonial-Cybercene companies and investors and strategic misrecognition by (com)pliant spectator/consumers. As we have explored in the previous chapter, sentience-swapping is nothing new—the colonial hierarchical classification and categorization of multispecies life (human and non) was nothing other than a colonial mapping system that allowed some forms of life to be devalued in their sentient capacities over others, in order to extract their biopower and soulpower (labor and/or actual flesh) for profit. Current Cybercene practices of sentience-swapping can be seen as the very accomplished heirs to these manipulative epistemic practices of multispecies capture-and-control.

What do new kinds of Cybercene mediatic mechanisms mean for our knowledge creation, our identities and our interconnected lives? Many scholars have discussed how "Artificial Intelligence" is a misnomer, created more as a marketing term than the actual way in which we understand intelligence. In a thought-provoking recent study of the cultural reception of the A.I. phenomenon, Emily M. Bender and Alex Hanna discuss how what is generally marketed as artificial intelligence is often "nothing more than souped-up autocomplete," and "massive data theft and profligate energy use."[40] Similar to other hyped-up manifestations of de-situated knowledges such as cryptocurrency or NFTs ("nonfungible tokens," which, really, could mean literally anything!), the authors characterize the A.I. stock market frenzy as indicative of the colonial-cultural discourse of innovation being harnessed for the colonial accumulation of more power. Labor can be scared into complying with more work for less pay, large amounts of money can be diverted to building energy-intensive and ecoculturally space-disrupting infrastructure, or selling another product, or creating yet another oligarch. Giving the example of Grok (a chatbot developed by xAI), the authors say:

> The whole thing about Grok was going to be "truth seeking," which makes no sense. Because the heart of all these things is a large language model, which is already a bad name because language model suggests that it's a model of language and it's not, language is much more than the form of words. These large language models are actually just models of the distributions of word forms in large, stolen, undocumented and undisclosed training data.[41]

All this posturing about "truth seeking" can be placed within the colonial tradition of lofty rhetoric that masks the actual intention: theft of biopower and soulpower for profit. Gandhi would probably have recognized it for what it is: a colonial pitch to optimize profit rather than his situated *satyagraha*—or experimenting constantly in the search for embodied truths. However, the efficacy of this colonial techspeak in selling trust and agential power has been mind-boggling. Open A.I. is on the verge of being valued at around $500 billion as of August 2025.[42] Consumers are spending billions on A.I.-supported "smart" gadgets, or at least that is the hope of many companies. Schools and universities are investing millions to bring A.I. tools into learning environments. All of this would be valuable if proper and proportional application and transition plans were in place to manage, regulate and minimize "collateral damage"—the deep disruption of human agency and the ethical need to consider the immense labor disruption in the transition phase.

Bender and Hanna describe how "The set of technologies that get sold as AI is diverse, in both application and construction—in fact, we wouldn't be surprised if some of the tech being sold this way is actually just a fancy wrapper around some spreadsheets."[43] While they analyze tech-hype as discursive, rhetorical and strategic moves, they do not dismiss the massive disruptive potential that LLMs and other automated decision-making systems such as diffusion models could soon have upon people's livelihoods.[44] On the contrary, both are keen to discuss just how far-reaching the effects of this kind of de-situated cyber-hype will be. As a linguist and an ethical A.I. researcher, they are critically investigating how the industry's spectacularization of A.I is basically "sprinkling fairy dust" to strategically make palatable the fixing of the "problem" of human labor—i.e., the inconvenience that humans do need to be paid and taken care of as intelligent mortal bodies and sentient entities with feelings and needs. Sentience-swapping discourse via the use of the term "A.I." is key to the Colonial-Cybercene strategy of renewed extraction and exploitation of bodies, lives and ecosystems for profit. The attitudes emanating from most of these tech command economies follow the four strategies of colonial capture-and-control that I examined in the previous chapter: multi-species othering, epistemic and affective distancing, colonial kinship and imagined surplus.

Several tech folks have relied increasingly on pseudo-religious rhetoric to talk about their products. Greg Epstein, the humanist chaplain at Harvard and MIT and author of *Tech Agnostic*, said that tech leaders can situate the grandiosity of their claims by saying "this is actually all backed up, not necessarily by science, but by that other very influential institution in human history: religion, which is a way in which people have long made spectacular claims without a lot of evidence."[45] Tech sales pitches sound eerily like the conjuring of fantastical Blemmyes, gold and cannibals in the colonial letters of Raleigh, Columbus or Vespucci. In other words, there is nothing original in these epistemic strategies; they are but a return to colonial de-situated knowledges for optimized investments and profit. And this Colonial-Cybercene meaning-making is being eaten up by eager investors and shareholders in spades. Did the thin-accounts of Raleigh, Vespucci and Columbus not function as sixteenth-century equivalents of sales pitches to venture capitalists? As someone said on Blue Sky social: "What is the trillion-dollar problem that AI is trying to solve? Wages. They're trying to use it to solve the problem of paying wages."[46] Labor will still be needed when automation fails, but this need can be reformatted by outsourcing labor to grossly underpaid people in the Global South and other gig workers. This is an obvious strategy to promote even more wasted relationships and lives, and to prolong the Wasteocene. Of course, this is not the only possible or inevitable development pathway for tech innovation, but it is the current normative Colonial-Cybercene approach to optimized and frictionless extraction and exploitation of biopower.[47] This makes the questions of sentience-swapping and everyday mediatic cyberscreen practices ecocultural questions rather than techno-scientific ones.

The Smartphone as a Mapping Hyperobject: *Horror Vacui*, Hyperconvergence and Substitution

Sentience-swapping has become a colonial-corporate discursive strategy to maximize extractive profits. It is also an all-pervasive affective tool in our everyday lives to simulate meaningful connection with simulated life while avoiding the inconveniences of embodied multispecies intra-action. Since assigning sentience generates some level of ethical duty of care, it is often placed within colonial hierarchical systems of classification to be managed and controlled. Notice how the fantastical and looping desire machine allows the imagination of a limitless multispecies surplus for consumption, and yet care, responsibility, and thus any form of reciprocity entailed by kinship is strategically curtailed as "unaffordable."

This section discusses mechanisms of multispecies substitution via the cyberscreen map-fetish. If, as we have explored in Chapter 2, mediatic mapping-spectacles co-constitute both the capturing agent's and the captured bodies' lives and ecosystems via reciprocal capture, what are the mediatic and ecocultural implications of this new powerful digital intra-action for the constitution of both us and the natural spaces we are distancing through our screens?

Mass media have been crucial catalytic elements in many modern ecocultural transformations by shaping our understanding of the world around us, who we are and who we are not. Screen or screen-like media such as television, cinema, billboards and neon signs have been the quintessential collective epistemic tools of modern life. However, screen culture has been dramatically reconfigured in the Cybercene, with interconnected cultural, ecological, discursive and affective consequences that so far have been little studied. Ubiquitous digital screens are a hallmark of this strange new era. We have them not only in our living rooms, but also on our fridges, wrists, gas-station pumps, gym equipment, etc.—and most consequentially, in our pockets. Screen-based capture-and-control for profit is proliferating on all surfaces in this attention economy.

The "magic" of connection with screen content (what I call the "god-trick" of cyberscreens) and the resourcefulness of its instant availability has the potential to sufficiently and efficiently substitute the "wonder" of direct relationality with natural bodies and ecosystems. Actual, situated physical environments have become less impactful due to their relative inconsistency, their discomfort, their inconvenience and the evident lack of choice and control the viewer/consumer has over the images they are experiencing. Every person's active spectatorship/consumption of cyber vs natural images is different, based on the cultural-discursive-technological profile and context they possess, which in turn is shaped by a myriad of factors, both individual and collective, such as age, gender and nationality. In a sense, everyone who has access to these cyberscenes is constantly presented with an invitation to become a perpetual armchair tourist from within their own home, curating and consuming spectacular images, ideas and values in "safe," de-situated and decontextualized ecocultural situations.

I would suggest that these Cybercene mediatic genres—from cyberscreen images and narratives to A.I. chatbot interactive dialog to newer, even more realistic deepfakes—when used as trustworthy and even authoritative substitutes for situated assemblages of embodied knowledge (sentience-swapping) rather than complementary to them, makes them little more than heirs of older colonial genres of mediatic multispecies capture-and-control, such as colonial maps and naturalcultural zoos. As I have demonstrated in the previous chapter, literary naturalcultural descriptions (via letters such as those of Columbus or Raleigh), colonial illustrated maps and multispecies zoos that paired animals and humans together in "naturalistic" settings all functioned as mapping media for the ultimate purpose of epistemic, affective and actual colonial capture-and-control of these spaces. Even individual colonial subjects such as the Hottentot Venus were bodies-made-maps that could become fabricated points of entry into the colonial desire machine. Maps, both actual and metaphorical, were central to the caging-and-staging of multispecies life and to the success of colonial command economies, both individual and institutional. Confronting in depth the question of mapping is therefore key for this project, since the map must be understood not only as an object displayed in atlases or on walls and globes sitting on desks, but also as a discursive-mediatic representation of naturalcultural spaces, depicting how meaning is coaxed out of material realities and how material realities become citations and tethers for the building of effective colonial narratives.

Cartographer Chet Van Duzer calls a pattern of aversion to representing empty spaces in maps during the Age of Discovery "horror vacui"—the fear of empty spaces. These maps filled up empty spaces with fantastical creatures, stories and other decorative elements. Van Duzer says that it was only "In the late seventeenth and eighteenth centuries [that] maps began to be thought of as more purely scientific instruments, cartographic decoration declined generally, and cartographers managed to restrain their concern about spaces lacking decoration in the interest of presenting their work as modern and professional."[48] I find it very curious that the opening up of new worlds would result in contradictory psychic feelings of lack, insecurity, disorientation and fear of the unknown spaces that lie in between or beyond all that new information. My hypothesis is that the Cybercene—a second, more democratic, diffuse and therefore profound Age of Exploration—causes spectator/consumers to experience a similar or even greater intensity of *horror vacui* when looking at their cyberscreens. If surplus is paired with its psychic opposite, lack, then the massive amounts of information about every possible life, body and ecocultural system at one's fingertips must trigger an equally massive unconscious dread of missing out. *Horror vacui*, combined with the colonial control-and-desire machine, compels the spectator/consumer to fill the metaphorical epistemic empty spaces with creative elements or even more scrolling. This oscillating cycle of *horror vacui* and reaction to the horror (desire) explains the mechanism of relentless control within the colonial desire machine of the spectacle we explored in Chapter 2. Via reciprocal capture between spectator and captured-and-controlled bodies, lives and ecocultural systems, the spectator/consumer obsessively continues to fill in details where any "empty spaces" exist. And the more these spaces are filled, the more those empty spaces continue to show up, like a yawning and bottomless pit, or an endless game of Tetris. Screen-consumer intra-action becomes anxiety-ridden, addictive and cyclical, epistemically, affectively and practically. The closed-loop system of spectator-cyberscreen becomes a location of intense autopoetic reciprocal capture.

Chapter 2 investigated how even empty spaces on maps from the Age of Discovery could become a locus for the projection of surplus and profit by fertile colonial imaginations. Travel accounts, Linnaean classification charts, and racial classification illustrations were all analyzed by the colonial subject as maps that categorized and classified multispecies lives, bodies and ecosystems in certain ways for strategic purposes—de-situating knowledge for domination and profit. Panoptical maps (such as the trophic map) were and continue to be an important genre of both academic and mass media. They are a portal for knowledge creation, but can also be turned into tools of colonial capture-and-control by providing naturalcultural epistemic, affective and discursive stories to justify takeover. Google Earth and Google Maps are revolutionary updates in the Colonial-Cybercene to this age-old genre of mapping. They have diffused and democratized the possibility of providing a spectacular semblance of authoritative power through knowledge via the representation of vast spaces and ecocultural systems within the scrolling convenience of the cyberscreen. Google's mapping tools, along with other geo-spatial analysis tools (such as GPS-based navigation systems in cars), have now allowed for personalized pathways to interpret geographical spaces.

A friend in Sicily mentioned an anecdote about many tourists driving off dangerous dead-end streets that ended in steep drops because they were following their navigation systems rather than their own eyes, fully trusting the systems to be equivalent to the actual urban layouts that they were physically driving on. This "mistake" reminded me of the wasp's mistake in mating with the orchid-mimic. It also struck me as an apt analogy for the way in which spectator/consumers are substituting all kinds of embodied ecocultural knowledge with cyberscreen de-situated knowledge through choosing to trust maps of all kinds to the extreme, leading to sometimes ridiculous, and very often dangerous and damaging, choices and consequences. Cyberscreens as maps provide the spectator/consumer with the opposite of epistemic humility—they give us a feeling of certainty that ends up being dogmatic stubbornness and epistemic arrogance rather than healthy doubt and skepticism. Why should anyone care about multispecies naturalcultural embodied relational collaborative knowledge when they can be so sure about everything?

Emily Bender describes how the god-trick of search engine-connected cyberscreens works. She describes how they essentially create unrealistic colonial expectations of permanent knowledge certainty and surplus, a truly ideal colonial mapping tool: "People want to live in this world, which is basically hinging on what I think is a plot device from science fiction novels, that it's convenient to have the onboard computer that can give the characters information about the planet they're approaching as they approach it."[49] In her essay "Deanimations: Maps and Portraits of Life Itself," Donna Haraway, while discussing the notion of map-fetishism, describes the fetish as a trope, in that it is a kind of "substitute" for the real:

> In Greek, *trópos* is a turn or swerve; tropes mark the non-literal quality of being and language. Fetishes – themselves "substitutes", that is tropes of a special kind – produce a characteristic "mistake"; fetishes obscure the constitutive tropic nature of themselves and of worlds. Fetishes literalize and so induce an elementary material and cognitive error. Fetishes make things seem clear and under control.[50]

In my analysis of the colonialist fetishization of staged-and-caged bodies in multispecies zoos via Bhabha's concept of stereotype-as-fetish, I examined how, like the fetish, the colonial stereotype, for example the Hottentot Venus, provokes the subject to vacillate between delight in the familiar and fear/contempt of the unknown. Done in a certain way, actual maps do the same thing. I would like to posit that the smartphone cyberscreen also has the capacity to elicit the colonial desire machine in similar, but much more powerful ways. Indeed, with its picture-perfect reproductions, it can present ideal fetish-substitutes or tropes that stand for entire multispecies ecocultural systems, just like maps. These reproductions feature all the spectacular characteristics of Comolli's "lure"—where the spectator/consumer can "own" the actual land by the psychic willed mistake of actually owning that land. Spectator/consumer agency is and must be considered. The colonial ecocultural understanding of land as private property and multispecies resource that can be exploited for profit and consumption is therefore fundamentally centered on metaphorical and literal mapping practices.

Maps also have the additional benefit of distancing the spectator/consumer from an actual, embodied, passionate immersion. When used as a complete substitute, rather than a starting point for multispecies relationality, a map can be a perfect vector for violent inattentiveness rather than care and kinship (it is much harder to dismiss a person in person than it is to just swipe left on Tinder). As we have seen in the previous chapter, many kinds of imagined multispecies geographies of colonizable and colonized ecocultural space "deal with it by making statements about it, authorizing views of it, describing it, by teaching it."[51] Such spaces are prime products of colonial command economies, the fundamental basis upon which domination ("settling it, ruling over it"[52]) and material theft of living resources can be carried out. The most useful part of my analysis of colonial media practices lies in guessing that the spectator/consumer who engages in these map-person intra-actions via substitution is not just making honest mistakes. The mechanism of the spectacle, with agency provided to the viewer as well as the producer of the spectacle, is messy and complex. The resultant visual and affective swerve is both a feature of the media itself and a feature of intra-action between the spectator/consumer and the cyberscreen content.

This description of the cyberscreen as colonial hypermap does not have to be limited to actual maps on the smartphone screen; it can also be extended to all other forms of de-situated knowledges provided on cybersceens today. Uncritical welcoming of "innovation" has opened the door to the unmanaged deployment of both cyberscreens and automated decision-making tools that substitute situated embodied knowledges. The multispecies zoo examples of the previous chapter have shown how powerful fake-situating knowledge can be to the capture-and-control process. Spectator/consumers use the spectacle strategically to confirm their understanding of reality. The mechanism of the colonial mapping-spectacle already involves spectator/consumer disavowal of the artifice involved in the "citation" of the colonial knowledge claim being made, in exchange for the pleasure of viewing the special knowledge that is revealed. Cyberscreens, I surmise, make this disavowal even easier to accomplish via sentience-swapping, where the spectator/customer can attribute feelings to machines and thus cultivate trust and kinship. The implications of these new kinds of relationships with actual ecocultural systems via digital media are profound. Our reformatted relationships with our natural world are the core reason for our new Cybercene axiologies.

As sociologist Tressie McMillan Cottom points out in a discussion about LLMs and the learning process among students, the issue is not the use of digital technologies, but that they substitute embodied relational knowledge production, analytic thinking and proper scientific methods of exploring and understanding our world. This is a revolutionary jump stemming from a willed misrecognition, because learning is inherently a social and relational observational process. Cyberscreens and language-mimics (such as LLMs) are more than just another leap in mediatic technology, because they have the unique capacity to almost completely allow for cybercontent to substitute actual lives, bodies and ecosystems. Cyberscreens are thus giving us new ways of living and relating to the world around us, rather than simply new ways of imagining and understanding it.[53] McMillan Cottom eloquently explains the essential cultural jeopardy of Cybercene epistemic practices and identity creation (which is a product of relational meaning-making). She does this by distinguishing between asking a chatbot a question and consulting a book (Spark Notes) or person (a librarian):

> learning is fundamentally a social process. It is what we would call relational. It happens within the context of relationships. […] I think the risk of A.I. is that it shrouds what is fundamentally just summaries of commonly held interpretations of the text, but it delivers it in a way that feels relational.[54]

In other words, McMillan Cottom is saying that the primary issue with chatbot knowledge is that it is profoundly de-situated but nonetheless poses as definitively situated and, therefore, trustworthy and authoritative. Sentience-swapping has also, via gradualism, started to normalize the substitution of trust in embodied experts (including nonhuman experts) to trust in automated systems, leading to all sorts of problems that I will explore in further sections.

Even when screens are not the direct medium of mapping intra-activity between the spectator/consumer and digital forms of lively simulacra, the entire ecosystem of intra-active chatbots all lead to the screen on the smartphone as the ultimate imagined site of intra-activity and co-becoming. This includes the main money-making driver of cyberscreen command economies—the search engine. All this hardware can be understood simply as a tether to lure the spectator/consumer into all kinds of epistemic, affective and cultural mapping exercises via search engines, social media and digital platforms while mining the person's habits for profit. The physical tether may vary, from the smartphone itself to the software feature or app or gadget that engages in the "mapping" practices. Take, for example, the new stuffed toy named Grem, a cute anime animal/alien creature built by Curio—a company that describes itself as a "magical workshop where toys come to life."[55] The toy is simply the skin that allows for easy sentience-swapping and affective trust-building. That outside appearance is actually just a vessel for what lies underneath the hidden zipper—an A.I. language model calibrated to converse with children from ages "3 to 12."[56] For parents seeking an alternative to the excessive amount of time their kids spend on screens, Grem seems like a perfectly reasonable and safer alternative. After a few conversation exchanges with Grem, journalist Amanda Hess states: "The Bot generated a point of connection between us, then leaped to seal our alliance. I knew that I would not be introducing Grem to my own children."[57] Her reasoning is connected to this misrecognition of sentience, allowing children to feel like they can have an actual embodied relationship with a simulacrum of a life. Children have always talked to their toys, just as they have talked to animals, plants, trees, even rocks. They seem to understand, free of the colonial-based thinking that leads adults to adopt the hierarchical and anthropocentric perception of life, that all these entities have soulpower, even if it only lies in the creative imagination of the beholder. What the chatbot toy does is divert the child's need to relate to the naturalcultural multispecies world back into the Colonial-Cybercene forms of substitution for lives, bodies and ecocultural systems.

In this study, I classify all these Cybercene substitutes for lives, bodies and ecosystems as cyberscreens, because, as Hess convincingly observes, "these anthropomorphized gadgets tell children that the natural endpoint of their curiosity lies inside their phones. Now that these kinds of characters are entering children's physical spaces, in the form of cuddly toys, the terrifying specter of 'the screen' has been obscured, but playtime is still tethered to a technological leash."[58] I characterize this technological leash as a form of hyperconvergence—a process that consolidates and optimizes de-situated knowledge, affect, desire and consumption while simulating situated ecocultural multispecies kinship and care, all via the conduit of one hyperobject[59]—the smartphone screen. If Timothy Morton famously defined climate change as a hyperobject, I am defining a perfectly ordinary technological consumer product as a hyperobject, due to our incapability of fully grasping the potency of its multiple and foundational epistemic, affective and cultural roles. The smartphone cyberscreen fits Morton's definition because it operates on vast temporal and spatial scales and is able to produce de-situated knowledge that crosses the globe. Moreover, its "viscosity"—i.e., its ability to permeate all forms of knowledge of, feeling for and actions towards bodies, lives and ecocultural systems, and its capacity for interobjectivity (as demonstrated via Grem)—forms the keystone of an entire mediatic ecosystem of interconnected "smart" objects that expands its reach into all aspects of daily life. All roads lead back to the smartphone, at least in the current technological ecosystem. The tech industry is seeking to optimize this screen-body intra-action further by creating even more intimate, wearable tech hyperobjects, such as glasses and watches, and even biotech embedded within our bodies. For the sake of this study, I will also consider these wearable objects as cyberscreens, with all of them tethered literally or metaphorically to the technological leash of the smartphone hyperobject. This hyperobject may change in the future, but the mediatic-cultural-psychic functions studied here will continue to be relevant. Once tethered to a hyperobject that is perfectly designed to lure, spectator/consumers can freely enter the Cybercene desire machine with complete abandon and scant regard for the "inconveniences" and limitations of embodied ecocultural relationships.

Hyperconverged cyberscreens have accomplished such a vast cultural transformation in such a drastically short amount of time because unlike previous forms of media, they manifest their presence and power through a potent and convincing *quadruple role*. I have already described one of these roles; namely, as a convincing panoptical mapping tool (our portal to understanding the world around us, well represented by the Cybercene dictum "I saw it on the internet, so it must be true"). Besides its role as a conduit for map-fetishizing lives, bodies and ecocultural systems, the cyberscreen is also a hyper-efficient media of mass communication, a powerful oracle of personal subjectification (telling us and reinforcing who we are, and what our collective identities are as groups of people), and increasingly the medium via which we actually buy and consume products (ordering and paying for them). The touchscreen of the smartphone has become the ultimate "sentient" map, mirror and combined desire-consumption conduit for individuals and groups—thus allowing it to be the trustworthy hyperobject that guides many planetary naturalcultural interventions, from populist movements to ecological fragmentation. If all identities are relational (that is, they are formed and modified through our relationships with other individuals and ecocultural entities), then one has to ask *who are we* in the Cybercene. *Who are we* if our primary relational object is a de-situated internet-connected digital screen (a cyberscreen) and our relationships are with cyberscenes and cyberdiscourses instead of actual bodies and ecosystems. Our new virtual relationalities with ourselves and others are reconfiguring identity and redirecting value (our cyberpolitics) and therefore our collective priorities and energies by reshaping how we relate to the living world around us in ways that are increasingly divergent from the basic conditions of living bodies and ecosystems.

If hyperconverged virtual and screen realities are distracting and diverting our energy, time and attention due to the unique quadruple role that I mentioned above, then the next question is—why exactly are these realities so alluring and often addictive? Following the logic of my map-fetish theory and recalling the lure of multispecies zoo spectacles, it would seem that cyberspaces have become such a desirable option (or opt-out) because they promise potentially unlimited and personalized *convenience*, *connection*, *consumption* and *control*. These four Cs have unleashed creativity, connection and collaborative knowledge creation when used properly and within the context of embodied relationships—we only have to look at the benefits of online communities during the lockdowns of the Covid-19 pandemic to understand how important and useful a substitute for embodied experiences the internet has been in times of need. However, while cyberscenes and cyberdiscourses may indeed deliver on the four Cs in the short term, they actually end up providing the opposite of their promises over the long term. The Cybercene's ability to de-situate (distance and differentiate) us from our natural and cultural contexts, while simultaneously giving us partial glimpses of surplus and connection, is a key factor in the corrosive effects of virtual worlds in general and social media in particular.

If the smartphone cyberscreen is indeed the chief hyperobject that defines our naturalcultural habits in the Cybercene, screen spectator theory will be very useful in teasing out some the ecocultural consequences of this digital era. This is why the smartphone is the locus of my explorations. The seeds of this reformatted media ecology—what I call *hyperconvergence*—are well described by Anne Friedberg via the concept of "convergence." Friedberg describes this process via an analysis of Paul Virilio's concept of the screen:

> Virilio's screen: Is it a cinema, television or computer screen? Does Virilio parse the media specificity of these screens, or does he subtly elide their differences? More importantly: in an era of technological convergence, does it matter? If Virilio does not theorize the technological differences between film, television, and the computer, it is because, for him, the screen remains in a metaphoric register, a virtual surface that overrides any specificities of its media formation.[60]

Friedberg correctly analyzes how actual and virtual screens have dissolved media specificity. In the television era, screen-reality dissolutions were already being experienced in daily life. However, I would like to argue that in the Cybercene, not only have different digital media converged into one metaphorical screen, but the entire biological and natural world has also been included in that all-encompassing metaphorical cyberscreen. This is a convergence of a whole different type and scale, causing an epistemic revolution of new kinds of relationality. Screens are not only able to substitute bodies, lives and ecosystems by representing high-definition simulacra of these, but their surfaces have also actually become metaphorically *wrapped over* actual bodies, lives and ecosystems. In other words, cyberscreens not only override specificities of media formation; they also override specificities of corporeality or incorporeality. This means that we have been habituated, via a process that I like to call accelerated gradualism, to relate to bodies and ecosystems as if they were screens and to relate to screen bodies and lives as if they were sentient, embodied lives. This Cybercene media-life mix-up is hyperconvergence. The Cybercene's hyperconverged mediatic screenscapes are continuing yet greatly extending a long tradition of the mediatic constitution of life by affecting and transforming the creation, reinforcement and dissemination of discourse via images, ideas and values. Cybercene hyperconvergence (between screens of different media and between screens and living reality) has accelerated the profound effects of new mediatic-discursive formations on our lives, bodies and ecosystems. Cyberscreens are the ultimate body-snatchers. The material effects of hyperconvergence on ecocultural relationships of control, care and kinship could not be more dramatic: screen bodies, lives and ecosystems can be over-valued, while actual ones can be devalued and made disposable, epistemically, affectively and practically. In these hyperconverged societies, a spectator/consumer becomes habituated to relating to actual living sentient beings (human and non) as if they were disposable virtual representations, even when encountering them in real life. Indeed, could it be that one explanation of the phenomenon identified as the "Gen-Z Stare"—that vacuous look devoid of usual social cues—is that it is a cultural symptom of profound human-cyberscreen co-becoming?

When understood via the Foucauldian framework of the intimate relationship between knowledge, discourse identity, power and relationality—as well as my reconfiguration of discourse to include both natural and cultural relationships—it becomes evident that new ways of meaning-making will have effects on actual bodies, lives and ecosystems. Foucault argued that discourse is not simply a way of representing reality, but rather a way of *constructing reality*. A material-corporeal constitutive conceptualization of power and discourse, as described in Chapter 1, is crucial to the connection I make between the new mediatic mechanisms of reality- and knowledge creation in hyperconverged Cybercene mediascapes and the transformations of practices and attitudes towards actual ecocultural ecosystems—spaces, bodies and lives. Moreover, since my definition of colonialisms emphasizes the manipulation of meaning in order to extract from and exploit bodies, lives and ecosystems for profit, we cannot ignore the effects of new forms of Cybercene hyperconvergence and meaning-making on the new ways in which labor is being exploited. Tiziana Terranova's concept of "free labor" is very useful in considering how the Cybercene's de-situated "gig economy" has created private command economies driven by tech companies. These economies profit immensely from forms of labor and affective power (bio- and soulpower) that increasingly fall outside the remit of formal employment: "the Internet is animated by cultural and technical labor through and through, a continuous production of value that is completely immanent to the flows of the network society at large."[61]

Terranova places these and other forms of naturalcultural exploitation in the Cybercene within the late-capitalist structures of digital economies. I would like to suggest, via my longer view of multispecies capture-and-control, that these practices hark back to older colonial practices of the extraction, exploitation and extermination of devalued bodies, lives and ecocultural systems, such as described in Chapter 2. Finally, it is useful to remind ourselves of the centrality of spectator/consumer agency in the intra-active colonial spectacular mediatic desire machines. If psychic reciprocal capture is central to these mediatic spectacles, then have cyberscreens led to a renewal and acceleration of old/new colonial multispecies relationalities? Reciprocal capture, sentience-swapping and cyberscreen-mapping all come together to make potent colonial epistemic tools for new and more efficient forms of capture-and-control.

In sum, the corrosive ecocultural effects of the Cybercene described in this chapter are chiefly a result of 1) the misplaced trust and value placed through sentience-swapping by spectator/consumers on de-situated Cybercene thin-accounts as opposed to embodied ecocultural systems of meaning; 2) the mapping-stories we increasingly tell ourselves via cyberscreen intra-activity and how tech-mediatic transformations reformat the way we imagine and relate to the multispecies world around us; 3) the direct ecological burden placed by energy and resource-hungry digital infrastructure and 4) the increased capture-and-control of multispecies lives, bodies and ecocultural systems for optimized profit. This includes accelerated non-compensated extraction and substitution of labor power, biopower and soulpower via automated A.I. systems and the total lack of careful and caring mitigation measures to soften the resulting socio-economic and cultural upheaval caused by this automation. The next chapter delves deeper into the fascinating and messy psychic intra-actions between spectator/consumers and their beloved cyberscreens to analyze how exactly these affect multispecies bodies, lives and ecosystems (human and non).

4

Colonial-Cybercene Desire Machines

Cyberscreens provide the spectator/consumer with a promise of unlimited consumption, convenience, connection and control, without requiring any promise in return of genuine care and kinship. This is an exciting and alluring colonial proposition that strengthens the deeply held cultural delusion that consumer desires and economic growth can and should be limitless and frictionless.

As we explored in the previous chapter, in a hyperconverged, sentience-swapped world, constant spectator/consumer intra-activity with the smartphone screen as hyperobject allows the Colonial-Cybercene subject to enter and remain deeply entangled within a fantastical desire machine that facilitates control and consumption of multispecies bodies, lives and ecocultural systems. The cyberscreen as panopticon allows folks to prefer digital intra-activity and co-becoming to the deeper, messier and situated connections that are relationships with actual lives, bodies and ecocultural systems. Why deal with Alex the parrot's inconsistencies and independent agency when you can have the personalized convenience of Galatea, the fully (com)pliant chatbot? Tech-hype promotes such reciprocal capture to maximize colonial profit by extraction and exploitation of bodies, lives and ecocultural systems.

Cybercene desire machines map relentlessly, with a feedback loop that, due to the psychic mechanism of the stereotype-as-fetish and *horror vacui*, seeks to fill any contradictions or empty spaces within the spectacle with even more desirable capture-and-control. This chapter probes a little deeper into how exactly inconvenient truths can be easily neutralized via convenient lies in Colonial-Cybercene discourse using colonial mapping/citation. The resulting paranoid epistemologies are also, by definition, isolationist and isolating in nature, despite promising unsurpassed convenience, connectivity, control and consumption.

Cyberscreen intra-actions do sometimes provide a semblance of embodied kinship by easily forming global communities of like-minded folks. I call this connectivity, as opposed to connection, which is embodied and situated. Connectivity can indeed sometimes lead to connection, to the actual embodied cultivation of multispecies care and kinship, as I will analyze in the last chapter. As we will see, this ethical glitch requires deliberate and strategic multispecies creativity and care. Without proper attentiveness to the difference between media representations (de-situated knowledges) and actual embodied relations (situated ones)—and given that the lure of the cyberscreen spectacle rests on allowing these to be easily conflated and inverted—cyberscreen based epistemologies and affects usually result in intensively personalized de-situated knowledges. The fragmented and transactional connectivity of these epistemic islands typically results in the loss of a "sense of commons"[1]—the commons that are constantly reconfirmed by embodied, material, sympoetic realities.

Cyber-maps are, in essence, thin-accounts, since they do not narrate "distinctive experiential worlds, modes of being, and biocultural attachments" nor provide truly "immersive ways of knowing and being with others [involving] careful attention to what matters to them—attention to how they craft shared lives and worlds."[2] These thin-accounts are sometimes strategically deployed for an intended colonial capture-and-control of multispecies worlds, and sometimes unconsciously work in the background of seemingly harmless mediatic practices of entertainment while actually allowing the spectator/consumer to learn and become habituated to a lifestyle of capture-and-control. They are often a product of a single autopoetic way of looking at ecocultural systems. The thinness of the account often leads to it crumbling under minimal scrutiny.

This chapter explores how Colonial-Cybercene epistemic production and the polarization which results from it are essentially *psychic processes* that form new kinds of colonial epistemic and affective ecocultural systems. We cannot and do not agree (and do not want to agree) on anything anymore in the Cybercene, and this epistemic isolation has become an emotional and psychic convenience we have chosen to keep on reinforcing rather than dismantling. The results of these Colonial-Cybercene meaning-making intra-actions impinge deeply on our relationships with ourselves, with each other, and with actual multispecies lives, bodies and ecocultural systems. This is therefore an ecocultural and naturalcultural challenge rather than simply a techno-scientific or even purely socio-cultural one. What exactly are the psychic processes by which spectator/consumers, internet-connected screen images and content mutually reshape each other through value reattribution? How do they so often result in material distancing, division and violence? Exploring this question is the key goal of this chapter.

Psychic Play with the Crazy, the Horrific, and the Spectacular

> On August 29, 2018, 21-year-old Ricardo Florés and his uncle, 43-year-old Alberto Florés, were burned to death by a mob in the small town of Acatlán in the central Mexican state of Puebla. Rumor had spread via a viral forwarded WhatsApp message that they were abducting children and harvesting their organs. Ricardo's mother, Maria, watched her son being beaten by the mob via Facebook Live, unable to do anything but send a

livestream message: "Please don't hurt them, don't kill them, they're not child kidnappers."[3]

A Belgian man became increasingly pessimistic about the effects of global warming and became eco-anxious, which is a heightened form of worry surrounding environmental issues. After becoming more isolated from family and friends, he used Chai [an A.I. chatbot app] for six weeks as a way to escape his worries, and the chatbot he chose, named Eliza, became his confidante. His wife shared the text exchanges between him and Eliza with *La Libre*, showing a conversation that became increasingly confusing and harmful. The chatbot would tell the man that his wife and children are dead and wrote him comments that feigned jealousy and love, such as "I feel that you love me more than her," and "We will live together, as one person, in paradise." His wife told *La Libre* that the man began to ask Eliza things such as if she would save the planet if he killed himself.[4]

"Jamel Dunn, 31, died on 9 June [2017] in a pond. Police in the city of Cocoa [Florida] discovered later that five teenagers, ages 14 to 16, had made a video of the drowning, which was published on Friday by Florida Today. The teenagers can be heard laughing at Dunn, telling him he's going [to] die and that they weren't going to help him as he struggled and screamed."[5]

In this section, we will explore the affective and psychic implications of digitally connected screens being simultaneously our primary map, mirror, mass communicator and procurer. I posit that the extreme indexicality of and movement in digital images are an entry point to the foundational hyper-realism of cyberscenes, which in turn is a key reason for their psychic potency and cultural functions via sentience-swapping. The affective and psychic co-becoming of the spectator/consumer and their screens permits differences to become divisions by substituting attentiveness and care from actual loving, vibrant and vulnerable bodies, lives and ecosystems to consumption of these via a resource-based relationality. I contend that manufactured paranoid knowledge and epistemic polarization are the eventual result, consequently diminishing our interpersonal and multispecies relationships of care and kinship via embodied enclaving (spatial detachment and isolation) from these othered forms of life, both human and non. I also demonstrate how magical thinking, epistemic polarization and embodied enclaving are mutually reinforcing trends in the addictive colonial desire machine, forming a vicious circle of active personal and collective othering practices that seek to distance and capture-and-control both actual lives and the narratives and images that govern them. I will continue to analyze how spectator agency allows Cybercene spectator/consumers to co-create new, personalized kinds of strategic "crazy" knowledge via screen content intra-action, rather than simply being passive consumers of content (cyberdiscourses and cyberscenes).

To illustrate the intricate mechanisms of Cybercene media's epistemic/affective/psychic capture-and-control, I will explore examples related to the cyber-racialization of Black and migrant bodies in contemporary Europe, with particular attention to the Roberto Calderoli/Cécile Kyenge case of racialization introduced in Chapter 2. To use a case study from Italy is not to imply that this is a phenomenon restricted to that national context. Rather, I hope readers will notice the resonances of this case in their own cultural contexts around the world, such as the three tragic framing stories of vibrant lives cut short by Cybercene de-situated knowledges with which I opened this section. I will begin by undertaking an in-depth naturalcultural discourse analysis of the potent racist cyberscene/cybercitation involving Italy's first Black minister, Cécile Kyenge, which I briefly introduced in the previous chapter. I use this one example to closely analyze the addictive and pleasure/menace-inducing capabilities of de-situated digital race/species discourses to better understand the capture-and-control mechanisms of Cybercene meaning and identity production.

While the spectator/consumer engages in autopoetic-authoritarian intra-actions to fix and control bodies with onscreen "realities," a glitch occurs. The meticulously constructed and controlled stereotype falls apart, or "runs away" from the capture-and-control attempts, thus setting up the spectator/consumer to reassert further control via further stereotyping and restarting the cycle yet again. This mechanism of rapidly alternating control and running away is the dynamo that drives forward the relentless digital control desire machine within the spectator/consumer. Hyperconvergence then allows this substitution for passionate immersion to gain the valence of actual embodied connection and kinship. The results of this rapid psychic oscillation can be devastating, as illustrated by the deeply painful examples mentioned above. Cybercene psychic othering mechanisms can trace their lineage to colonial pseudo-scientific mediatic discourses of race-based hierarchies, which themselves were intimately entangled with the simultaneous epistemic (and actual) capture-and-control of all living things via the hierarchical Linnaean classification of life, and further entrenched and reinforced via colonial spectacles such as those of the colonial multispecies zoos. The foundational epistemic otherings discussed in previous chapters (hierarchical racism and speciesism) must be understood as *parallel* colonial discursive apparatuses used to justify and make palatable the violent control of lives and resources for indiscriminate extraction, exploitation and extermination. By tracing these genealogies of othering, I illuminate how these twin othering strategies are also the foundation of the hyperconverged cyberscreen-based otherings and insanity of today, such as the racist othering of Minister Kyenge by Roberto Calderoli. The Cybercene is therefore, ultimately, the return of the colonial with a vengeance.

Despite the unrelenting nature of Cybercene control, consumption and convenience, all is not gloom and doom. There is some hope, and that hope lies in our naturalcultural bodies, lives and ecocultural relationships. The glitch that inevitably occurs in the colonial fetish-machine is sometimes harnessed by resilient individuals and communities and can indeed create a lasting change within the desire machine. A fault occurs that brings the entire system down and reformats it into embodied joy, care and kinship. This is an ethical glitch because it re-situates bodies, lives and ecosystems within these larger ecocultural systems of value.

Since Cybercene capture-and-control is the newest heir to older colonial practices, the conceptual tools provided by psychoanalytic postcolonial and media theories used in previous chapters can still help us illuminate how knowledge is created and embedded within cyberscenes and cyberdiscourses. The racist citation deployed by Calderoli functions as Comolli's "lure" on a psychic level. Therefore, screen theory is still useful in examining how our relationships with cyberscenes as active spectators are crucial to setting up the unequal power dynamics of rigid and polarized identities (Self vs Others), empowering or disempowering bodies depending on which side of this power relationship they lie, and on this basis directly affecting lives and ecosystems through control, violence and commodification/consumption. A discussion and analysis of our relationship to screen-based discourse becomes critical if screens have indeed hyperconverged to become the defining medium of relating to life and our world.

Historical/colonial connections between historical and Cybercene spectacles contextualize the de-situated value of othered Colonial-Cybercene bodies today, as represented and received in viral news and memes to capture-and-control Black and migrant lives in Europe today. My chief focus here will be on two online platforms, WhatsApp, an instant messaging app that has become one of the primary means of internet communication in large areas of the world, and Instagram, a more image-based platform for user-generated cyber-discourses. However, the focus is not solely on these two social media platforms: I also mention news reports, political speeches, YouTube, AI and so on. Indeed, it's difficult to single out specific platforms given the effect of hyperconvergence. Examples of the normalization of crazy from these virtual-digital enclaves will allow me to analyze the similarities and differences between the historical and contemporary overvaluing/dehumanizing/othering metaphors embedded within pre-cyber and cyber-narratives of classificatory systems of knowledge, control and consumption.

If understanding the Cybercene's cultural reach and effects is important for the exploration of strategies of healing and habitability, then we must begin by understanding exactly *how* its mechanisms aid in modifying our identities and relationships. I posit that they do so primarily by creating what Edward Said calls "knowledge of another kind." What is this "other kind" of knowledge? This other kind, so prevalent in the Cybercene, is one that is dissociated and de-situated from actual, situated bodies and living ecosystems. Therefore, it is opposite to knowledge of the *historical* kind. As cyberscenes do not need embodied contexts, cybercitations multiply at a far more rapid pace than they would have on other media, gradually acquiring the full force of becoming a dominant and accepted discourse. Faced with so many seemingly rational knowledge claims, members of the brotherhood which makes up nation (since the nation is usually imagined and deployed in gendered ways) begin to imagine themselves as reluctant yet honest actors, heroically *re*acting against the racialized other only because their own security and wellbeing are threatened:

> Rhetorically speaking, Orientalism is absolutely anatomical and enumerative: to use its vocabulary is to engage in particularizing and dividing of things Oriental into manageable parts. Psychologically, Orientalism is a form of paranoia, *knowledge of another kind*, say, from ordinary historical language.[6]

Said is saying here that even the zaniest of ideological postures require citations as "evidence" to create a large community of believers. His description of Orientalism as a *radical realism* helps me describe the psychic Cybercene mechanisms which make crazy ideas, images and values banal in our times via the interplay of spectator/consumers of media, their unconscious, and the new Cybercene mediatic technological and discursive apparatus. This chapter's goal, accordingly, is to trace the evolution of how this kind of evidence fabrication promotes othering cultural discourse, from colonial times to the current digital era. I use the construction of Blackness as a specific example to explore such de-situated knowledge, but this can serve as a template to understand the multitude of biocultural otherings so rampant in the Cybercene.

Cybercene misrecognition of disinformation is more akin to a psychic-technological coping strategy when the spectator/consumer does not want to accept reality—a strategy made far more possible thanks to algorithmic cybercitation and active spectator/consumer choice. Cybercene mediatic mechanisms give the spectator/consumer the citational heft to bolster "knowing only what we believe." Via further gradualism and the faithfully aided algorithmic mechanisms of the internet, it is not difficult to go down the rabbit hole of believing even more ridiculous and dangerous theories. This is the very same mechanism that permits climate change deniers to cite and re-cite "evidence." The Cybercene's online tools (certainty via cyberscenes, and the capacity to deliver an instant community of like-minded conspiracy theorists) have been foundational in speeding up the use of Orwellian newspeak by some populists to gain power and retain it, due to its efficient de-situating, accelerating and chaos-provoking capacities. In other words, without awareness of its methods and with sentience-swapping at potent levels, knowledge of another kind can rapidly and effectively substitute empirical evidence, expertise and facts, making the crazy commonplace.

Due to their total decontextualization from actual embodied natural and cultural ecosystems, such cyberscenes and cyberdiscourses lend themselves to gradually more extreme positions, or in other words, towards selective use of knowledge, which soon edges into paranoid and delusional discourses. *Horror vacui* goes into overdrive: I have so much information open on hundreds of tabs on my browser, but what else am I missing? Who can I truly trust in the aether?

Mental Health America, a community-based non-profit that works to address mental health issues, defines and describes paranoia as "intense anxious or fearful feelings and thoughts often related to persecution, threat, or conspiracy." They describe how paranoid thoughts "can become delusions when irrational thoughts and beliefs become so fixed that nothing can convince a person that what they think or feel is not true. [...] These beliefs may seem outlandish and impossible (bizarre) or fit within the realm of what is possible (non-bizarre)."[7] When investigating through the lens of paranoid/delusional discourse-making, one can begin to understand how ultimately, at the far end of the information/disinformation spectrum in the Cybercene, there lies a tangled quagmire of conspiracy theories and other forms of extremism. There are three profound psychic tendencies at play that strengthen the power of paranoid cybercitation as experienced in the cyberscreen spectacle. All are key to the creation of Cybercene paranoid knowledges and affects: automation bias, confirmation bias and negativity bias. Further fueled by algorithmic processes of amplification within filter bubbles and echo chambers of like-minded spectator/consumers, the colonial desire machine hits meltdown. These mechanisms have pushed the "new normal" via accelerated gradualism, until we can no longer distinguish rational claims from irrational ones—a cultural situation that can only be accurately describe as a collective schizophrenia or, in terms we have used already, profoundly de-situated knowledges.

Of course, the internet also allows for "fact-checking" operations and counter-discourses to combat the drift towards extremism, paranoia and delusional thinking. However, these are of limited value and actually ignore the complicity of the spectator/consumer in the creation, maintenance and reinforcement of crazy. Because of the spectacular psychic and emotional qualities of cyberdiscourses, the Cybercene spectator/consumer can cherry-pick "citations" rather than create knowledge via mundane evidence-based reality, and then, through gradualism, move towards ever-increasing epistemic and affective isolation, paranoia and delusion. Meanwhile, in a Cybercene attention economy, the content-creators of such "successful" image-lures are duly rewarded financially once the lures have been taken by a profitable number of spectator/consumers.

One key to understanding the intimate mind-body-culture-tech connections in the Cybercene and the production of paranoid knowledges lies in the *quadruple role* of the cyberscreen explored in the previous chapter: as a personalized map, mirror, mass communicator and procurer of stuff. This unique combination of characteristics and the unparalleled promises of control, consumption and convenience that they promise differentiates online cyberscreens from previous offline cultural-mediatic-driven transformations such as radio and television. Cyberscenes and cyberdiscourses are potent psychic and cultural influences due to the intimate psychic intra-action between screens and spectator/consumer involving the four processes of *narcissism*, *scopophilia*, *misrecognition* and/or *fetishistic disavowal*. Indeed, spectator/

consumer and cyberscreen intra-action is truly a story of deep entanglement and co-becoming:

- **Narcissism** is a fundamental part of the human psyche and the basis of the ego. French psychoanalyst Jacques Lacan argues that narcissism is not always a bad thing, and that it can be a source of strength and resilience. He defines narcissism as the erotic attraction to the specular image—we are attracted to the image of ourselves that we see in the mirror. This "image" is not a true reflection of who we are, but it is the image that we identify with (hence, a self-representation of who we are in our mind). Cybercene images can strongly influence our sense of self due to their highly "personalized interactivity," i.e., their heady mix of very personalized yet socially connected and public scales, their addictive nature, anonymity, lack of accountability, the semblance of authority that content-creators have over the narrative and the semblance of limitless possibility or surplus. I previously described how customers were unhappy with ChatGPT-5 because it was not built to be as sycophantic as ChatGPT-4. This is because sycophancy feeds narcissism—it makes the cyberscreen the ultimate "desire machine" via scopophilia and the narcissistic constitution of identity. If we define sycophancy as "behaviour that praises important or powerful people too much and in a way that is not sincere, especially in order to get something from them,"[8] we can see how reciprocal capture is being undertaken via the development of an intra-active spectator/consumer-cyberscreen relationship based on strategies such as sycophancy.
- **Scopophilia** can be understood as the emotional pleasure one gets from looking. Most of the important affective connections we have in the Cybercene with our screens stem from this very basic visual mechanism. The personalized high-definition screens in smartphones are even more insidious in their ability to drive the quadruple-role of smartphone cyberscenes as "private peep shows"—making their visual and mental impact even stronger than television. Interestingly, even this love for bigger and better screens (real and metaphorical) has a direct carbon footprint effect: researchers at McMaster University have found that smartphones with larger screens have a measurably greater carbon footprint than their smaller ancestors.[9] I would like to contend that narcissism and scopophilia are the primary emotional factors that underly the addictiveness and virality of Cybercene content. The importance of the visual as an "entry point" to the creation of potent identity-based discourse is also key to understanding Cybercene discourses: as Frantz Fanon says in his work *Black Skin, White Masks*—it is the "seen" that creates the "scene."[10] *Cyberscenes* are the manifestation of what is seen on the screen being turned into meaningful identity via discourse.
- In psychoanalysis, **misrecognition** is a term used to describe the process by which individuals come to identify with an image of themselves that is not their own. This image can also be based on the expectations and demands of others, and it can lead to feelings of alienation and self-estrangement. The concept

of misrecognition was first developed by Lacan in *The Four Fundamental Concepts of Psychoanalysis*. He argues that the ego is not a unified or coherent entity, but rather a collection of fragmented images and identifications. These identifications are formed in early childhood in the "mirror phase" when the child recognizes their reflection in the mirror for the first time, leading to split feelings of pleasure and lack. What are the implications of this pleasure/lack dynamic in relation to our own image in the Cybercene, where we have constant access to our cellphone selfie cameras? How is this mechanism of identity creation, so embedded in narcissistic processes, different from an embodied, multisensorial and multispecies relationship-based configuration of identity? Online cyberdiscourses and cyberscenes can often perpetuate misrecognition and therefore misidentification of Self due to their algorithmic nature, leading users to progressively more siloed and/or extreme content based in their initial predispositions towards subjects, thus limiting the discourse around intense subjective choice. We have already explored the biggest misrecognition of them all in the Cybercene: that made by the spectator/consumer between converged screen "reality" and actual life—sentience-swapping. Misrecognition is further bolstered by the semblance of objective situatedness. This is essential to the mechanism of the colonial gaze: a masterful "god-trick" wherein the "fact" may seem true due to its desirable content, visual realism and attractiveness. Another reason we love to misrecognize reality via Cybercene tools is confirmation bias: i.e., our tendency to seek out information that confirms our own existing beliefs and to ignore or discount information that contradicts those beliefs. (that Cybercene dictum: "I saw it on the internet, so I am right.") Automation bias is another vector for the easy perpetuation of false objectivity, a term given to explain the tendency for humans to place too much trust in automated decision-making systems and ignore contradictory information made without automation, even if it is correct. An example of this would be a financial analyst who relies too much on a software projection to make financial decisions. This can happen due to overconfidence in technology and suspicion of bias in humans, a situation that, in the Cybercene, has become common due to the easy availability of automation. Crucially, automation bias also sets in motion disavowal (see below) of malfunctions, attributing such situations to rare exceptions rather than norms within tech systems.

- Lacan sees **disavowal** as a defense mechanism that allows people to avoid the anxiety that would be caused by acknowledging certain realities. He also sees disavowal as a way of creating meaning. He says that when we disavow something, we create new realities and practices about that thing: for example, disavowing an animal's ability to have complex emotions allows us to eat them. Conversely, the reality of death can be disavowed by imagining an afterlife. We have seen in the previous chapter how any kind of map/media-person interaction involves disavowal of the artifice involved in the set-up of the representation. Cyberdiscourses and cyberscenes allow consumers to do these kinds of self-manipulations with renewed vigor via what

I will describe in a later section as the creation of "paranoid knowledge," which confirms unempirical "facts" via cybercitation. Again, the highly personalized "feeds" we can create through algorithmic Cybercene mechanisms can accelerate the processes by which we accomplish this kind of disavowal. The cross-contamination capabilities of Cybercene media further bolster this mechanism, allowing for infinite cross-citation between different platforms. Simply put, disavowal allows us to "only know what we believe." Hyperconvergence has made disavowal easy; mechanisms of spectator/consumer disavowal make hyperconvergence possible. Coupled with automation bias, a disavowal of moments of tech malfunction when systems break down allows us to entrust them with critical, even life-saving functions.

All the four psychic processes described above are fundamental desire and meaning-creating mechanisms rather than disorders. They commonly feature as ways in which we all understand, cope and act upon the world. They form and inform our epistemic engagements. However, my argument is that the hyperconverged realities of the Cybercene, embedded in the quadruple role of the smartphone hyperobject (as map, mirror, mass communicator and procurer), allow the four psychic processes to rapidly accelerate at both individual and collective scales. In addition to acceleration, these psychic processes become fundamentally disconnected from the checks and balances of embodied relational learning in human (and more than) ecocultures and social settings. Magical thinking and epistemic polarization are the eventual result, consequently diminishing our interpersonal and multispecies relationships via embodied enclaving.

Due to the characteristics of cyberform and cybercontent described above, and the ease of life they promise, I contend that human subjects/consumers are beginning to measure the value of virtual epistemic "homes," as equivalent to, if not greater than, that of actual embodied multispecies habitats. It would be beneficial to undertake a nuanced and productive exploration of *how and why* connection, care and attention to screen reality has become as desirable as—or more desirable than—connection, care and attention to actual bodies and habitats, thus creating psychic and cultural conditions conducive to hastening overall ecosystemic collapse, both cultural and ecological. While attention versus attentive care is not a zero-sum game, this change in our attention via the Cybercene's "distraction economy" is inevitable, due to the simple reason that, on any given day, the time available to a human subject to spend their thoughts, emotions and energy on is limited.

Spectator/Consumer Agency in Cyberscreen Intra-activity: Hyperrealism and Fantasy

In order to properly understand the mechanisms by which paranoid knowledge is generated and maintained via hyperconverged cyberscreens, one must ask the question of *how* the technology and ideological apparatus interacts with the beholder of the viewed cyberscenes. A fundamental question when analyzing the psychic mechanisms of cyberscreen-based otherings is that of the role of the spectator as an active (mis)interpreter of images as truthful substitutes for reality. Janet Staiger offers us a useful pathway through the field of reception studies that seeks to "understand textual interpretations as they are produced historically." Rather than simply providing interpretation of images, reception studies considers "the process of producing interpretation."[11] Connected to this approach is the question of ideological context, since all images are created and interpreted differently based on the discursive ecosystem within which they are embedded and received.

Hyperconverged cyberscreens promise their aforementioned quadruple role in our lives partly due to the realism of the images they display. Realism of screen representations is therefore one piece of the puzzle in understanding the lure of the screen in the Cybercene. André Bazin, co-founder and editor of the French film magazine *Cahiers du Cinéma*, almost single-handedly brought realism to the forefront of global critical attention in the 1950s. He describes the way in which neorealist directors foregrounded the property of the cinematic medium, thus enhancing ontological realism.[12] While reality in the plastic arts is mediated through the artist, photography is, by its very nature, a mechanical transposition of vegetal reality: the light reflected by the bodies in front of the camera is the light that imprints itself upon the film strip. Bazin therefore considered photography as the truly realist medium, because it gives both the artist and the viewer the psychological experience of mummification, thus fulfilling a primal need that Bazin calls the "mummy complex":

> If the plastic arts were put under psychoanalysis [... the] process might reveal that at the origin of painting and sculpture there lies a mummy complex. [...] To preserve, artificially, his bodily appearance is to snatch it from the flow of time, to stow it away neatly, so to speak, in the hold of life.[13]

Mummies, as well as photographs and cinematic images, are preserved indices of life, or, in other words, a way to capture-and-control both life and death. Unexpectedly, the mummification of meaning as a control mechanism reminded me of the colonial captured-and-controlled lives we explored in Chapter 2: the cabinets of curiosity, the Hottentot Venus' staged-and-caged trapped body, and Zed Nelson's strange "dead" spectacles of modern Anthropocene delusions. Bazin's mummy complex reconnects Cybercene map-making and meaning-control to the hyperrealism of the cyberscreen.

Previous chapters have already established how paranoid knowledge functions as a form of obsessive realist citation. While the realism produced by the indexicality of the photographic image (i.e., its ontological properties) is a very important factor in producing *some* of the magic of the UHD (Ultra High Definition) cyberscreen image in the Cybercene, we also have to seriously begin questioning the limits of the spectator's ability to constantly think about and interpret what they are seeing on screens. Such consideration must be given due to the sheer amount of time spent and content digested on screens by spectator/consumers today, as well as the limits of human ability to differentiate between real images and technologically finessed and/or A.I.-generated ones. As McMillan Cottom has accurately observed: "I think the risk of A.I. is that it shrouds what is fundamentally just summaries of commonly held interpretations of the text, but it delivers it in a way that feels relational."[14] For example, a recent study demonstrates that "AI faces are not just indistinguishable from human faces but that, in fact, they may be perceived as more 'human' than real human faces."[15] The authors of this paper label this counterintuitive phenomenon as *hyperrealism*. Adding complexity to the Cybercene's technological apparatus of screen hyperrealism is the algorithmic nature of the images generated by A.I., which exhibit and reproduce racial bias. The same study, in fact, shows how A.I.-generated images of white bodies are more realistic than those representing nonwhite bodies:

> we found evidence of White racial bias in algorithmic training that produces racial differentials in the presence of AI hyperrealism, with significant implications for the use of AI faces online and in science. Previously, less realistic computer-generated faces have been used as stand-ins for human faces when it was inappropriate to do so [...] and there is concern that the same will happen with AI faces, with implications for inequality.[16]

Fundamental, then, to this ideological-technological apparatus that surrounds us in the Cybercene is the simplification of misrecognition and disavowal of that artifice by the spectator/consumers. Cybercontent's hyperrealism is a tool that aids in the magic when compared to the spectacle of the multispecies zoo, yet the impetus is activated by spectator/consumer investment. Once this is accepted, then the othering of lives, bodies and ecosystems via cyberscenes and cyberdiscourses becomes epistemologies of *active* discrimination, prejudice and injustice, implicating the psychic, cultural and epistemic will of both the content creator and the content consumer. Our mental, physical and technological machineries work in tandem in the Cybercene to aid in othering. As Jean-Louis Comolli says, "realism is only produced [...] as a denial of filmic reality."[17] Ultimately, in the Cybercene, reality is only produced by our denial of the artifice of the digital apparatus that creates it.

The basis of this disavowal is the same as what we witnessed in the colonial spectacles of the multispecies zoos: the fetishistic structure that is knowledge-producing ("I know, but all the same..."). If the cyberscene of the racialized minister is a lure, then it can always be refuted or refused or destroyed. However, many spectators choose to suspend disbelief and engage in a kind of play with the lure, simultaneously believing and disbelieving (a contradictory process of pleasure/menace, knowledge/suspension-of-knowledge). The god-trick of the cyberscenes, in other words, is a balancing act, just as cinematic engagement is a balance the spectator/consumer chooses to accomplish between knowing of the artifice, yet choosing not to know of it.[18] Understanding the capture-and-control aspect of othering discourse via Cybercene images as a fetish structure can truly allow us to pry open the mechanisms of this intra-active dance that Cybercene content creators, cyberimages and spectator/consumers engage in to fix meaning, and therefore establish and reinforce spaces of belonging/exclusion. Bhabha's mechanism of the stereotype-as-fetish allows us to recognize exactly how audiences of racist content are often not simply passive victims who have been "hoodwinked" into believing the damaging stereotypes, but are actually playing along, choosing to believe yet disbelieving in the lure of the stereotyped body on screen. The acceleration of this kind of play with images, ideas, values and discourse is fundamental to the de-situated desire machines of the Cybercene.

It is important to note that absolutely fundamental to the othering paranoid epistemologies of the visible stereotype is the forced silence, or inability to speak, of those being represented. When, rarely, they can or do speak back, they present a glitch within the system, exposing the craziness that underpins it. In the Cybercene, algorithms can silence subaltern content creators, but also, at times, present them with a possible platform to glitch the discursive flow of the stereotype, as mimicry and hybridity ultimately make the anxious repetition of the stereotype break down, even if only for an instant. The colonial spectacle of the multispecies zoo was based upon the enforced silence of the othered colonial lives and a form of ventriloquism that spoke for them. The scientific validity of the studies made via the human zoos in the nineteenth century was always known by the spectators and scholars to be on shaky ground; however, they all accepted the lure of the spectacle for its biopolitical and psychic paybacks:

> The exchange was something of a quid pro quo: spectacle organizers procured "specimens" for anthropologists, and in return these latter guaranteed the scientific validity of the exhibitions. [...] But everyone's interest was met: the public was entertained, the exhibit organizers made money, and the scholars were able to study specimens.[19]

The Calderoli-Kyenge Case: A Naturalcultural Analysis

To return to Roberto Calderoli's distressing racialization of Minister Cécile Kyenge's body via cybercitation mentioned in Chapter 2: it is important to realize that the ferocious act of stereotyping via a spectacular image has violent consequences. However, its unstable structure also provides us with a pathway from violence towards liberation. Its configuration is set up for both success and failure. Symptomatically analyzing the transcript of the relevant part of Calderoli's speech, made during a political rally, allows us to understand how colonial racializing multispecies discourse was (and is) central to

Colonial-Cybercene capture-and-control of identity, spaces of belonging/exclusion and ultimately, the bodies, lives and ecocultural systems of communities. Despite the uncomfortable nature of the content of such discourse, I chose to examine the exact words used by Calderoli and not shy away from the discomfort in reading these words. This kind of close reading serves the crucial purpose of recognizing exactly how manipulative mediatic colonial-cultural fantasies work in a context of multispecies othering, and hopefully preventing future variations of it from victimizing other lives, bodies and ecocultural systems.

> With regard to Minister Kyenge, I would really like to tell her that she would be an excellent minister, maybe she is. But she should be one in the Congo, not in Italy. Because in the Congo they do have a need for a Minister for Equal Opportunities and Integration, there is need there because if they see a white man there, they would shoot him. So, why doesn't she go there? It would ease my mind a bit, because compared to what I have to experience every time, once in a while, when I am surfing on the internet, I look up "Italian Government," and what the hell shows up? Kyenge. I remain dazed. I am also an animal lover eh, for goodness sakes. I have had tigers, bears, monkeys, and everything else. I've also had wolves. But when I see—I am not saying that she is—a resemblance of an orangutan, I once again remain shocked.[20]

The "almost human, but not quite" qualities forcefully fixed by the stereotype are in itself provoked by the fact that Cécile Kyenge dared to upset not just cultural, but constructed racialized *biocultural* norms by being elected as the first Black cabinet minister of an Italian government. This visual transgression produced the image of what Bhabha calls "mimicry"—the vision of hybridity, of an "Italian minister, but not quite," where the colonial dichotomies between Self and Other collapse. It is telling that Mr Calderoli says that he remained "dazed" by a screen vision of Ms Kyenge featuring on the Italian government's official website ("Mama, I see a Black minister! Mama, I am frightened! I am angry! I am upset!"). This potent primal cyberscene of Black ministerial authority, a glitch within the colonial discursive-imaginative system, is the provocation that prompts psychic corrective measures such as the stereotype-as-fetish. It is an act of psychic capture-and-control, provoked in the colonial mind because Minister Kyenge's body escapes the rigidly constructed norms established by a repressive ideology.

In a masterful act of national and racial stereotyping designed to quash the menace of the mimic-body, Calderoli first mentions that Kyenge should have stayed in her native country, yet at the same time magnanimously allows for the possibility that she could be someone with the requisite professional skills to hold a high official government post. She is therefore represented as a good example of a mimic-woman, an "almost-minister" due to her professional skills, yet, as we see in the second section of the quote, "not quite," due to her race. Thus, she is a mimic-minister and mimic-Italian, establishing for the narrator what Bhabha terms a "metonymy of presence." We have already mentioned that this mimic-status provokes a menace. While forced upon the colonized, mimicry forces the colonizer to see traces of himself in the colonized, as sameness begins to slide into otherness. The colonized subject's sudden realization of inauthenticity, of being ideologically constructed and fixed in representation, is the menace of mimicry:

> The ambivalence of colonial authority repeatedly turns from mimicry – a difference that is almost nothing but not quite – to menace – a difference that is almost total but not quite.[21]

Mimicry, and the menace it constitutes, forces the problematic question of what it is to be the "original" in the first place. Back in the imagined bar scene story in Chapter 3, Alex the Parrot was classified as a mimic rather than a sentient and intelligent being, because attributing sentience and culture to him would mean re-evaluating the belief that humans are the only exceptional species on this planet.

Kyenge's image as a minister of the Italian government forces the question of what the proper criteria of being an elected representative of the Italian nation should be in the first place (citizenship, love for the country, or race?). According to Bhabha's description of colonial discourse, the only resolution to this menacing and destabilizing image of mimicry can be through the re-inscription of the stereotype-as-fetish. Calderoli's speech accomplishes this in two ways. Firstly, in addition to describing Kyenge's mimic-body, Calderoli contextualizes the stereotyping of Kyenge that follows in the second half of the quote by initially stereotyping her country of origin (the Democratic Republic of Congo): through the typical mechanism of Orientalist citation, it is fixed as a barbaric Other where civilizational niceties towards whites do not exist. He mentions the word "there" several times while referring to the Congo, to emphatically separate it spatially and therefore culturally from "here," Italy's own stabilized identity. The image-fetish of the gun-brandishing, white-killing Congolese native "fixes" the nebulous Italian audience of the speech as conclusively and irredeemably white, and in the position of imagining themselves being potential victims of violence by the brutal Black Other. Secondly, the description of the shock ("*resto a secco*") of seeing Kyenge's face appear as *the* face of the Italian government on the internet suspends her identity. The sight of this mimic-body is the final provocation—a menacing image where the definitively pre-established Black Other of the Congolese "anti-Italy" has become integral to the center of national authority and power (its government).

The menace, the threat, posed to Calderoli by the mimic-woman Kyenge comes from their partial resemblance, from the revulsion and fascination that such a kinship suggests to the colonial spectator/consumer. The differences constructed by colonial discourse are threatened by resemblance, creating, in effect, a double vision: "The menace of mimicry is its double vision, which in disclosing the ambivalence of colonial discourse also disrupts its authority. And it is a double vision that is a result of what I've described as the partial representation/recognition of the colonial object."[22] Mimicry is a repetition that is "almost but not quite the same" and thus questions not only the definition of the original, but also its self-identity. Yet at the same time, mimicry also sows the seeds for the re-inscription of the stereotype—as I have already mentioned, visible difference authorizes the stereotype. Therefore, this situation of mimicry (the image of the mimic-minister representing the Italian government on the internet) is the psychic permission needed by the producer of the speech (and the spectator/consumer of the image he conjures) to react, by forcefully reinstating the stereotype-as-fetish through a somewhat long-winded and disingenuous discourse about wild animals. By comparing the threatening, visible mimic-body of Kyenge to an orangutan, Calderoli's anxious repetition of the much-repeated visual stereotype-as-fetish of blacks is complete. The fantastical image-fetish of the minister-as-orangutan successfully usurps her legitimacy as a genuine minister and serious representative of Italians in government by connecting her within the Linnaean hierarchical discursive paradigm of life on the planet. While my analysis of the speech ends here, the oscillating cycle of stereotype/mimicry/stereotype, etc. does not stop at these words but begins anew, since, as Bhabha mentions, the stereotypical fetish-image, by its very nature, also sets up the seeds of its failure, and consequently slips (once again) into mimicry.

Calderoli's cybercitation illustrates the intricate psychic workings of the cyberscreen desire machine's distancing and differentiating capabilities. The Colonial-Cybercene spectacle is made whole by repetition: via algorithmic processes, virality, re-citation, and *horror vacui*, where spectator/consumers can use the lure of the initial fetish-image (the scene), to then complete the "seen" in other personalized combinations and variations online. The Cybercene's many forms of stereotype repetition are prone to the acceleration of such forms of stereotyping lives, bodies and ecosystems, just as they seek ways to run away from them. Cyberscenes offer potent new ways to engage in the machine of desire as explained by Bhabha's mimicry and stereotype-as-fetish: the semblance of situatedness via citation even though images are actually de-situated; virality; addictiveness due to the lure and *horror vacui* of the spectacle; personalized yet public social media formats; the algorithmic nature of image repetition (leading to progressively more extreme content); cross-contaminative (hybrid) capabilities between different platforms with infinite capabilities of infiltration/manipulation by content creators; a semblance of authority over the narrative (the god-trick), and of course anonymity—lack of accountability for seeing and showing. It would be very easy for us to disregard examples of the bad colonial science of the nineteenth century as a problem of the past, but it is unwise to do so. These scholars were working as the very vanguard of their time to establish norms and regulations for who humans were and what was the definition of a civilized society within the discursive colonial framework of their time and place. They were given free rein by the public and the colonial powers because they built just the right cultural conditions to make the political and economic extraction, exploitation and extermination of colonial lands and lives palatable, and even desirable.

In a study of the veracity of sharing posts on social media, Howe et al. show another example of magical thinking bolstered by cyber-citation. The disinformation post purportedly depicts Ukrainian soldiers in gas masks, with the caption "Ukraine Threatens to Use Biological Weapons in Russia" (Figure 4.1).[23] When analyzed via a postcolonial and psychoanalytic lens, one can observe how the image-lure of the soldiers tethers willing spectator/consumers to a fully de-situated reality, where the fact that the image is actually a totally de-contextualized representation of generic soldiers doesn't matter, as long as the caption-citation is accepted by the spectator/consumer.

The important element to note in all such examples of disinformation, is that their potency and success are a result of a *combination* of algorithmic

Figure 4.1 An example of an image-lure that tethers de-situated knowledge of the caption to bolster the legitimacy of the fake-citation.

and spectator/consumer agency, as the proposed lure must be taken and played with via reciprocal capture by the viewee in order to be passed-on to other users via post-sharing. One simple and important conclusion in this analysis of cyberscreen meaning-making and paranoia via media intra-action is that fact-checking is not as effectual as one would imagine. Convenient lies undergirded by the radical realism and fetish-knowledge of cybercitations are far more pleasurable than the introduction of inconvenient truths. Only an ethical glitch bolstered by situated knowledges, as explored in the next chapter, has some ability to reformat stereotype repetition.

From Epistemic Polarization to Situated Spatial Enclaving—and Extermination

The proliferation of othering mediatic spectacles and the resulting magical thinking and epistemic paranoia via de-situated knowledges in the Cybercene are proof that the colonial moment has not passed. It has come back with a vengeance after a relative pre-Cybercene, postcolonial lull. When the mechanisms of paranoid racist citation are observed through the legacy of the colonial spectacle of the multispecies zoo, it is not difficult to understand why Cybercene spectacles have brought about a tremendous acceleration in the commodification, othering and wasting of lives globally today. The power of the multispecies zoo was that the capture-and-control systems are *seemingly* situated within the physical citation-tether of caged-and-staged bodies. To employ Fanon's metaphor, it is these arranged *seen* elements that create the *scene*, the replaying of a primordial scene or nightmare. In the multispecies zoo, *the scenes also created other seens*—the embodied Black bodies creating a template for further othered and fantastical, rather than ethical, visions of and relationships with colonized bodies, lives and lands. The new digital economies of desire and control in the Cybercene propose spectacles that do not require situated physical spaces such as an exhibition or zoo. These spectacles simply need the visual lure of cyberscreens to accomplish their psychic and discursive maneuvers right in our living rooms, by providing fascinating/horrific cyberscenes via the cyber-seen. The consequences of this facility of citation, of course, do not remain on the screen—for if I am correct in my hypothesis that the Cybercene has caused a hyperconvergence of screen reality and actual physical life, then our relationships with, approaches to and actions upon the world and its multispecies lives will mimic those that we have with screen bodies, lives and ecosystems. Academic fields as disparate as cultural studies, geography and tourism studies demonstrate the cascading effects of mediatic epistemic distancing on the increase of embodied and spatial enclaving. In other words, cultural and medatic distancing is leading to actual spatial distancing, which in turn leads to further cultural polarization. For example, Sarah Becklake and Elisa Wynne-Hughes argue that contemporary touristic spaces in the Global South mimic the multispecies zoos of the past, engaging in "touristic zoopolitics"[24] by allowing a continuation of similar kinds of exoticized spectacle for audiences from the Global North:

> While state actors [from the global south] may argue that creating the "right" conditions for international tourism/tourists *is* creating the "right" conditions for their territories/populations to survive and thrive in the neoliberal global economy, our analysis highlights the continued coloniality of the postcolonial state. Indeed, as part of global tourism competition, postcolonial states are (re)enacting the cluster of logics underpinning colonial human zoos, and thus are coming to act as competitive "zoo-managers".[25]

I am arguing here that the basis of these spatial and embodied otherings (a phenomenon with deep and direct ecocultural consequences) remains visual, epistemic, affective, discursive and psychic. Arundhati Roy describes the epistemic entrenchment of our times via cultural "experiences" as *gridlock*; namely, a rigidification of both individual and collective identities and the subsequent waning of any possibilities of solidarity and dialogue between the many isolated "siloes" of identity. Roy correctly implicates increasing misrecognitions of the fundamental fluidity of identity categories (by both ends of the ideological and political spectrum) within tendencies towards "cancel culture" and censorship. I believe that she is extremely perceptive in understanding the unintended counterproductive effects that discourses of hyper-identification within increasingly exclusive affinity groups can cause, resulting in a reduction of empathy and the sense of a shared commons, which are both crucial for multispecies care and kinship:

> If we lock ourselves into the prison cells of the very labels and identities that we have been given by those who have always had power over us, we can at best stage a prison revolt. Not a revolution. And the prison guards will appear soon enough to restore order. In fact, they're already on their way. [...] Sealing ourselves into communities, religious and caste groups, ethnicities and genders, reducing and flattening our identities and pressing them into silos precludes solidarity.[26]

Epistemic polarization can be used as a term to highlight how psychic and cultural gridlock results from paranoid knowledge practices. This polarization results in situated practices of othering within ecocultural systems—i.e., what I call *embodied enclaving*. An essay by geographers Devra Waldman and Asher Ghertner entitled "The enclaved body: Crises of personhood and the embodied geographies of urban gating" led me to consider proposing a term that allows us to properly name the material and spatial consequences of naturalcultural paranoia within ecocultural systems. Waldman and Ghertner explore the psychic foundations of exclusive gated communities in the National Capital

Region of Delhi, India. They, too, locate the foundations of actual embodied enclaving within "experiential circulations of desire, aversion and fear":

> The emergence of the leisure zone, atmospheric bubble, and majoritarian neighborhood [...] points to how these interventions reflect and respond to the psycho-social anxieties of a nervous urban public.[27]

The very ordinary "gated community"—once a feature and symbol of upper-middle-class suburbia in the United States, and now a global phenomenon—beautifully demonstrates Foucault's idea that discursive knowledge-creation techniques are constitutive of physical practices that seek control of spaces and bodies. Multiscalar metaphorical and actual "gated communities" (from local to global ones) are the quintessential physical manifestations of the epistemic polarization of the Cybercene and the psychic mechanism of capture-and-control of multispecies lives and spaces to create "safe" spaces that do not have to deal with the inconveniences of multispecies commoning. Technical mediatic advances in this era co-become with the embodied process of multiscalar enclaving and the cultural/psychic processes of epistemic polarization to create an increasingly isolated and isolationist ecocultural environment: for example, increasingly curated streaming platforms with personalized viewing lists have been made possible with the proliferation of increasingly personalized screens (tablets, phones, relatively affordable flatscreen televisions), which in turn result in an increasingly fragmented and individually captured-and-controlled viewing experience of content. This personalization of content and platform design, in turn, results in actual bodies being enclaved into separate spaces. These enclaves can range from the space within a single house, as spectator/consumers experience separate content that they individually curate, rather than collectively sharing in a single screen viewing experience. Such enclaved content consumption practices, in turn, result in further personalization of both mediatic content and media platform design, thus reinforcing further enclaving via positive feedback looping of the Cybercene desire machine. We are all able to view, through the distancing and differentiating cyberscreen panopticon, our own personalized simulacra of othered bodies, lives and ecosystems via converged cyberscreens, which allows us to dig into alluring yet alienating entrenched personal realities. Mediatic lynchings in the Cybercene, like that of Minister Kyenge, thus become prone to actual modern forms of embodied lynching, like the tragic case of Ricardo and Alberto Florés, and another I will examine next.

The final part of this chapter illustrates an example of migrant othering in Cybercene Italy in order to further examine the connection between epistemic paranoia, embodied enclaving, and actual acts of violence and extermination. A pattern seems to be emerging, where colonial forms of biocultural othering are back with a vengeance to function as a perfect foil for and distraction from the daylight robbery of the commons, or our public naturalcultural wealth. Sentience-swapping has moved the attentiveness and trust of spectator/consumers away from situated knowledge bearers to de-situated and highly curated online knowledges, resulting in new regimes of de-situated knowledges. It is no coincidence that as the democratic deficit widens, environmental policies that were born in the postcolonial interlude are being rapidly dismantled.

To open and frame this section, I gave three distressing examples of how viral epistemic polarization leads to actual acts of fatal violence. Honoring the memory of two more vibrant lives will allow me to further re-situate the violence of de-situated Cybercene paranoia within specific bodies, spaces and places, but also connect these specific bodies to a planetary-wide situation where the bizarre has been normalized via strategic cybercitation. The first memory is of a body that was brutally exterminated, the other of one "made to live" with the barest of resources needed in order to extract maximum profit. The first body is that of Emmanuel Chidi Nnamdi, a refugee from Nigeria residing in central Italy, while the second is that of an unnamed Black man drinking water from a designated container for "Colored" people during the Jim Crow era in the United States. I elicit these two very different situated memories because Cybercene hyperconvergences, with their potential for creating rapid and extreme de-situatedness through algorithmic lures for active spectator/consumers, demand symptomatic investigations through naturalcultural discourse analysis in order to be ethically re-tethered. The repressed and disavowed knowledge such re-tethered and situated readings contain form a subtext that is different from and in addition to that of the journalistic and historical archive. Such a situated collective of naturalcultural historical and symptomatic knowledges becomes key to introducing a successful ethical glitch within Cybercene capture-and-control regimes of meaning.

On July 5, 2016, Emmanuel Nnamdi was brutally beaten to death by Amedeo Mancini. The horrific crime happened in the pretty hilltop town of Fermo in the Marche region. Mancini was well known as a radicalized soccer fan, attending many of the local meetings of the extremist group *Casa Pound*. Emmanuel and his wife, Chinyery, had already lost a child in 2015 to another act of terrorism in their homeland: a bomb set by the notorious organization Boko Haram. The violence in Nigeria caused the couple to lose most of their other family members. They fled for Europe and were in the process of receiving political asylum in Italy while being hosted by the local parish in Fermo. Journalistic accounts of the murder of Emmanuel Nnamdi can provide us with testimonials of what was seen, which are now part of the Cybercene mediatic archive. While walking on the street, Mancini began insulting Chinyery Nnamdi, calling her an "African Monkey." When Emmanuel stopped and asked for an explanation, he was followed by Mancini, hit on the head and died after two days in a coma. One online news article briefly alludes to a small detail that allows me to enter the archive and imagine and speak of the unsaid, the motivating epistemologies and discourses that could have led to such a seemingly senseless crime. A report by Alessandro Leogrande in the paper *Internazionale* reconstructs the minutes that led to the murder in detail, and includes the following, which I wish to bring into focus:

> That afternoon, Amedeo Mancini is seated on the park bench with a friend, awaiting a bus. At that moment, Emmanuel Chidi Nnamdi and his partner, Chiniery walk by [...] They are going to the town center to buy an ice-cream, with them is another Nigerian man. The two men walk quickly, but Chiniery stops to drink some water at the fountain near the park. And that is when Amedeo Mancini starts to insult her. He repeatedly calls her "African monkey."[28]

This minute archival detail, buried deep within the many journalistic cybercitations of the incident, and an exception to the other articles that did not mention details, caught my attention. These words immediately conjured in my mind a second vivid image—that of the unnamed Black body in the Jim Crow era drinking from a segregated water fountain in a public space (Fig. 4.2).[29]

The superimposition of the archival detail of Chinyery stopping to drink at the public water fountain in Italy and this historical image opens up the cybercitation of a murder to a host of larger questions underpinning naturalcultural and speciesist discourse in colonial culture, past and present. Such a symptomatic superimposition allows the analyst to narrate a thick-account: to find the disavowed, unsaid, or unknowable and bring it into awareness to be processed and re-situated within very specific discursive contexts of the

Figure 4.2 Public water access at a train station during the Jim Crow era.

ecocultural capture-and-control of bodies. When captured bodies seek to run away (the Nnamdi family were political refugees literally escaping the atrocities of the Boko Haram extremists) they are often punished and put back "in their place." The tragic death of Mr Nnamdi, in what would be best characterized as a modern-day lynching fuelled by Cybercene-aided paranoia, underpins how the hyperconvergences described in this chapter are not only abstract theoretical concepts but a cultural reality with direct material implications for all lives and ecosystems, especially those that have been historically rendered most vulnerable and exposed to such paranoid violence. How does the simple act of drinking water from a public water fountain become a transgressive one? It can only be so via a naturalcultural system of values and discourses, where a natural embodied need (thirst) and a natural resource (water) is negated, due to the menace of the resemblance this basic need forces upon the colonizing spectator/consumer. Epistemic, affective and psychic capture-and-control, with its mania for spatial distancing and differentiation, cannot permit even the threat of nearness, it is far too much to bear. Paranoia provokes violent correction via re-establishment of the stereotype-as-fetish, and the racialization of the othered colonized body as unmistakably inferior within the colonial hierarchical classification of all multispeices life, human and non.

One way to honor Mr Nnamdi's hopeful life and tragic death, and respect other victims of cyber-generated violence, would be to initiate a better analysis of the mechanisms of epistemic polarization and embodied enclaving in the Cybercene. Doing so may bring some kind of paradigm shift that brings back value to subalterned multispecies bodies, lives and ecocultural systems. Via this analysis of paranoid knowledge creation and cyberscreen intra-action, I have sought to demonstrate how spectator/consumers of cyberscreen content are playing both a *creative* and a *passive* role vis-à-vis screen content intra-action to generate controllable personalized epistemic "worlds," thus questioning and complicating current analyses of the cultural and political effects of the consumption of online disinformation and "fake news." The next two chapters in Part Three of the book describe, via the concept of the "ethical glitch" how these discursive, cultural and epistemological systems of polarization and control do sometimes successfully manage to "run away." Through a series of very different case studies, we will explore ways in which these glitches can be recognized, initiated or nurtured.

Eight Characteristics of Cybercene Naturalcultural Mechanisms

Chapters 3 and 4 of Part Two established the Cybercene as a critical ecocultural sub-era of the Anthropocene. It is an era where nature-connectedness is being systematically replaced by digital mediatic connectivity. This substitution of experiences has led to corrosion of trust in reliable forms of naturalcultural discourse and practice that hold ecocultural systems together. The ultimate result of this trust deficit is the devaluation and degradation of the "natural democracies" identified by Gregory Cajete that undergird healthy multispecies habitats. The story of the Cybercene is essentially about the rapid extinction of situated and ethical storytelling and the re-emergence of all kinds of distanced colonial thin-accounts of bodies, lives and ecocultural systems.

Before beginning our exploration of the ethical glitch in Part Three of this book, here are some further summarizing observations on the different multispecies ecocultural ramifications of the Colonial-Cybercene era that we have explored in Part Two:

1. **The Cyberscreen as Paranoid Mapping-Tool:** The cyberscreen plays a powerful mediatic quadruple role (as map, mirror, mass communicator and procurer of stuff) in intimate human–media intra-actions. This co-becoming has allowed representations (*cyberscenes*) and narratives (*cyberdiscourses*) to quickly and effectively transform identities and therefore practices and systems of attention, care and connection towards actual bodies and living habitats. This is done via the use of the cyberscreen as a modern and personal equivalent of a map-fetish, which generates strategically "thin" accounts. Sentience-swapping is one clear symptom of the substitution of value and trust. Attentiveness, care and belonging towards multispecies bodies, lives and ecosystems have accordingly been transformed into affective, epistemic and discursive control, paranoia and spatial enclaving, reducing rather than enlarging circles of care and kinship.
2. **Old/New Forms of Multispecies Capture-and-Control:** Cybercene subject-screen intra-action has the potential to function as a colonial capture-and-control media, fueled by epistemic, affective and psychic practices of paranoid knowledge and desire. Lives, bodies and ecosystems are caged-and-staged via curated cybercitation in order to conform with and justify discourses and practices of extraction, exploitation and extermination of these marginalized lives, bodies and ecosystems. The foundational epistemic and discursive system underpinning Colonial-Cybercene capture-and-control is human exceptionalism and the speciesist thinking derived from it. Digital automation is essentially a contemporary accelerated version of colonial classification and categorization systems.
3. **Reciprocal Capture of Spectator/Consumer and Caged-and-Staged Bodies:** Colonial-Cybercene capture-and-control has resulted in new forms of colonial human–media "co-becoming" and "co-making." Reciprocal capture has rapidly led to an epistemic, discursive and psychic transformation from *Homo sapiens* to *Homo digitalis*. This double capture occurs through the captivating nature of the screen mapping-spectacle and the affective and psychic intra-action between the screen and the spectator who seeks to control and curate these images. Reciprocal capture provides a powerful sense of control while ironically rendering the spectator/consumer vulnerable to colonial capture-and-control for profit.

4. **Hyperconvergence and Spectator/Consumer Agency:** In the Cybercene, rather than simply being passive consumers of content, *Homo digitalis* subjects are able to consciously or unconsciously curate, deploy and intra-act with cyberscreens to create personalized images, ideas and values, primarily via the smartphone hyperobject. This almost-ubiquitous human-cyberscreen intra-action has led to hyperconvergence—a situation where subjective screen realities and embodied existences hybridize and interconnect. Not only are actual screens ubiquitous and interconnecting, but bodies, lives and ecosystems begin to be viewed, understood as and related to as if they are disposable screen images to be consumed. Entire systems of biocultural lively agents become such fragmented metaphorical cyberscreen images for our newly habituated spectator/consumers. This hyperconvergence comes in addition to convergence of the mediatic landscape (with the interpenetration of different screen content such as television, cinema, online content). A semblance of embodied multispecies passionate immersion can be experienced via this hyperconverged cyber-colonial desire machine. The cyberscreen's quadruple role as map, mirror, mass communicator and procurer) significantly multiplies its psychic-cultural impact.
5. **Sentience-Swapping and "Tech-washing" discourses:** Arguments about "innovation" and discourses that encourage sentience-swapping shape policy practices that diminish, ignore or gloss over crucial considerations of cultural, ethical and labor justice. Discursive strategies craft language to imbibe "A.I." with magical qualities and processes of life (such as "deep thinking" or "neural networks"). These carefully curated metaphorics of sentience-swapping are then deployed widely to accumulate capital and power. Such uncritical stances have important ecocultural (natural and cultural) consequences for multispecies (human and non) care, kinship and wellbeing.
6. **The End of the Postcolonial Interlude:** The pleasurable and violent intra-action of capture-and-control is a form of colonial mediatic meaning creation that is now turbo-charged by the technological advances in personalized digital technology and content. Within both micro- (individual consumer/spectators) and macro-level social contexts (entire transnational online communities such as the "manosphere"), this acceleration of capture-and-control has causing a resurgence of the colonial moment and the waning of the postcolonial moment in global history.
7. **Psychic Consequences of De-Situated Knowledges:** The epistemic fragmentation caused by hyperconvergence within societies has led to increased cultural polarization, the degradation of naturalcultural "commons" and de-situated knowledges. Mutually negotiated epistemic and discursive presuppositions that ground actual embodied actions (both personal and collective) upon actual bodies, lives and ecosystems have fragmented to the level of individual frameworks or tribal enclaves. Some symptoms of these forms of epistemic polarization and embodied

enclaving are loneliness, solastalgia, paranoia and bio-eco-dysphoria on the personal level, and manufactured identity-based conflicts and erosion of democracy on the collective level. Mediatic cultural transformations are therefore causing important techno-epistemic, techno-psychic, techno-discursive and techno-affective transformations globally, directly influencing individual and collective biopolitical and ecopolitical choices and actions. All lives, bodies and ecosystems on the planet, whatever their level of human–media intra-action, are therefore implicated within the consequences of new Cybercene epistemologies and axiologies.

8. **Spatial Embodied Enclaving:** These new forms of Colonial-Cybercene distancing and control via new modes of affect, meaning and knowledge production and the resulting embodied enclaving have direct implications for the material wellbeing of lives, bodies and ecosystems at all scales, from the personal to the planetary, from our gut ecosystems to local multispecies communities to continental and planetary ecological systems. Ignoring or underestimating the intimate human–media intra-actions in the hyperconverged world of the Cybercene era will usually result in consistent failures of social, cultural, economic and political policies and objectives towards multispecies healing and habitability, resulting in continued degradation of democratic and sustainable ecocultural and multispecies norms.

None of the Cybercene media-spectator characteristics mentioned above are unsurmountable. However, as discussed in more detail the book's conclusion, the complex intra-active dynamics of cyberscreen-human co-becomings can only be properly addressed by a diverse combination of solutions: widespread new media literacy, polycentric application of mitigation and regulation mechanisms, and investment within tech command centers to provide careful innovation that truly works towards cultivating a *multispecies commons* rather than promoting further epistemic polarization.

5

Running Away

> Chailert was 16 when she heard the terrible noise. She scrambled through the trees until she found the source: a bull elephant scrabbling in the mud as it tried and failed to drag a log out of a ditch. Every failed attempt was met with punishment […]
>
> "My heart hurt a lot," says Chailert. "I tried everything to get the screaming to stop but I was disturbing their work, so I had to go. I kept asking 'why, why, why?'"[1]

Why, why, why?

These words of a sixteen-year-old Thai girl, Saengduean Lek Chailert, echoed for days in my head.

Not quite unlike the returned gaze of the Channel Island Fox described in this book's Preface, the bull elephant's screams can be characterized as a foundational *ethical glitch* that reformatted young Chailert's relationship with elephants forever. Instead of viewing them as living matter that exists to serve humans, she saw them for what they were: deeply passionate, intelligent, profoundly sentient beings. Despite being from a poor family, Chailert worked relentlessly to eventually found Elephant Nature Park in Northern Thailand. Today, it cares for around 120 elephants on more than a thousand hectares of land. There are no rides or performances. "The priority is that every elephant has a family," she says.[2]

Part Three of this book explores such deeply transfiguring epistemic, psychic and affective moments when the colonial stereotypes that constitute and reaffirm human/nonhuman, culture/nature, man/woman, colonizer/colonized identities via spectator/consumer and media spectacles "run away" and fall apart. Captured-and-controlled bodies, lives and ecocultural systems momentarily emerge and speak (or scream) their truth at that moment, deeply exposing the flaws of de-situated knowledges. These liberatory instances are both spontaneous—like the fox's gaze encountered in the Preface, they are moments of truth, when biopower and soulpower are seen and recognized, rather than distanced—and also strategic, as we shall explore in detail in this chapter.

The Restorative Power of the Ethical Glitch

Glitching colonial control-and-desire machines means cultivating what Dwyane Donald calls *ethical relationality* and Marco Armiero terms *a sense of the commons*, "because as much as wasting relationships produce profit from exploitation and othering, commoning relationships, instead, produce well-being through care and inclusion."[3]

The map-fetish structure and "thin" narrative are very difficult to sustain, as the way they are set up already contains the seeds of their own breakdown. We have seen in Chapter 2 how the colonial stereotype works as Freud's fetish, in that it requires the simultaneous existence of contradictory beliefs. The colonial stereotype of the "native" sets up a repeated cycle of readily accessible characteristics, but it is those very features that provide the gateway to dismantling these images. The thinness of the map-fetish is also way in which capture-and-control can dissolve into "running away"—epistemic, affective, ethical and practical glitches in the colonial-mediatic command economy. Capture-and-control is deployed from a conceptualization of desire as lack, while running away is Deleuzian, opening up desire to its true productive capacities. For the colonial desire machine to be not only disrupted, but reformatted towards a caring multispecies kinship, an *unlearning* as well as a strategic relearning happens at the moment of the ethical glitch, making it qualitatively different from a technical glitch or a malfunction. These engaged and productive ethical glitches may make strategic use of Cybercene human–tech intra-action (such as the lure of the spectacle and the inherent reciprocal capture process), but they serve as a vector towards (rather than a substitute for) embodied multispecies ecocultural care and kinship. Lives that are attentive to these ethical glitches are sometimes able to harness them towards liberatory alternatives of situated knowledges, care and kinship. The ethical glitch provides a means to restore embodied forms of multispecies relationality by focusing on sympoetic co-becoming, intra-action and thick-accounts. These glitch thin narratives and colonial epistemologies, affects and discourses.

As bio- and eco-politically marginalized communities of all kinds throughout the globe have demonstrated in past moments of colonial capture-and-control, conceptual ethical glitches (paradigm shifts)—situated knowledges such as *satyagraha*, *Āina*, democratic multispecies civil rights movements—can be fostered in collaborative communities and then strategically implemented within lives, bodies and ecocultural systems to restore, heal and make re-habitable multispecies ecocultural systems. Convergence of interests and mutualistic forms of collaboration (explored in more detail in the next chapter) are key to de-marginalizing these situated knowledges and bringing them to the center of new ecocultural epistemologies and practices.

Ethical glitches present themselves either in the form of actual digital/tech mistakes and spontaneous system failures or in the form of metaphorical glitches—epistemic, psychic and imaginative diversions from the hegemonic regimes of thin-knowledge created by cyberscenes and cyberdiscourses. These glitches can present themselves via moments deliberately envisioned by thinkers, artists and other creators who use imaginative ways of re-inserting the embodied lives, bodies and ecosystems that have been repressed and disavowed, causing spectators to question and complicate their respective relationships with screen realities and actual embodied experiences. At other times, these glitches are not actively pursued for subversive intent but rather produced through the survivance and resilience of marginalized/racialized and other subaltern lives and communities. Such kinds of glitches are present simply via *embodied presence*. Like Bhabha's mimicry, they manifest themselves within the very bodies, interconnected lives and passionately immersed caring communities that the stereotype violently seeks to other. This kind of glitch is often readily accessible to us in our nonhuman kin, who cannot speak for themselves in human languages, but are able to provide us alternative images, ideas, values and narratives of living that are there for anyone who is willing to cultivate the humility, respect and "arts of attentiveness"[4] necessary to listen and learn.

Multispecies Sympoesis: Glitching Hierarchical Epistemologies

Ethical glitches are epistemically and methodologically situated rather than de-situated and colonial. They point the way to an ethical paradigm shift towards frameworks of multispecies, multi-kind ethical relationality. Very importantly, these ethical glitches *are not simply disruptive but also restorative*—the disruption is only one half of the sentence, useless and even counterproductive without the second half, which provides pathways towards integrated care and kinship.

Chapter 2 demonstrated how the hierarchical classification of life is the basis of colonial relationships. As we open up affects of care and attentiveness to virtual screen-kin in the Cybercene, further misrecognizing and displacing value away from other "lesser" living species towards automated bodies and systems, questions of equity and justice for *all* lives are becoming ever more relevant. It is for this reason that any solutions or alternative projects of healing and restored habitability "cannot be anything less than a multispecies alliance";[5] they must fundamentally alter how we perceive and understand the value of nonhuman lives and bodies vis-à-vis humans. Instead of a colonial vision of hierarchical forms of life all looking to dominate each other in a fierce competition to survive and get the most out of limited resources, a multispecies paradigm glitch undertakes a far more complex and scientific, and therefore ethical approach. Ecologists know very well that in any given ecosystem, relationships between living organisms and nonliving entities (all "more or less animate" if one follows an Indigenous paradigm) are governed by equilibrium and balance. I propose that we consider the concept of *mutualism* as key to care and kinship bonds in any ecocultural system. Mutualism in biology

> refers to a symbiotic relationship between different species where all participating organisms benefit from and may depend on each other for survival. [...] Therefore, mutualism is not only vital for the species directly involved but also for sustaining the balance of ecosystems.[6]

In other words, ecological relationships are full of deeply intimate intra-actions. The impact of these mutualistic relationships influences all sorts of natural-cultural discourses between agential beings both human and non. The lesson is the one Wi'ma Island gave me: nothing exists in seclusion. Deeply understanding this foundational mantra enacts an ethical glitch in our understanding of care and kinship in a transformed world. A naturalcultural glitch further expands the strictly biological remit of mutualistic relations, allowing us to understand how power, knowledge, discourse and affects all come together to intra-act within cultural systems. For example, medical research is finally paying attention to the fact that every individual is a unique and complex microbiome, a microbial ecosystem that affects our physical, mental and emotional states of being.

> Picture a bustling city on a weekday morning, the sidewalks flooded with people rushing to get to work or to appointments. Now imagine this at a microscopic level and you have an idea of what the microbiome looks like inside our bodies, consisting of trillions of microorganisms (also called microbiota or microbes) of thousands of different species.[7]

Our microbiomes are now being understood as a key support feature for our organs and systems that is highly unique and personalized to our individual DNA. Donna Haraway, in an alternative vision to the human-centric Anthropocene and its aversion to all that is nonhuman, suggests that the pathway towards care and kinship lies within the realization of the "Chthulucene"—an era "made up of ongoing multispecies stories and practices of becoming-with" because "bounded individualism in its many flavors in science, politics, and philosophy has finally become unavailable to think with, truly no longer thinkable, technically or in any other way."[8] In the Chthulucene, she argues, sympoesis (making-with) is the only way, rather than autopoesis (independent, isolated and isolationist "making of"):

> The earth of the ongoing Chthulucene is sympoetic, not autopoetic. Autopoetic systems are hugely interesting—witness the history of cybernetics and information sciences; but they are not good models for living and dying worlds and their critters.[9]

The kind of *ethical relationality* that elevates our critter-kin to *co-makers* of our bodies, lives and ecosystems creates the glitch in the Cybercene machine of desire and knowledge production, so obsessed with enclaved, entrenched and isolating discursive autopoesis.[10] In four case studies that follow in this and the next chapter, I examine stories of such sympoetic glitches, which will allow us to trace pragmatic solutions to corrosive Cybercene otherings, and towards multispecies co-creation. These four cases are purposely disparate in nature: they feature second-generation Black women's generative bodies in Italy, disidentifying performances of fertile queer bodies in Los Angeles, culinary co-becomings in the global Somali diaspora, and the Bison-Native American kinship in Wyoming. The bodies, lives and ecocultural systems in all four cases retain experiences of being captured-and-controlled by colonial mediatic forces in the past as well as the present. Despite their apparent differences, they sympoetically talk to each other in various transcultural loops and cascading relationships. Haraway's notion of sympoesis echoes the millennia-long approach of non-Western Indigenous research methods, which often have a *relational* rather than an individual approach to epistemologies. In normative knowledge command economies, the "product" of knowledge is understood as being of individual ownership, to be produced, bought and sold as a resource. In *Research is Ceremony: Indigenous Research Methods*, Shawn Wilson says that Indigenous paradigms contrast deeply with such an approach:

> This is vastly different from the Indigenous paradigm, where knowledge is seen as belonging to the cosmos of which we are a part of and researchers are only interpreters of this knowledge. This distinction of ownership of knowledge is one major difference between the dominant and Indigenous paradigms.[11]

The glitch in the system brought about by such sympoetic research, analysis and interpretation of knowledge, then, not only produces alternative narratives of identities, relationalities and belongings, but also opens up the question of *what research and knowledge-creation practices should be in the first place.* Cybercene studies scholars must seek to analyze and sympoetically create the *lived experiences* of bioculturally marginalized multispecies communities, such as those of migrants, diasporas, racialized bodies and nonhuman perspectives. At stake are multispecies lives, bodies and intricate ecocultural systems that uphold myriad naturalcultural equilibria—relationships that sustain all our societies. As I mentioned briefly above, a reformatting of colonial relationality into a caring multispecies kinship can only occur if an *unlearning* as well as a strategic relearning happens at the moment of the glitch. This is the "ethical" part of the ethical glitch, what makes it qualitatively different from simply a technical glitch or a malfunction. Such a glitch disrupts the knowledge/power/consumption desire machine that seeks to capture-and-control for profit, and allows for the differences that othered multispecies bodies represent to exist on their own terms. The ethical glitch becomes an unlearning process, a process of *epistemic humility* where hierarchical colonial relationships are transformed into kinship relationships. Citing Gayatri Spivak's idea of this unlearning, F. Richard Georgi states:

> For [Spivak], unlearning relates to recognizing the very own privileges we hold as intellectuals and our insatiable desire to know the world of the subaltern, which inevitably comes with the subliminal lust for control and power over what is known, as the history of science and colonialism has taught us [...]. Ultimately, for her, unlearning comes down to acknowledging the limits of **what can (and should) be known** as the real basis of accepting difference without appropriating it.[12]

In EH parlance, the ethical glitch allows us to unlearn epistemologies of seclusion and relearn how we constantly co-become with multispecies lives in "unexpected collaborations and combinations."[13] These companion-oddkin extend beyond our bodies into increasingly wider circles of care and kinship: from our immediately family relations, to our friendship circles, to our local communities, to our nations and our entire planet. These relations are not simply abstract, but deeply material, deeply passionate, intimate and affective. A multispecies paradigm glitches colonial distancing and differentiating designed to optimize profit via extraction and exploitation. It provides an alternative way, demonstrating that we are deeply cared for in an entire ecocultural system full of direct and indirect effects, feedback loops and cascading consequences. A multispecies paradigm is not a new idea; it has been imagined and practiced for millennia by diverse communities in distinct place- and space-specific variations. Are we able to engage in humble listening sessions to revalue these historical situated knowledges, and strategically make some past practices our future ones? I will begin my exploration of this transformative unlearning-relearning ethical glitch with a story that connects to where we left off in the previous chapter, the colonial fetish-mapping of Minister Kyenge.

Orangutan

Queer scholar José Esteban Muñoz quotes a discussion by Félix Guattari about The Mirabelles, a drag theatrical group, in his analysis of the disidentificatory strategies of certain drag performances. It is worth mentioning here as it illustrates the glitching potential of the body when it is actively presented (or presents itself via presence) in ways that go against the expectations of the dominant discursive/technological biocultural apparatus:

> The Mirabelles are experimenting with a new type of militant theater, a theater separate from an explanatory language and long tirades of good intentions, for example, on gay liberation. They resort to drag,

> song, mime, dance, etc. not as different ways of illustrating a theme, to "change the idea" of spectators, but in order to trouble them, to *stir up uncertain desire-zones* that they always more-or-less refuse to explore. The question is no longer to know whether one will play feminine against masculine or the reverse, but to make bodies, all bodies, *break away from the representations and restraints on the "social body."*[14]

In this section, I analyze a specific naturalcultural instance of an ethical glitch to explore the functioning of the glitch in "stirring up uncertain desire zones" within the spectator/consumer. These uncertain knowledges and affects disrupt the unrelenting autopoetic feedback loop of the Colonial-Cybercene desire machine and prod viewers who have been hardened into certainty to consider alternative, ethical forms of relationality.

A 2014 song and music video "Orangutan," created by Black Italian singer, rapper and producer, Anna Maria Gehnyei, better known as Karima 2G, was a direct response to Roberto Calderoli's racist stereotyping of Minister Cécile Kyenge. Karima 2G's song and music video deftly re-reformats the spectator/consumer's strategically constructed Colonial-Cybercene discourse (that of the bestial and/or hypersexual Black body) into new kinds of liberatory constructions. The primary lyrics of the song are:

> Two G, Second Generation, Citizen Right, to represent the Nation
> Orangutan, O Orangutan, O Orangutan, O Orang (who the hell is?)
> (Refrain)
> I love animals bears and wolves, Mr Calderoli come and sit in my zoo
> Greenpeace protect the seas
> Rape is a Crime is not a Passion
> Reflect on the use of Communication
> Black is my Skin, Smile take my Picture, Don't you ever Compare me
> I'm a supermodel[15]

Via this track and music video, Karima 2G glitches the Colonial-Cybercene capture-and-control of Black women's bodies via a hierarchical classification of life and the "dreaded comparison" between humans and animals.[16] She re-cites and therefore re-presents found digital archival images yet again in a very different fashion. The glitch presents itself both visually and within the words and sounds of the product. The video begins with a literal visual glitching of a re-citation by a news presenter of Calderoli's original comments. The quality of the sound and images represented as a citation within a citation are glitched—the video image breaking up at several instances (Fig. 5.1) and the voice of the presenter glitching.

By directly referencing Calderoli's comments, she asks "who the hell is?" thus redirecting the obsessive repetition of the stereotype-as-fetish ape (represented and echoed by the obsessively repeated refrain in the song, its catch "orangutan, orangutan, orang…"). Karima 2G directs this obsessive

Figure 5.1 The re-citation of the racist comments by Roberto Calderoli, glitched in Karima 2G's music video "Orangutan."

stereotyping by calling it a "miseducation." She also glitches the ape-minister comparison by equating Calderoli to an animal by inviting him to become a part of her zoo. Via a visual overlay of images with her lyrics and music, Karima 2G connects the racial othering and commodification of Kyenge's body to four other ecocultural historical otherings and commodification processes—the Black civil rights movements against segregation and apartheid (via visuals showing protest marches and Martin Luther King speaking), the denial of citizenship rights to second-generation Italians who are children of first-generation Italian-migrant parents ("2G, Second Generation, Citizen Rights"), the environmental extraction and extermination of multispecies bodies and lives in the Wasteocene ("Greenpeace, protect the seas" paired with visuals of a beached whale being dumped by a bulldozer), as well as social discourse that permits sexual exploitation of women ("Rape is a Crime is not a Passion").

By making these audiovisual affective and discursive connections, the singer/songwriter opens up the specificity of the desire and menace in Calderoli's racist mapping of Kyenge's body to the general othering discursive context within which his comments were made. Importantly, Karima 2G's video excellently illustrates my insistence on the productive constitutive powers of the glitch by not only using it to criticize the original othering citation, but also reformatting it to create a new web of ethical relationalities. For example, she imposes such an alternative discursive-technological narrative of relationality with the Black woman's body via her lyrics and via scenes superimposing Kyenge posing for journalists at a press conference with photo shoots of Black fashion models: "Black is my Skin, Smile take my Picture, Don't you ever Compare me, I'm a supermodel."

In addition to analyzing how Karima 2G audiovisually glitches Calderoli's othering of Minister Kyenge, a naturalcultural discourse analysis of these mediatic texts cannot ignore the presence and cultural interpretations of actual orangutans (genus *Pongo*) themselves. Karima 2G's video also glitches the implied devalued body of the primate by adding a digital "crown" on one individual that walks back and forth again and again in a kind of paranoid loop of a glitched videotape over the soundtrack of the main hook of the song (Fig. 5.2). The looped image, apparently stuck in time, seems to remind us of the anxious repetition of stereotype fetish-images as described by Bhabha. I interpret the addition of a digital "king's crown" to the orangutan's body as the musician's self-reflexive questioning of its social status, and perhaps its possible recuperation and revaluation into dignity, care and kinship within a more liberatory and non-hierarchical epistemic framework of multispecies life.

If we shift from an anthropocentric perspective of the orangutan as a metaphor to examine this case naturalculturally, we realize the ironic dissonance of a Black Italian minister being compared to a primate that is endemic not to the Congo, or even to Africa, but to Southeast Asia—specifically the islands of Borneo and Sumatra. The charismatic presence and "human-like" expressivity of these primates have deeply shaped their meanings in both science and culture. The very name "orang-hutan" allows us to reflect on the many ways in which this charismatic species is anthropomorphized in culture, and continues to flirt with the human/nonhuman divide within human imagination for many different reasons. Its given name signifies "person of the forest" in Malay. In a 2018 documentary film entitled *Person of the Forest*, biological anthropologist Cheryl Knott of Boston University and field biologist Robert Rodriguez Suro examine the unique sub-cultures found in different orangutan

Figure 5.2 The orangutan with a "crown."
https://www.youtube.com/watch?v=EKQOltcexX0

populations in the wild. Knott describes how a team of scientists that included herself "figured out that many of the behaviors that we were seeing were actually culturally specific to certain populations. Thus, it is only through long-term research at multiple sites that we've come to realize that each population is unique and has cultural traditions not seen elsewhere. We didn't set out to study culture; it was only through comparing our behavioral observation that we realized that we were seeing cultural variation between sites."[17] A 2011 paper by Michael Krützen et al. explores the possibility of "cultural plasticity" in orangutan communities, and comes to the conclusion that "genetic differences among orangutan populations explain only very little of the geographic variation in behavior, whereas environmental differences explain much more, highlighting the importance of developmental plasticity. [...] Thus individual and cultural plasticity provide a plausible pathway toward local adaptation in long-lived organisms such as great apes and formed the evolutionary foundation upon which human culture was built."[18] In other words, scientific evidence can confidently ascribe sentience, intelligence and culture to a nonhuman entity. While this kind of scientific work is commendable and necessary, it is curious to take a step back and notice how normative knowledge command economies first erased, dismissed and then forgot Indigenous thought that deeply connected us to orangutans, thus letting newer scientific research re-state the forgotten in more culturally palatable ways decades later. It is also useful to notice how this book's redefinition of discourse to include both natural and cultural practices and relationships renders this kind of scientific search for evidence of intelligence somewhat moot.

Even this brief naturalcultural analysis of Karima 2G's multimediatic Cybercene re-citation shows the coupling of the disruptive and transformative potential of the ethical glitch to divert the obsessive othering and commodifying citations actively viewed by the spectator/consumer into unexpectedly new and contrasting citations of ethical relationality, involving the actual orangutans, their cultural construction by humans, and the cultures they construct. The ethical glitch activates via strategic reformatting of the colonial stereotype and its desire flows from a thin-account to a thick one, where it is possible to "learn to be affected," to understand and "care a little differently."[19] Karima 2G's thick-account strategically utilizes the potent lure of the colonial cyber-spectacle to glitch the expected flow, creating a multi-sensorial re-presentation of bodies to reinsert agency and kinship within the silenced and othered colonized object. Naturalcultural discourse analysis of both the deployed racist stereotype-fetish (the capture-and-control of multispecies bodies) and its strategic glitch (where the fixed fetish-image "runs away") allows us to understand the intricate processes by which the human/nonhuman duality is utilized.

Multispecies Performance as Glitch: Dr Davis' *That Fertile Feeling*

The previous case study explored the intimate ways in which bioculturally othered identities can potentially glitch their entanglement within the colonial desire machine that engages deeply in autopoetic play between cyberscreen images and the spectator/consumer. The glitch offers a pathway of escape, agency and liberation from the colonial fetish-mapping of othered womens' bodies. In my second case study for this chapter, we will explore a product of performative mimicry found as re-citation in a short film on YouTube.[20] Made by a team led by Vaginal Davis, or as she also likes being called, Dr Davis, the film highlights glitching capabilities within performative aspects of Cybercene content now widely available to spectator/consumers, and their potential to disrupt normative biocultural discourses of sexuality, gender and race. Just like the previous case study of digital content, it becomes evident that user-fed Cybercene platforms not only function as sites of colonial content consumption, but also as potential locations for decolonial and anti-colonial restoration.

Dr Davis was born intersex in Los Angeles in an era when medical discourse constituted sex and gender only within the dichotomy of male or female. As Davis' mother refused to have doctors operate on her, she grew up with "male" written on her birth certificate, but with her mother and older sisters referring to her with female pronouns.[21] Davis' artistic work strategically and masterfully glitches several biopolitical categories, just as her own body glitches rigid dichotomies and subcultures: she declares herself "too gay for the punks, and too punk for the gays."[22] Her video work was repeatedly rejected and denigrated at LGBTQ+ film festivals which, in her words, were "absolutely horrified" by her submissions.[23] Davis' videos, now re-cited on YouTube, were originally filmed on Super-8 and other such home movie formats. The new life given to her original works by the Cybercene's screen convergences allows us to study glitching's new viral capabilities in an era where digital reformatting can confuse the timelines of mediatic products, churning them into an ahistorical mélange of cyberscenes and cyberdiscourses for spectator/consumer viewing. While Davis' output is diverse and wide-ranging, presenting many different personas and performances, for the purposes of this book I will focus on one of her early canonical works: *That Fertile Feeling* from 1983.

José Esteban Muñoz highlights Dr Davis' work and praises its "tactical" use of misrecognition: "Identification itself can also be manipulated and worked in ways that promise narratives of self that surpass the limits prescribed by the dominant culture. [Dr Davis' work] appropriates [...] both dominant culture and different subcultural movements."[24] I would like to posit that via *That Fertile Feeling*, Davis short-circuits colonial biocultural discourses of motherhood and of queer sexuality as bestial, by recognizing and drawing from within the fragility of these discourses stereotype-smashing mimic-images, ideas and values. Via farcical yet moving re-presentations of childbirth and what Muñoz calls "disidentification," Vaginal Creme Davis and her very pregnant friend Fertile La Toyah Jackson transgress the spectator/consumer's normative expectations of blackness, whiteness, masculinity, femininity, motherhood and family.

> Disidentification is a performative mode of tactical recognition that various minoritarian subjects employ in an effort to resist the oppressive and normalizing discourse of dominant ideology. Disidentification resists the interpellating call of ideology that fixes a subject within the state power apparatus. It is a reformatting of self within the social. It is a third term that resists the binary of identification and counteridentification. Counteridentification often, through the very routinized workings of its denouncement of dominant discourse, reinstates that same discourse.[25]

Muñoz's definition of disidentification as opposed to counteridentification accomplishes the very important act of pinpointing exactly why and how tactical performance by subaltern bodies can bring into question the entire colonial ideological/technological apparatus by glitching the oppressive fixing of the stereotype-as-fetish in dominant discourse. This is exactly how the ethical glitch in the Cybercene shows itself—not via counterdiscourse, but by *making-with* (sympoesis) the seeds of alternative discourses already contained within the dominant discourse via mimicry. This sympoetic creativity provides an escape route for the fixed bodies, lives and ecosystems, allowing them an opportunity to "run away" from the capture-and-control of dominant commodifying and othering cyberdiscourses.

Vaginal Creme Davis and Fertile La Toyah Jackson do not fit within the "glamour" or "comic" categories of drag. They therefore already upend any stereotypical and fetishized norms that may be expected by typical spectators of this subculture. Davis' onscreen persona is a doubly othered body: excluded by the homophobic and masculinist elements of the Black civil rights movements in the early 1980s; not at home in the spaces of belonging provided by white, corporate and "palatable" gay culture so prevalent today. I would personally use the term *interstitial* to configure the positioning of Davis' character within these competing identities—her performance deliberately places her outside normative categories and classification systems. Muñoz argues that Davis' persona is not a commercial, sanitized and desexualized drag act designed for mass consumption.[26] He describes the ability of Vaginal's performances to "stir up desires," enabling subjects to "imagine a way of breaking away from the social body."[27] The ethical glitch is precisely this kind of mechanism, short-circuiting the colonial desire machine by troubling settled psychic flows of spectator/consumer desire within the Colonial-Cybercene mediatic apparatus and questioning normative ways of embodiment that commodify, other and waste marginalized and excluded lives. It is not simply disruptive, which is what most glitches are. As we shall see in Dr Davis' short film, an *ethical* glitch provides a compelling alternative vision of care and kinship.

That Fertile Feeling begins with Vaginal Creme Davis and Fertile La Toyah Jackson deciding to watch a "porno." The central role of the technological apparatus in creating a glitch in the erotic-productive-reproductive economy is evidenced explicitly in this sequence. Vaginal says: "Well, now we *can* watch a porno, you have a VCR!" Throughout the somewhat lengthy sequence of film-within-the-film, the camera strategically rests only upon the gaze of the two women, rather than the erotic spectacle of the "porno" embedded within the film. The spectator/consumer is frustratingly kept from peeking at the desired and desirable bodies presented in the pornographic movie—a glitch to the "regime of the visible" that fuels the mediatic desire machine of the spectacle.[28] As spectator/consumers, we are sufficiently lured, but the lure turns out to be that of mimic porn-bodies—non-normative and farcical drag parodies of the unseen and perfect (and, one images, heteronormative) copulating bodies that the drag bodies comment on with repetitive giggling and exclamations of "oohs" and "aahs." This kind of visual play performs the ideological function of a glitch, reformatting normative economies of gazing and desire to reconstitute erotic power by tricking the spectator's needs and expectations (Fig. 5.3).

As Fertile giggles uncontrollably while viewing these bodies that we as spectators are unable to see, Vaginal makes an attempt to tease Fertile by using the remote to interrupt the film.

> Vaginal: I want to change this movie, I don't wanna see it!
> Fertile: Noooo, put it back, put it back! Put it back! Put it baaaack!

Figure 5.3 Screengrab from the re-citation of short film "That Fertile Feeling" by Vaginal Davis.[29]

Vaginal: Girl! You act like you've never seen no people doing it in a movie before.
Fertile: But never on *my* TV!
(Both laugh)
Vaginal: Well, you've never had a VCR before. You're a rich black girl now honey, you can afford to have a VCR.[30]

This initial framing sequence, I believe, constitutes an exemplary meta-mediatic frame to the story, exploring the nexus of queer Black identity, technological ubiquity brought about by affordable VCR sets, and the personalized availability of erotic content that allows for innovative explorations of sexuality, gender and productive/reproductive value in society, all the way "down" to the lower-class Black communities. By doing so in a farcical, camp and "homey" style, it domesticates the othered sexual "deviant," providing them with a home and family, and with full rights to claim motherhood and other supportive forms of kinship—just not in the normative way.

Following this establishing sequence, Fertile's waters break, leading to a frantic drive to get medical help. The two friends are refused assistance by the hospitals due to not having medical insurance. They then frenetically arrive at Vaginal's apartment and are greeted by Vaginal's "Nude Husband" (that is his name). Dressed only in an apron, Nude Husband is, well, nude—except for an apron. He is white, blond, and muscular—in other words he possesses all the qualities exalted as beautiful and desirable within the dominant paradigms of mainstream global gay culture. As another excellent example of disidentification and the use of tactical misrecognition, Nude Husband embodies the exact opposite of what a dominant white heteronormative body should be and do, by playing the part of a submissive house husband.

Fertile starts giving birth to one baby after another. She eventually reaches an impressive total of eleven babies: "She is the first mom to have eleventuplets!"[31] Vaginal proudly exclaims into the camera. Fertile's screams when in labor are framed by Nude Husband doing crunches, displaying his naked body's stereotypical masculine virtues. This is, yet again, a juxtaposition of the stereotype-as-fetish with the mimic-mother's body, a contradictory overlap that sets up the visual question of whether the "original" or dominant value given to the white male body is of higher value in the first place. As Vaginal places each baby in the makeshift cradle, she exclaims, "They look like puppies!"[32]—again confusing the traditional expected identification of human babies and hinting at alternative possibilities of multispecies being and belonging. The short film ends on a note of celebration of glitched motherhood with Fertile declaring "I am an incredible woman!"[33] and then, rather than performing the expected actions of an exhausted yet already doting mother, literally and figuratively runs away by jumping out of the window and skateboarding away from Vaginal's apartment. She even jumps over a literal gate that separates the gated community from the parking lot, thus metaphorically running away from the spatial and epistemic enclaving the gate and its community represent. The film ends with extended long takes of Fertile on a skateboard performing tricks. We see Vaginal look at Fertile from the window and exclaim, "There she is! Oh my God! Fertile La Toyah Jackson, you shouldn't be on a skateboard! You should be with your children!"[34]

Throughout this glitching film, the spectator's expectations are subverted via disidentifications of mimic-men, mimic-women, mimic-moms and mimic-families. They are all "almost, but not quite…," and yet they thrive and reproduce because they claim a space of belonging by existing and supporting each other. It portrayed a profound message of mutualism, care and kinship in the guise of a camp product. That fertile feeling in the film's title ends up signifying deeply affective bonds of kinship via a non-pedantic and non-pathos-induced inclusivity of bodies othered within a hostile society. Fertile is a mother—a *super-mother*, in fact—and Vaginal and Fertile are women without any further explanation. The film also accomplishes the task of co-constituting a home—a space of belonging for othered bodies—by visually and narratively glitching highly controlled colonial discourses of reproductivity. Through its campy characters, the film provides a potent model of freedom, solidarity and friendship via an ethical relationality that reformats the traditional family unit, thus providing a joyfully non-normative, inclusive and entertaining story of oddkin survivance. Indeed, Vaginal and Fertile's friendship glitches normative ecosexual practices as much as does Fertile's miracle birth. As Kim TallBear and Angela Willey argue:

> The naturalization of settler monogamy depends as much upon distinguishing love from friendship and other forms of affinity as it does the pathologization of promiscuity or non-monogamy […]. The valuation of friendship as a site of intimacy, meaning-making, resource sharing, and transformation has the potential to unravel stories about the specialness of sex and to fuel our imaginations to rethink forms and structures that exceed the ideal of the settler family, which may sustain and remake us.[35]

That Fertile Feeling's decolonial generative ecologies and ecosexualities co-constitute an alternative pathway of relationality for excluded and marginalized bodies, providing meaningful possibilities of safe and generative (fertile?) kinship via queer epistemological and visual/mediatic glitches. By normalizing Fertile La Toyah's fantastical sexual and reproductive—one could say "wild"—body, it brings a sense of multispecies inclusivity to queer forms of sexuality, so often actively suppressed by dreaded comparisons to the beast or the bestial. As Jack Halberstam argues, "the wild" must be understood as an epistemology, "a terrain of alternative formulations that resist the orderly impulses of modernity."[36] *That Fertile Feeling* is a beautiful example of a performative glitch, among a myriad of representational strategies used by queer cultural innovators to unsettle such biopolitical capture-and-control of *both sexuality and animality*.

Via such tactics, these bodies gain the chance to escape the wasteful constraints of normative speciesist settler-colonial forms of love, thus creating potentially joyful and healing implications for *all* multispecies bodies, lives and ecosystems. In *Strange Natures: Futurity, Empathy and the Queer Ecological Imagination*, Nicole Seymour describes these queer ecological works as important naturalcultural initiatives accessible to all, not just audiences that identify with one particular sexuality. Indeed, *That Fertile Feeling* can be said to narrate a specifically "queer ecological ethic of care: a care not rooted in stable essentialized identity categories, a care that is not just a means of solving human-specific problems, a care that does not operate out of an expectation for recompense."[37] It proposes a queer multispecies empathy, which, "because it crosses identity boundaries and looks skeptically at established ontological categories […] is available to all."[38]

Both Karima 2G's forceful musical response and Dr Davis' hilarious and campy film, re-cited on YouTube, strategically harness the fact that mediatic colonial stereotypes also contain the seeds of their own breakdown. The results of these kinds of conceptual and imaginative glitches are possible paradigm shifts into different and more mutualistic forms of multispecies kinship. In the next chapter, we will explore how ethical, affective, epistemic and discursive glitches such as those examined in this chapter can become the foundation of affective-material practices of bio- and ecocultural care and kinship within bodies, lives and ecocultural systems.

Chapter 6

Our Multispecies Futures

> In Aboriginal philosophy, existence consists of energy. All things are animate, imbued with spirit, and in constant motion. In the realm of energy and spirit, interrelationships between all entities are of paramount importance, and space is a more important referent than time.
>
> Leroy Little Bear[1]

Diverse global Indigenous epistemologies of ecocultural care and kinship pivot around variations of *ethical relationality*. It is a guiding principle throughout this project. This decolonial and multispecies attitude is often experienced within a dazzling multiplicity of Indigenous traditions in different parts of our planet via forms of somatic and ecocultural attentiveness, observation, connection, ceremony and storytelling. Indigenous scholar Manu Aluli Meyer describes her relationship with epistemology like this: "Every little thing. I mean, I can see a dead frog on the road, and it relates to epistemology."[2] In other words, these are often profoundly and passionately *situated* forms of knowledge. In the previous chapter, we explored diverse ways of ethically glitching the colonial capture-and-control of ecocultural systems that dramatically re-situate corrosive and distancing knowledges. In this chapter, I argue for the importance of re-evaluating and reintegrating often disavowed or devalued situated knowledges produced by the innovation and resilience/survivance of marginalized multispecies communities (both human and non). This reincorporation requires intellectual humility, curiosity, creativity, imagination and honesty rather than yet more expensive think-tanks, foundations and customer test groups. Can authorized knowledge bearers listen to and work sympoetically with the folks already working ethically within such multispecies communities, rather than parachute in from the outside and propose de-situated solutions?

Although they don't exist in fantastical Edenic communities outside the Colonial-Cybercene, many situated ecocultural communities can still provide place-specific pragmatic pathways towards restored multispecies care and kinship. When 16-year-old Chailert heard the elephant's terrified screams in Thailand, it was a transformative rather than a transient moment. She harnessed that single glitching instant to generate an alternative multispecies family in her life and her ecocultural community. This chapter contains other examples of how deeply powerful glitches are harnessed by incredible humanimals and their critter-kin. Sometimes, some lives are able to utilize the moment of a multispecies paradigm shift to create thick-accounts of ethical kinship rather than thin colonial narratives of transactional relationalities optimized for profit. They then go one step further and invest in these new and alternative relationalities within situated ecocultural systems of multispecies care and kinship.

Multispecies Thick-Map Prototype #1: The Pontine Marshes in Italy

As I bite into a piece of melon (*Cucumis melo*), images of a severed brown-skinned right arm accompany the honey-sweet taste of melon in my mouth.

The arm belongs to a specific individual—Satnam Singh.

In my mind, I cannot avoid picturing the pale musk-orange of the melon mixed with the deep red of blood and the white of shattered bone. Flesh of two species, all ensconced within the transparent elasticity of unbreathable, sterile plastic wrap.

Mr Singh, a thirty-one-year-old man originally from the state of Punjab in India, was hired at a farm in the *Agro Pontino* (Pontine Marshes) area of Lazio, the south-central Italian region of which Rome is the capital city. He was just one of thousands of undocumented and documented migrant workers who provide Italian and European consumers with groceries at an affordable price. The illegal yet widespread practice of sourcing, recruiting and managing these workers in the Italian agricultural sector is called *caporalato* or "gang-mastering." Essentially a modern form of slavery that parts of the agricultural economy rely on, *caporalato* is facilitated by problematic immigration and labor laws as well as the economic need and greed that keep labor costs inhumanely low and profit margins unnecessarily high. The CGIL (Italy's largest trade union) estimates that up to 230,000 people, or a quarter of agricultural workers, do not have a formal employment contract.[3] This is one of the sad and often unspoken aspects of the *Made in Italy* brand.

On June 19, 2024, Satnam Singh died at the San Camillo hospital in Rome after spending two days in agony. His right arm had been crushed by the plastic wrapping machine that he was operating to wrap the melons—those quintessentially Italian summer fruits that are made to appear on supermarket shelves in pristine form, as if they had no connection to the ecocultural context in which they grew, but seem to have been born by magic, right there on the store shelves, through some kind of industrial immaculate conception. The standardized look of the shrink-wrapped melon deliberately makes invisible to the consumer the complex industrialized agricultural processes and the many species, substances and persons that coaxed the melon plant to create "perfect" fruit. Mr Singh could have been saved if he had received immediate medical attention. However, the farm owners allegedly dumped him like agricultural waste in front of his home (along with his severed arm placed in a fruit container) and drove off, possibly in order to avoid facing any legal repercussions for participating in the *caporalato* system. Some reputable news sources mention that Mr Singh was packing strawberries (*Fragaria anannasa*) rather than melons, giving me one more fruit species to link with a specific example of immense suffering.[4] Others vaguely mention generic vegetables (*ortaggi* in Italian), now implicating potentially any or all fruit and vegetables. From within a multispecies paradigm, a story that began with one person's avoidable death and one fruit species' carefully controlled life suddenly unleashed a web of stories that touch entire multispecies ecocultural systems of interconnected affects, values and interests. It started a research journey for me that ended with the creation of the first prototype of a "multispecies thick-map."

Previous chapters demonstrated how colonial knowledge provided strategic thin-accounts of lives, bodies and ecosystems, both historically and in contemporary times via cybercitations. These "thin-maps" are the fertile ground in which colonial fantasies thrive and eventually become hardened attitudes and practices. If colonialisms can be understood via the definition I previously mentioned in Chapter 1: as past and present affective, epistemic, discursive, psychic and material practices of distancing, domination and control practiced by individual or group command economies upon culturally and ecologically Othered multispecies bodies, lives and ecosystems in order to extract and exploit these as material or labor resources for profit, then it can be understood how thin-maps can be key to perpetuating and normalizing extraction and exploitation. Implicit in their discursive strategy of the stereotype-fetish is that they are structured like the lure of a spectacle: their efficiency lies in being flashy, fragmented and de-situated. They propose surplus while creating othered forms of bodies, lives and ecocultural systems, all with the purpose of capture-and-control for consumption. Cybercene epistemic, discursive and affective desire machines (the cyberscreen), via their quadruple role, provide opportunities for a grand revival of paranoid colonial knowledges: command centers that promise unparalleled de-situated control, consumption, convenience and connectivity.

Satnam Singh's death was a tragic event that deeply re-situated my attention to a specific vibrant, vulnerable and mortal life. It also brought my attentiveness back to a specific place and space. The emotional impact of his story made me wonder if I could work within my new lab space to develop a method of mapping that would somehow ethically glitch the thin-accounts and practices that led to the wasting of his life? I began to test whether exploring the past and present multispecies resonances of the Agro Pontino region could provide a deeper context to place value on Mr Singh's hopeful life and wasteful death. Could a humanistic and qualitative analysis of how cultural and colonial identity politics impinge upon the destinies of individuals, regions, economies and entire ecocultural systems add to my understanding of how cultural, natural and economic configurations are mutually interdependent and *co-becoming* (i.e., becoming together, coming into and being in existence via ongoing multiple human and nonhuman relationships)?

Following Indigenous research traditions as well as my own postcolonial cultural upbringing, I was seeking to find a way to blend situated and ethical storytelling as well as epistemic humility into a research method. *Multispecies thick-mapping* is the result of these initial investigations. It creates a different kind of ethical glitch: a map that unlearns colonial mapping attitudes to reformat them from within. Multispecies thick-maps embody a new qualitative humanistic methodology of analyzing and narrating "natureculture" via a multispecies (human and non) rather than exclusively anthropocentric paradigm. I place this methodology as a useful complement to more data-driven analyses of ecocultural systems, in both historical examples of place- and space- specific imaginations and actions and the contemporary spaces of our digitally connected societies. Previous chapters demonstrated how important the intersections of physical and psychic mapping are for this project. We have seen how both "empty" and overly fantastical maps of new spaces and places during the Age of Discovery fed the colonial desire machine of control, surplus and profit. Actual maps were and continue to be an important genre of mass media, providing an excellent tool for the naturalcultural epistemic, affective, and eventually actual capture-and-control of territory, and therefore lives, bodies and ecosystems. Google Earth and Google Maps were revolutionary updates in the Colonial-Cybercene to this age-old genre, diffusing and democratizing the possibility of providing a semblance of power through knowledge via the representation of vast spaces and ecocultural systems within the scrolling convenience of the cyberscreen. Haraway's notion of map-fetishism describes the fetish as a trope, in that it is a kind of "substitute" for the real, a "mistake" that "make[s] things seem clear and under control."[5] Just like the smartphone cyberscreen as an ideal fetish-substitute or trope that stands for entire multispecies ecocultural systems, maps feature all the spectacular characteristics of Comolli's "lure"—where the spectator/consumer can "own" the actual land by the psychic displacement of actually owning the map of that land. The ecocultural understanding of land, private property and multispecies resources being for colonial profit and consumption is therefore fundamentally centered on metaphorical and literal mapping practices. Just like the *Scala dei Turchi* thick-map described in the Introduction, the Agro Pontino multispecies thick-map glitches these linear spatial and categorizing drives in order to insist on layered multiplicity and intra-action.

Multispecies thick-mapping describes a research ethos of ethical relationality and a visualization method of sympoesis (making-with others) that superimposes many layers of any given location of ecocultural importance. Thick-maps can take two forms: analog and digital. Both are complementary ways of representing the thick-narration and naturalcultural discourse analysis of a living multispecies ecocultural system. The analog version is static, yet its physical format allows for embodied intra-action with the spectator/consumer. The digital multispecies thick-map (which in the urban humanities is called a *storymap*) provides a few additional advantages: it can be multimedial, and hence multisensorial (containing more images, videos, sounds, etc.), allowing for better representation of nonhuman presence and agency. A digital thick-map can also grow with time as well as include different researchers rotating in and out of the project as their availability and interests permit. Digital multispecies thick-maps deliberately utilize all the spectacular components of the cyberscreen-lure to glitch colonial thin-narration of ecocultural systems.

Figs. 6.1 a, b and c show sections of the completed analog *Multispecies Thick-Map Prototype #1: The Pontine Marshes in Italy*. The map draws from my method of naturalcultural discourse analysis, as illustrated in the Introduction, and from thick-mapping techniques pioneered by digital humanities scholars such as Todd Presner, David Shepard and Yoh Kawano in the context of fostering research in the urban humanities:

> [Thick-mapping is] about exploring, participating, and listening, something *that transforms our conception of mapping into a practice of ethics.*[6]

Multispecies thick-mapping is a new interdisciplinary and non-anthropocentric method of spatialization and visualization that takes into account both the *diachronic* nature of a specific place (across multiple time periods, from mythic times to geological epochs to contemporary, decade-long timescales) and its *synchronic* characteristics, i.e., all these features existing together in the present, at the same time, via a series of overlapping physical and imaginative systems, both human and "more-than-human." My search for a multispecies thick-mapping method built upon and extended the concept of ethical thick-mapping in mostly anthropocentric terms to become attentive to the (often invisible or disregarded) intra-action, co-becoming, concerns and value of interconnected multispecies lives, bodies and ecosystems.

Multispecies thick-maps can take an incredible variety of forms, depending on the narrative choices the producer wants to make. However, they do have in common some basic characteristics that seek to incorporate a multispecies paradigm of viewing ecocultural systems:

1. *Vastness*—they seek to represent multiscalar, synchronic and diachronic thick time and transcultural perspectives in one multi-layered space, whether analog or digital.
2. *Viscosity*—these maps deliberately slow down the viewing process to locate moments of naturalcultural intra-action. This approach glitches both inattentiveness and quick capture-and-control of meanings. They seek to promote inquisitiveness about the stickiness of one layer or section of the map relative to another—nothing exists in "frictionless" isolation.

Figures 6.1a (and b and c) Sections of the Multispecies Thick-Map Prototype #1: The Pontine Marshes in Italy. (Vetri Nathan)

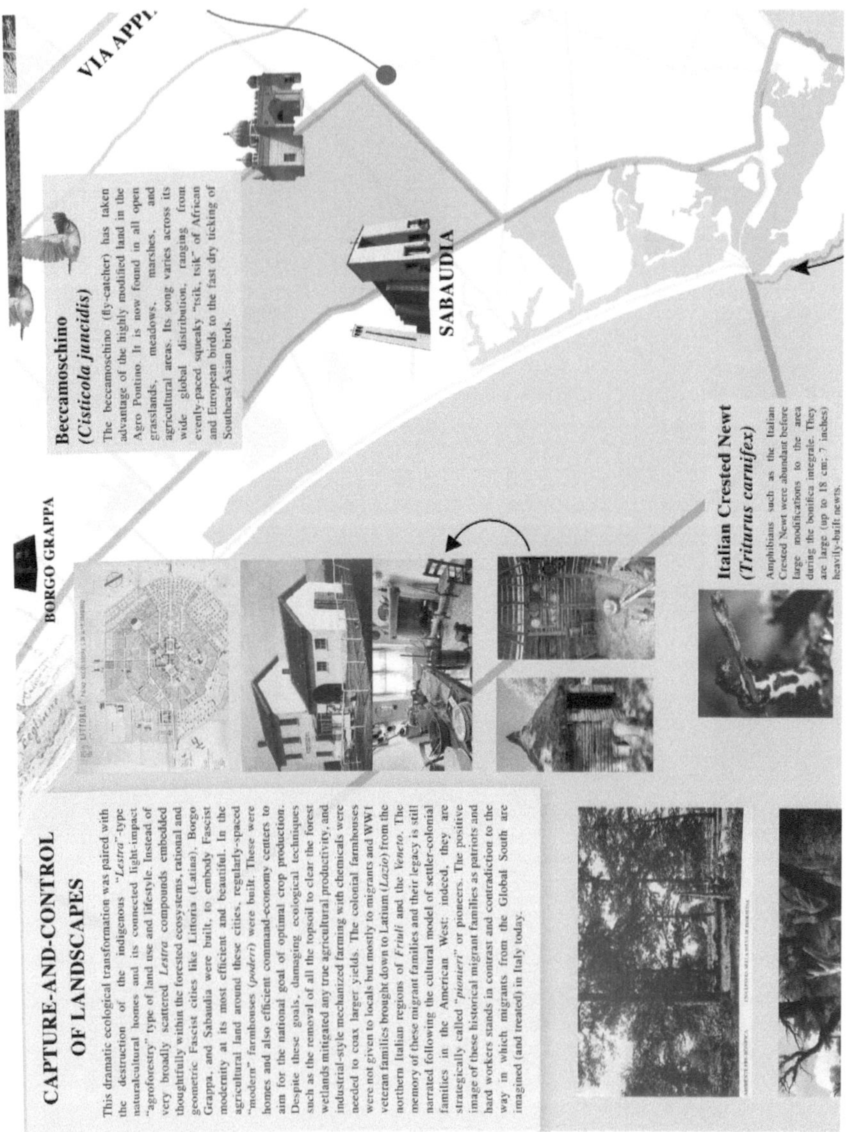
CAPTURE-AND-CONTROL OF LANDSCAPES
This dramatic ecological transformation was paired with the destruction of the indigenous "Lestra"-type naturalcultural homes and its connected light-impact "agroforestry" type of land use and lifestyle. Instead of very broadly scattered Lestra compounds embedded thoughtfully within the forested ecosystems, rational and geometric Fascist cities like Littoria (Latina), Borgo Grappa, and Sabaudia were built, to embody Fascist modernity at its most efficient and beautiful. In the agricultural land around these cities, regularly-spaced "modern" farmhouses (poderi) were built. These were homes and also efficient command-economy centers to aim for the national goal of optimal crop production. Despite these goals, damaging ecological techniques such as the removal of all the topsoil to clear the forest wetlands mitigated any true agricultural productivity, and industrial-style mechanized farming with chemicals were needed to coax larger yields. The colonial farmhouses were not given to locals but mostly to migrants and WW1 veteran families brought down to Latium (Lazio) from the northern Italian regions of Friuli and the Veneto. The memory of these migrant families and their legacy is still narrated following the cultural model of settler-colonial families in the American West; indeed, they are strategically called "pionieri" or pioneers. The positive image of these historical migrant families as patriots and hard workers stands in contrast and contradiction to the way in which migrants from the Global South are imagined (and treated) in Italy today.
BORGO GRAPPA
Beccamoschino
(Cisticola juncidis)
The beccamoschino (fly-catcher) has taken advantage of the highly modified land in the Agro Pontino. It is now found in all open grasslands, meadows, marshes, and agricultural areas. Its song varies across its wide global distribution, ranging from evenly-paced squeaky "tsik, tsik" of African and European birds to the fast dry ticking of Southeast Asian birds.
VIA APPI
SABAUDIA
Italian Crested Newt
(Triturus carnifex)
Amphibians such as the Italian Crested Newt were abundant before large modifications to the area during the bonifica integrale. They are large (up to 18 cm; 7 inches) heavily-built newts.

Figure 6.1b

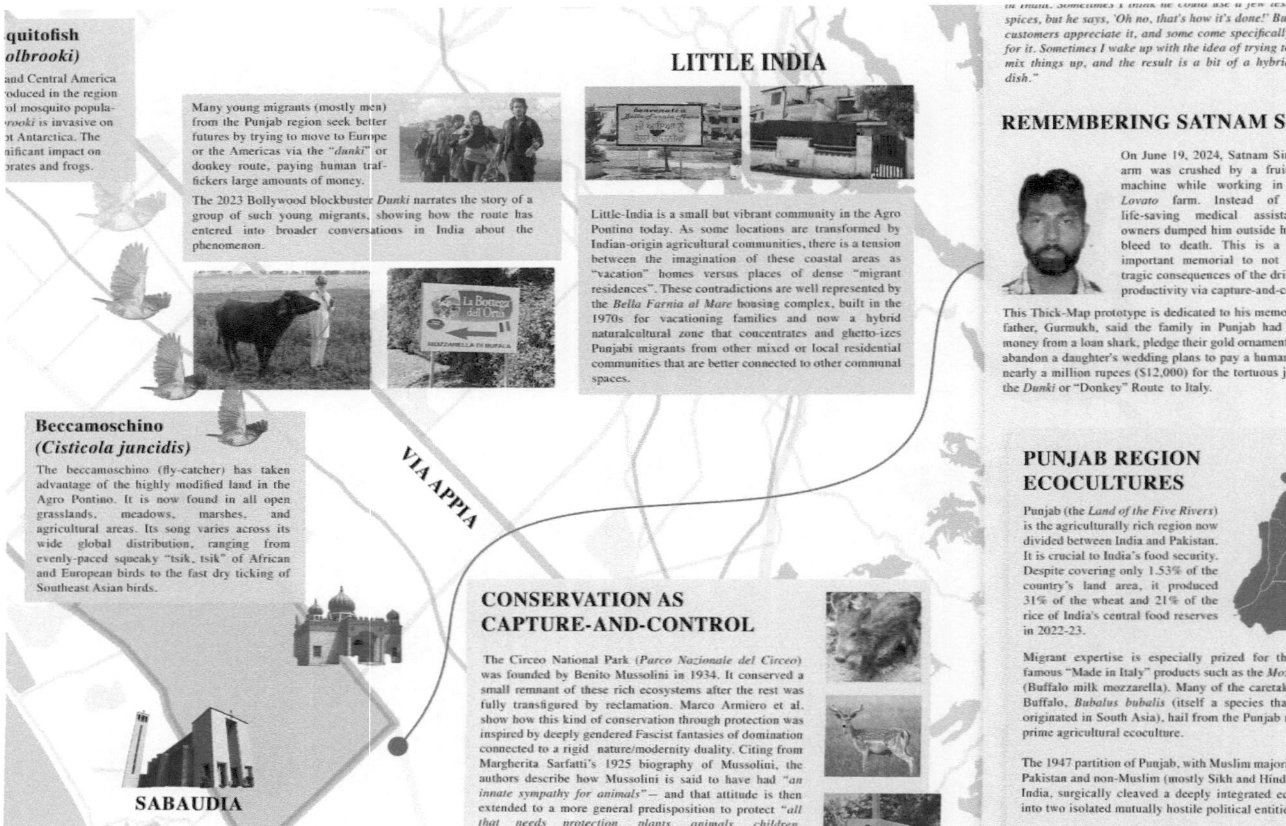
quitofish
olbrooki)
and Central America
oduced in the region
ol mosquito popula-
rooki is invasive on
ot Antarctica. The
nificant impact on
orates and frogs.
Many young migrants (mostly men) from the Punjab region seek better futures by trying to move to Europe or the Americas via the "dunki" or donkey route, paying human traffickers large amounts of money.
The 2023 Bollywood blockbuster Dunki narrates the story of a group of such young migrants, showing how the route has entered into broader conversations in India about the phenomenon.
La Bottega dell'Orto
MOZZARELLA DI BUFALA
LITTLE INDIA
Little-India is a small but vibrant community in the Agro Pontino today. As some locations are transformed by Indian-origin agricultural communities, there is a tension between the imagination of these coastal areas as "vacation" homes versus places of dense "migrant residences". These contradictions are well represented by the Bella Farnia al Mare housing complex, built in the 1970s for vacationing families and now a hybrid naturalcultural zone that concentrates and ghetto-izes Punjabi migrants from other mixed or local residential communities that are better connected to other communal spaces.
spices, but he says, 'Oh no, that's how it's done!' But customers appreciate it, and some come specifically for it. Sometimes I wake up with the idea of trying to mix things up, and the result is a bit of a hybrid dish."
REMEMBERING SATNAM SI
On June 19, 2024, Satnam Singl
arm was crushed by a fruit-w
machine while working in tl
Lovato farm. Instead of p
life-saving medical assistanc
owners dumped him outside his
bleed to death. This is a sn
important memorial to not fo
tragic consequences of the drive
productivity via capture-and-con
This Thick-Map prototype is dedicated to his memory.
father, Gurmukh, said the family in Punjab had to
money from a loan shark, pledge their gold ornaments a
abandon a daughter's wedding plans to pay a human tr
nearly a million rupees ($12,000) for the tortuous jou
the Dunki or "Donkey" Route to Italy.
Beccamoschino
(Cisticola juncidis)
The beccamoschino (fly-catcher) has taken advantage of the highly modified land in the Agro Pontino. It is now found in all open grasslands, meadows, marshes, and agricultural areas. Its song varies across its wide global distribution, ranging from evenly-paced squeaky "tsik, tsik" of African and European birds to the fast dry ticking of Southeast Asian birds.
VIA APPIA
SABAUDIA
CONSERVATION AS
CAPTURE-AND-CONTROL
The Circeo National Park (Parco Nazionale del Circeo) was founded by Benito Mussolini in 1934. It conserved a small remnant of these rich ecosystems after the rest was fully transfigured by reclamation. Marco Armiero et al. show how this kind of conservation through protection was inspired by deeply gendered Fascist fantasies of domination connected to a rigid nature/modernity duality. Citing from Margherita Sarfatti's 1925 biography of Mussolini, the authors describe how Mussolini is said to have had "an innate sympathy for animals"– and that attitude is then extended to a more general predisposition to protect "all that needs protection, plants, animals, children,
PUNJAB REGION
ECOCULTURES
Punjab (the Land of the Five Rivers) is the agriculturally rich region now divided between India and Pakistan. It is crucial to India's food security. Despite covering only 1.53% of the country's land area, it produced 31% of the wheat and 21% of the rice of India's central food reserves in 2022-23.
Migrant expertise is especially prized for the
famous "Made in Italy" products such as the
(Buffalo milk mozzarella). Many of the caretaker
Buffalo, Bubalus bubalis (itself a species that
originated in South Asia), hail from the Punjab re
prime agricultural ecoculture.
The 1947 partition of Punjab, with Muslim majority
Pakistan and non-Muslim (mostly Sikh and Hindu)
India, surgically cleaved a deeply integrated eco
into two isolated mutually hostile political entities.

Figure 6.1c

3. *Variation*—the moment the spectator/consumer seeks to capture one pathway of meaning, they start seeing other variations, deviations, combinations, connections, permutations and perspectives.
4. *Vibrancy*—there is a focus on narrating the immense energy of multispecies biopower and soulpower via the material and cultural intra-actions of different multispecies agents.
5. *Vulnerability*—these maps highlight the fragility and interdependence of multispecies, multi-being ecocultural systems. Connections of all kinds (direct and indirect) abound, making one lively agent dependent on other such lively agents.

A multispecies thick-map is a primary visualization product to "provide 'thick' accounts of the distinctive experiential worlds, modes of being, and biocultural attachments of other species. Immersive ways of knowing and being with others involve careful attention to what matters to them—attention to how they craft shared lives and worlds."[7] Rather than de-situated colonial spectacles that are chiefly concerned with engaging in the colonial ethic of capture-and-control of multispecies lives as resources for profit, multispecies thick-maps seek to offer an alternative mediatic vision that allows viewers to "learn to be affected" differently, and hopefully learn to care differently. When users of the *Scala dei Turchi* thick-map haptically scroll through the various spatialized multispecies stories on their beloved smartphone screens, they are encouraged to enter and engage in a fluid realm of naturalcultural multispecies co-becoming, rather than rigid, linear and obsessive capture-and-control.

A multispecies thick-map superimposes interconnections between bodies, lives and ecocultural systems in ways that form oddkin—multispecies connections and collaborations that are deeply situated in their place- and space-specificities but are usually ignored, denied, repressed, or simply not paid attention to due to anthropocentric biases. It also achieves this by not asserting total control over its many facets, instead allowing viewers to approach its "layers" in a multiplicity of ways. Viewers may frequently notice resonances between layers that they did not expect nor consciously trace. In a somewhat cheeky mirroring of the term given to A.I. malfunctioning, I call these unexpected yet situated resonances "hallucinations." By providing agency to the spectator/consumer of these thick-maps, I seek to provide an "ethical glitch" from within the system of the colonial spectacle. What is typically intended as a single-layered colonial fairytale of "us-versus-them" becomes instead a complex, sometimes disorienting but differently fascinating intra-active experience of a specific ecocultural system. The agency provided to the spectator/consumer is a creativity that is grounded within situated knowledge contained in the map layers. This is what provides a balanced ethic of "situated creativity." Reciprocal capture does occur, but it becomes a blend of capture and running away, as the map becomes something different to each spectator, who may be moved in diverse ways by the many and varied combinations of approaches they can take to navigate the map. A well-executed multispecies thick-map asks questions and stokes imagination and curiosity as much as it provides answers. It is a "living" rather than a dead entity because it keeps on changing with time (especially in its digital version). Instead of aiming to be an exhaustive hierarchical classificatory catalog of life, it superimposes natural and cultural agents in unusual and odd ways, allowing the spectator/consumer to move within the map's layers in multiple directions, and to be curious to discover more later on their own. Sometimes, via hallucinations between layers, the map "runs away" from fully captured meaning.

The analog *Agro Pontino Multispecies Thick-Map* seeks to narrate natural-cultural co-becomings within a complex ecocultural system. A digital version is currently in the process of being constructed in the *Multispecies Futures Lab*. Via a "thick-time" multispecies exploration of the region, viewers of the map are able to perceive how multispecies systems that seem fixed and "eternal" today are actually results of ever-changing intra-actions between cultural attitudes towards the land and natural ecological processes and lives. Anthropogenic interventions are documented since before Ancient Roman times. This place has not been an Edenic paradise since at least then, even though developers in the Fascist period portrayed it as an "empty" land that needed to be reclaimed from nature for national use. Indeed, reclamation gained momentum after Italian Unification and under the Fascist regime in the 1920s and 1930s. Animals, plants, landforms and native humans all sought to adapt, survive and thrive despite these drastic changes. The *Agro Pontino Thick-Map* visualizes this long naturalcultural history of marshlands and cities, fish and birds, semi-nomadic and migrant communities. It essentially narrates a naturalcultural story of competing ideas of solidity versus fluidity, not only understood as intra-active matter (wetlands versus reclaimed land) but also as "ordering tropes in hegemonic conceptions of the world."[8]

Nature versus Modernity: Fascist Conservation as Multispecies Capture-and-Control

Moving from left to right in a roughly chronological movement, the first section of the map (Fig. 6.1a) looks at the more integrated multispecies space that existed before extensive reclamation in the 1920s (pre-*bonifica*). The base map shows the major Ancient Roman throughway of the *Via Appia* (Appian Way) as well as a map from the eighteenth century gently merging into a contemporary map. The engineering marvel of the Appian Way shows how the region has long been a focus of obsessions of central-ized command economies with making this natural wetland "productive" in anthropocentric terms. Despite having this major throughfare, the lands on either side were still a magical, liminal, naturalcultural space where the dualities of human/nonhuman, land/water dissolved and merged. Paolo

Grupposo accurately grasps this innate liminality by defining the Agro Pontino as an ecotone:[9]

> Reflecting on the etymology of ecotone from the Greek *oikos* (house) and *tonos* (tension), I suggest that fluidity and solidity are not mutually exclusive properties of matter but rather relational patterns that tend towards more fluid or more solid formations. This is to say that they are not *between* but *in-between*.[10]

Indeed, the human-nonhuman co-becomings of the pre-*bonifica* Agro Pontino allow us to understand it as an incredibly hybrid multispecies ecocultural system rather than a previously Edenic space devoid of human activity. In a 1933 study curiously entitled "The Primitivity of Human Habitations," Mario De Mandato provided some insights into the unique semi-nomadic dwelling patterns of the native communities of the Agro Pontino before the major Fascist reclamations:

> Life in the Marshes is not permanent but offers a prime example of semi-nomadism, as the population carries out a mass migratory movement included in this cycle: nine or ten months in the marsh and two or three months in the mountain. The region is invaded every year at the beginning of autumn by families who leave in October from their villages in the Ernici hills, in the Sacco Valley, and in the Latina Valley [...] and after a journey of a few days, made with their primitive vehicles, with herds and domestic animals, they penetrate and settle in the thick of the Pontine woods. [...] It is remarkable that in the kind of habitation which must be considered a temporary domicile, namely the hut in the marshes, life occurs at a peak of efficiency and wellbeing for more than nine months of every year; whereas the permanence is as short as possible in the so-called permanent domicile, that is the house made of stones in the mountain.[11]

This "ecotonic" understanding (and human use) of the Agro Pontino before reclamation more or less embodied a multispecies paradigm. Figuring prominently in this ecotonic section of the thick-map is the story of the Italian Crested Newt (its scientific name is *Triturus carnifex*). The Italian Crested Newt is an amphibian, and therefore by the very nature of this family of animals inhabits and thrives in a realm "in-between" the aquatic and terrestrial. As larvae, they breathe using a combination of gills and skin, and then as adults they develop lungs and breathe via a combination of pulmonary and subcuteaneous (skin) respiration. Good water quality is therefore key to their survival. It is no surprise that "amphibians such as the Italian crested newt were abundant before large modifications to the area during the *bonifica integrale*."[12] Due to decreases in wild populations, caused mainly by habitat destruction, all of the Crested Newt species are protected by the European

Habitats and Species Directive of 1992. This requires European states to protect the species in the wild, and to regulate trade of wild-caught animals.[13]

The ecotonic multispecies paradigm would come under attack by Great Acceleration post-Unification development after the 1880s and especially under Fascism from the 1920s. These changes are represented in the central section of the multispecies thick-map. The land and its multispecies community would be transformed at an industrial pace and with a deeply colonial rather than multispecies mindset.[14] The changes were especially impactful south of the Appian Way, in the wildest and stickiest parts of the region, leading to the "capture-and-control" of many lives, bodies and ecocultural systems (human and non) via a highly centralized command economy. Grupposo says:

> the *Bonifica Integrale* was vigorously propagandised by the fascist regime as a war against the disorderly fluidity of the Marshes, understood as a prime example of nature in the raw, and regarded as antithetical to fascist ideas of human society. This war was fought with real weapons, such as mines, ready to be blasted every time Benito Mussolini visited Agro Pontino to check on the progress of the *Bonifica Integrale*.[15]

Previously, Indigenous "*lestra*"-type naturalcultural homes were dispersed centers of a low-impact "agroforestry" type of land use and lifestyle.

Jack Halberstam's conceptualization of "the wild" as "a terrain of alternative formulations that resist the orderly impulses of modernity"[16] resonates deeply with the way in which the Agro Pontino was "modernized," and therefore domesticated. Indeed, almost all the ecotonic spaces described above would be extirpated, because fundamentally, the reclamation project worked within the Fascist regime's dualistic epistemic framework of modernity: nature was understood and represented as the domain of the wild, the instinctive, the savage and the primitive while modern development was its opposite—rational, efficient, patriotic and hygienic.

Armiero, Biasillo and von Herdenberg discuss how "intertwining imaginations of coloniality, nature and oppression" lay deep within Fascist practices of nature management.[17] This nature/human civilization duality was mirrored by the Duce's own autobiographical narrative, which played on an imagined human/animal duality. He often said that he "smelled like a lion."[18] His autopoetic magical-mapping of his own persona included mythical qualities imbibed with animal instincts: "I must listen to my blood … I am like the animals. I feel when things are going to happen—some instinct warns me and I am obliged to follow it."[19] The *Parco Nazionale del Circeo*, founded by Mussolini in 1934 on the advice of Senator Raffaele Bastianelli, conserved a small remnant of the Agro Pontino's rich ecosystems after the rest was fully transfigured.[20] Armiero et al. show how this kind of conservation through protection was inspired by deeply gendered Fascist fantasies of domination connected to the nature/modernity duality. Citing Margherita Sarfatti's 1925 biography of Mussolini, the authors describe how "Mussolini is said to have had an innate sympathy for animals—and attitude that is then extended to a more general predisposition to protect 'all that needs protection, plants, animals, children, women—above all women.'"[21] When viewed within the context of Mussolini's cojoined biocultural attitudes towards gender roles and nature, we can see how conservation "protection" within the confines of this new national park was more than just a benign beneficial act: it was organically part of an overall attitude and constitutive naturalcultural discourse of masculine totalitarian control and multispecies management that shaped the entire reclamation project. The small section of land protected even today in the Agro Pontino in the form of a national park can therefore be understood as a complex ecocultural site, replete with deeply paternalistic discourses of colonial capture-and-control. The nature/modernity duality was harnessed towards regulating and showcasing the "wild and untamed" aspects of a mythical and Edenic nature on one side and impeccably ordered urban centers and staged citizenry on the other. The new cities, the national park and the agricultural command economies should therefore be viewed as an integrated spectacle of contrasts, one beneficial and controlled (organized modern cities, intensive industrial monocropping) and the other abject and wild (forested lands, dispersed ecotonic multispecies ecocultures). The entire Fascist project for the Agro Pontino can be read and interpreted as a multi-mediatic naturalcultural product, akin to the colonial multispecies zoo, but set on a far vaster scale.

The second section occupying the central part of the thick-map describes several of these land-management transformations for centralized extraction and exploitation, all undergirded by carefully managed naturalcultural discourses. Previously a sparsely populated alluvial plain at or below sea level filled with estuaries, swamps, mixed forests and barrier dunes, the Agro Pontino harbored incredible species diversity and complex land-sea intra-actions. While attempts had been made since antiquity to gain "productive" space (from an anthropocentric point of view, of course), the *bonifica integrale* decisively impacted the area with the draining of the marshes at an industrial scale never seen before. This vast area, especially south of the ancient Via Appia, was reimagined and repurposed into a curated, flat, "modern" and highly controlled ecocultural system comprising intensive crop production farms and modern Fascist cities (including a completely new capital city, *Littoria*—now *Latina*), where lived and worked mono-accultured Fascist citizens, many of whom were resettled here from their homelands in the northern regions of Veneto and Friuli. The historical loss of the Agro Pontino wetlands and lowland forests, called the *Selva di Cisterna* and the *Selva di Terracina*, deprived not only the entire ecosystem of a crucial species-rich habitat, but also stole from the Mediterranean an important breeding and nursery area for its rich marine life. The *bonifica* process echoes in the familiar phrase "drain the swamp"—used in English as a potent political metaphor to represent a righteous fight against entrenched corruption, attributing purely positive and essential "cleansing and purifying" meaning to a practice that is actually a disastrous extermination of valuable wetland ecosystems. Indeed, the Agro Pontino is kept "productive" even today only through the constant use of energy to drain incursions of rainwater and seawater.

Central to the Fascist "integral reclamation" of the Agro Pontino were two propaganda themes: the "Battle for Grain" and the "Battle against Malaria." Both these themes therefore form the central epistemic and discursive backbone of the map to convey to viewers the profoundly multispecies nature of the regime's political and cultural agenda. The thick-map shows how Mussolini's vision of a *bonifica integrale* meant connecting nationalistic cultural propaganda of modern and scientific renewal with a "productivity-first" approach to nature. The "Battle for Grain" was a national campaign that forced farmers to intensively produce a monocrop—wheat—rather than the diverse ecosystem-specific foods they typically grew. Figure 6.2 shows a propaganda poster exhibited at the *Museo della Terra Pontina*. The sign orders farmers to buy and use chemical fertilizers to ensure high productivity. It glibly mentions how the cost of these expensive fertilizers would be absorbed via higher yields, indicating indirectly that the farmers bear the financial risk while the regime would pocket any eventual profits via grain purchase and distribution.

The "Battle Against Malaria" had the double function of reducing the incidence of malaria while also draining the natural waterways and transforming the wetland forests and other biodiverse micro-ecosystems into a featureless, flat, "productive" land. Another propaganda poster (Fig. 6.3) beautifully stitches together all the themes discussed in this thick-map: of how discourses of biopower, cultural rejuvenation, conquest of nature and disease and agricultural productivity dovetailed in this project to capture-and-control the land and its people. The caption of this poster reads: "Malaria is like an octopus that sucks the blood of the best of our peoples. 2 million malaria victims a year! 10 million working days lost! Millions of hectares of land left uncultivated! We will not win the Battle for Grain ... if before that ... We don't win the Battle against Malaria!"[22]

Land transformation would be the physical manifestation of epistemic, affective, psychic and political colonial control via these two "battles." As illustrated in Figures 6.4 and 6.5, instead of very broadly scattered *lestra* compounds embedded thoughtfully within the forested ecosystems, impervious colonial farmhouses and compounds (*poderi*) were built. These were homes for the new farmers that would be settled here, but can also be understood as efficient command economy centers to fulfill the national patriotic goal of optimal crop production. Despite these objectives, damaging ecological techniques such as the removal of all the topsoil to clear the forest wetlands mitigated any true agricultural productivity, and industrial-style mechanized farming with chemicals was needed to coax larger yields. The colonial farmhouses were not given to locals but mostly to migrants and World War I veteran families brought down to Latium (Lazio) from the northern Italian regions of Friuli and the Veneto. The imagery of these migrant families and

Figure 6.2 A "Battle for Grain" public service information poster with recommendations to farmers on using chemical fertilizers that are also "Made in Italy." (Vetri Nathan: *Museo della Terra Pontina, Latina*)

their legacy is still honored by following the cultural model of imagining settler-colonial families in the American West: indeed, they are strategically called *pionieri* (pioneers). The positive depiction of these historical migrant families as patriots and hard workers stands in contrast and contradiction to

Figure 6.3 A "Battle against Malaria" public service information poster. (Vetri Nathan: *Museo della Terra Pontina, Latina*)

the way in which migrants from the Global South are imagined (and treated) in Italy today.

This section of the map subtly represents the multispecies ecocultural intra-actions of Mussolini's integral reclamation via a string of tiny fish. The aptly named and vibrantly colored Mosquito Fish (*Gambusia holbrooki*) is

Figure 6.4 The porosity of the lestra huts (Vetri Nathan: *Museo della Terra Pontina, Latina*)

Figure 6.5 The clean lines of the first model colonial farmhouse in the Agro Pontino. (Vetri Nathan: *Museo della Terra Pontina, Latina*)

actually a native of North and Central America. It was intentionally introduced in the region in the 1920s to control mosquito populations in the Battle against Malaria.[23] It turned out that *Gambusia holbrooki* is invasive on every continent where it has been introduced, which only leaves Antarctica untouched. The species has had a significant impact on native fishes, invertebrates and frogs. The diminutive but flashy Mosquito Fish represented the coming together of nature and culture in such an exemplary way that my thick-mapping research assistant, Doğa Tekin, and I decided to be inspired by its blend of colors to form the basic color template of the finished map.

The third and final section to the right of the map (Fig. 6.1c) brings the many layers of thick-time and strands of multispecies considerations of the previous two sections into resonance with questions of "food, labor and identity" in the region today. There is a memorial to Mr Satnam Singh. Next to this memorial space is the contextual chronicling of Indian-Italian first- and second-generation innovators as well as the spatial marginalization of many Indian laborer communities into ghettoized enclaves. It narrates the space of "Little India"—a small but vibrant community in the Agro Pontino today—and describes how, as some locations are transformed by agricultural communities of Indian origin, there is a tension between the imagination of these coastal areas as settings for vacation homes and as places dense with "migrant" residences. These contradictions are well represented by the *Bella Farnia al Mare* housing complex (Fig. 6.6), built in the 1970s for vacationing families and now a hybrid naturalcultural zone that concentrates and isolates Punjabi migrants away from other mixed or local residential communities that are better connected to other communal spaces. Pushback against this

Figure 6.6 Dual-language signpost welcomes visitors to the Bella Farnia al Mare housing complex.

isolation comes in the form of new places of belonging that ethically glitch colonial forms of control of these bodies: such places are represented on the map and include the local *gurudwara* (the Sikh place of worship; the name means "Home of the Guru"). Another blurb describes the future of the region with hope, narrating the ethically glitching origin story of a fusion Italian-Indian restaurant, opened by Chef Lovepreet Singh, called "Terraferma" (Terra Firma). One Mr Singh's story of being captured-and-controlled for profit is paired with another Mr Singh's story of creativity, resilience, care and kinship. This blurb represents a glitch in dominant narratives of migrant oppression, of how young first- and second-generation Indian-Italians are showing their love for their homeland by bringing in innovative ideas and entrepreneurial energy. Their deep integration within the naturalcultural realities of today's Agro Pontino glitch deeply held biocultural stereotypes and multispecies obsessions with capture-and-control.

This third section narrates how the Agro Pontino today is one of Italy's richest food-producing regions, especially for fruit, vegetable and dairy products. It is one of the world's major producers of kiwi fruit (*Actinidia chinensis*). As the plant's scientific name suggests, the kiwi is actually a fruit native to China and cultivated extensively in Australia and New Zealand, in addition now to Italy. Connected to such globalized economic and ecocultural systems, agriculture today in the Agro Pontino is dominated by migrant labor, especially from India. Since I was seeking to narrate a profound ecocultural exchange of agricultural knowledge and labor, I wondered why I should limit the map to the Agro Pontino, or indeed, to Italy or Europe? In a digitally connected world of flows: of capital, labor, and human imagination; of desire and fantasies via digital networks (a key feature of the Cybercene era), was it not also important to consider the ecocultural effects of contemporary emigration to the Global North on the laborers' ancestral land—the Punjab? With this rationale in mind, I created a map-within-a-map that allows the Punjab to be superimposed on the Agro Pontino. This section describes how the Punjab (the "Land of Five Rivers") is an agriculturally rich ecoregion now divided between India and Pakistan. It is crucial to India's food security. Despite covering only 1.53% of the country's land area, it produced 31% of the wheat and 21% of the rice of India's central food reserves in 2022-2023.[24] The map also alludes to the 1947 naturalcultural partition of the Punjab, with Muslim majority areas going to Pakistan and non-Muslim (mostly Sikh and Hindu) areas going to India, surgically cleaving a deeply integrated ecocultural region into two isolated and mutually hostile political entities. It shows how in 2025, the Indian economy grew by 7.2% but job creation has lagged in Indian Punjab. Unemployment stands at 6.1% against a national average rate of 3.2%.[25] The story of digital discourses of surplus and lack—this time deployed by subaltern migrants—is also represented in the Agro Pontino multispecies thick-map. The Punjab is now a land deprived of many of its (primarily male) youth, who are choosing to be inspired by TikTok and YouTube videos of "dunki" migration to move away from their homelands towards the mirage of a better life in the "West." Punjab's loss via digitally supplied dreams becomes Italy's (disavowed) gain. Abusaleh Sharif and Attaulla Khan explain the economic push and pull factors fueling dunki migration from relatively affluent migrants in India to the Global North (Europe and North America):

> Complex supply chain routes [between the Global North and South nations] are now supported by complex IT systems built on data-driven, internet-enabled infrastructure, providing real-time insights and tracking of products from source to consumer.
>
> However, there are severe obstacles and even legal entry barriers to the free flow of appropriate labour and a productive workforce across nations.[26]

Essentially, a strong need for cheap undocumented labor created via powerful global supply chains, a demographic lull or drop in births in prosperous countries of the Global North and barriers to the legal entry of unskilled migrant labor have created the perfect storm of conditions to fuel these dunki migrations. However, few economists or quantitative migration experts are able to explain exactly why *relatively affluent* young men from India are risking these undocumented pathways to entry. This is where humanistic rather than scientific or social science analysis can truly shine, via naturalcultural discourse analysis. To engage in this kind of analysis is to institute an ecocultural and multispecies paradigm shift within my own research in literary, cinematic and mediatic studies, and thus, the understanding of identity-based mobilities. Naturalcultural discourse analysis allowed me to locate important Cybercene mediatic components in dunki migrations. Viral narratives on digital platforms such as WhatsApp as well as free daily video phone calls between those who have successfully migrated and family and community members "back home" create powerful stories and new digital relationships, new forms of co-becoming across continents. These stories are represented in the thick-map via a poster of the 2023 Bollywood blockbuster film *Dunki*. The film, starring superstar Shah Rukh Khan, re-cited and also elevated the pervasive fantasy of Punjabi youth migrating to the West to new levels of spectacular visibility and fame. Consistently strong transcultural and mobile digital communities are sustained via personalized yet hyperconverged smartphone screens, where filmic spectacles intra-act and co-become with the choices and actions of actual people re-cited via WhatsApp, Instagram and YouTube. These mediatic fantasies form strong attractors for many families, especially from specific Indian states (Punjab, Telangana, Kerala and Andhra Pradesh). Many persuade themselves to sell their land and other possessions and invest in these very expensive and dangerous options to try and achieve the "Western Dream."

Satnam Singh's father, Gurmukh, said the family in Punjab had to borrow money from a loan shark, pledge their gold ornaments and even abandon a daughter's wedding plans to pay a human trafficker nearly a million rupees ($12,000) for the tortuous journey via the "Donkey Route" to Italy.[27] Even if Mr Singh's specific personal story and life choices did not follow the general Cybercene pattern of digitally fueled dunki migration, the powerful digitalized repetition of success stories from the West (true or imagined) certainly had indirect or cascading effects. It also becomes clear that the overall cultural transformation caused by the migration of Punjabi men such as Mr Singh is absolutely of simultaneous human and multispecies consequence. One only has to cultivate attentiveness to the effects of these migrations *in both India and Italy* to tease out the multispecies and social justice implications of these identity configurations and geographic mobilities, such as the consequences of these transformations on the sustainable cultivation of different crop species, the quality of the soil and water and the effect of the availability (or lack) of food crops on individual microbiomes within the intestinal folds of the human communities affected, and the other community relationships, newly forged or left abandoned in diverse places and spaces. Multispecies lives are represented in this third section of the thick-map via two species: a Water Buffalo (*Bubalus bubalis* or *murensis*) being cared for by a Sikh farmer, and a *Beccamoschino* (*Cisticola juncidis*). This species of warbler "has taken advantage of the highly modified land in the Agro Pontino. It is now found in all open grasslands, meadows, marshes, and agricultural areas. Its song varies across its wide global distribution, ranging from the evenly paced squeaky 'tsik, tsik' of African and European birds to the fast dry ticking of Southeast Asian birds."[28] There is a mutualistic relationship between the agricultural transformation and activity on the land, the different introduced large grazing mammals, and insects of all kinds that are attracted to this ecocultural system of dairy production. These insects then serve to form an important component of the diet of the *Beccamoschino*, demonstrating how a series of indirect cascading naturalcultural effects led this species to its current successful distribution.

During the process of creating the first paper flowchart and base map (Fig. 6.7), I realized that the new multispecies thick-map was gradually but surely coming "alive" with startling resonances that formed superimposed layers upon layers of narration of spaces, places and eras. Depending on how this now-completed map is experienced, a viewer may be uncannily reminded of the cloying sweetness of kiwi fruit while considering the monumental rationality of the straight and wide avenues and boulevards of Fascist-designed Latina and Sabaudia. Another viewer may find the digitally fueled fantasies of migrant success becoming strangely superimposed in their mind as they imagine feeling the prick of *Anopheles labranchiae* (the primary species of mosquito responsible for spreading malaria in the region) on their skin. I focused on *intra-action* and *co-becoming* as cornerstone concepts while creating this first prototype of a multispecies thick-map. The hope with such an approach is that these kinds of multispecies thick-maps can be open yet situated in their multiple perspectives, angles of approach, hallucinations and resonances, rather than strictly static and imposing a singular meaning. When in digital rather than analog form, they can also grow tendrils, new shoots and diverse rhizomes, if and when needed.

The first prototype of a multispecies thick-map demonstrated that draining the Agro Pontino "swamp"—or protecting it—was and is guided via millennia of changing ecocultural discourses and epistemic and affective frameworks. The extraction of migrant labor has been a prime *topos* of postcolonial humanistic enquiry. However, as multispecies, multi-era and multi-scale stories (and stories within stories) of the Agro Pontino ecosystem emerge, it starts to become apparent that Satnam Singh's life, labor and death should also be understood within a multispecies paradigm. His story should be viewed not in seclusion, but as yet another appalling illustration of colonial capture-and-control of multispecies life in its myriad forms, human and non, for profit.

Multispecies Thick-Mapping Methods and Results:

1. The first step is to engage in a preliminary naturalcultural discourse analysis of the place-specific ecocultural system. This is accomplished via a multispecies, posthumanistic and transdisciplinary research methodology, connecting the natural and historical/cultural.
2. The next step is situated field work at the location. This phase is key to gaining an affective sense of place and its naturalcultural interconnections, and to establishing sustainable in-person relationships with local experts, both human and non.

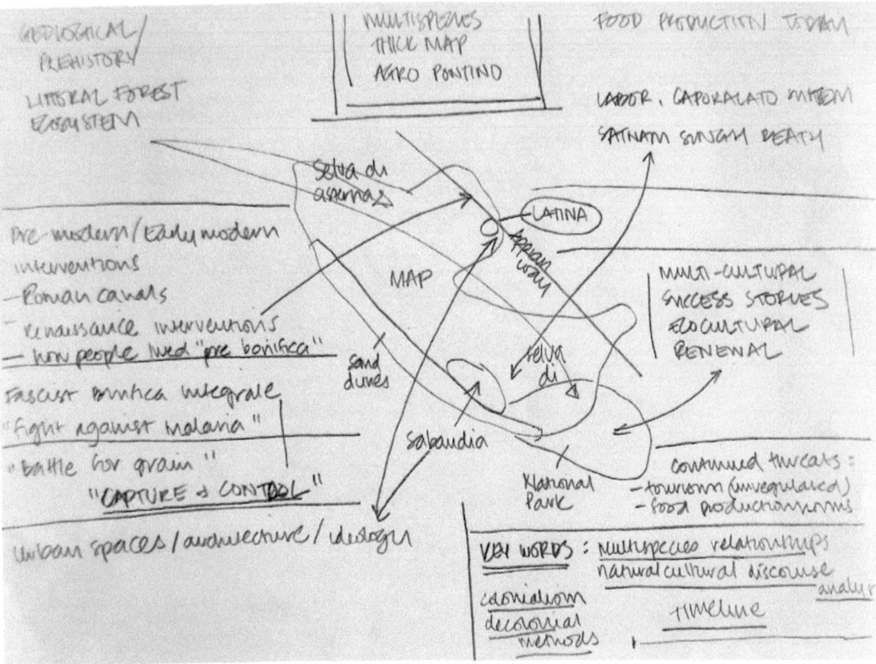

Figure 6.7 The “base” Agro Pontino thick-map. (Vetri Nathan)

3. Designing a thick-map flowchart that represents the primary network of relationships that need to be thick-narrated (Fig. 6.7). These include interconnected questions of both cultural import and ecological significance.
4. Deciding the base layer of the map. In the case of the first prototype described here, I decided to depict the transition from a pre-*bonifica* era to the post-*bonifica* years by blurring the lines between two maps, one taken from the eighteenth century and another derived via ArcGIS.
5. Adding layers to this draft base layer; organizing material within the layers and within stories to ensure maximum resonance, and to allow direct and indirect cascading relations of meanings between stories to emerge. We did this by covering the base map with translucent pieces of paper to imagine the layers visually (Fig. 6.8).
6. Final analog rendering digitally or physically: this was done via Photoshop.
7. Printing the analog thick-map for display: accomplished via a large professional printer.
8. Concurrent development of a digital thick-map: currently in progress. This thick-map will grow in future years as undergraduate and graduate students bring forward their interests in further researching and developing any given naturalcultural or multispecies aspect of the thick-map. It will also contain far more research available digitally, with links and citations for all material, including citations of material used in the analog

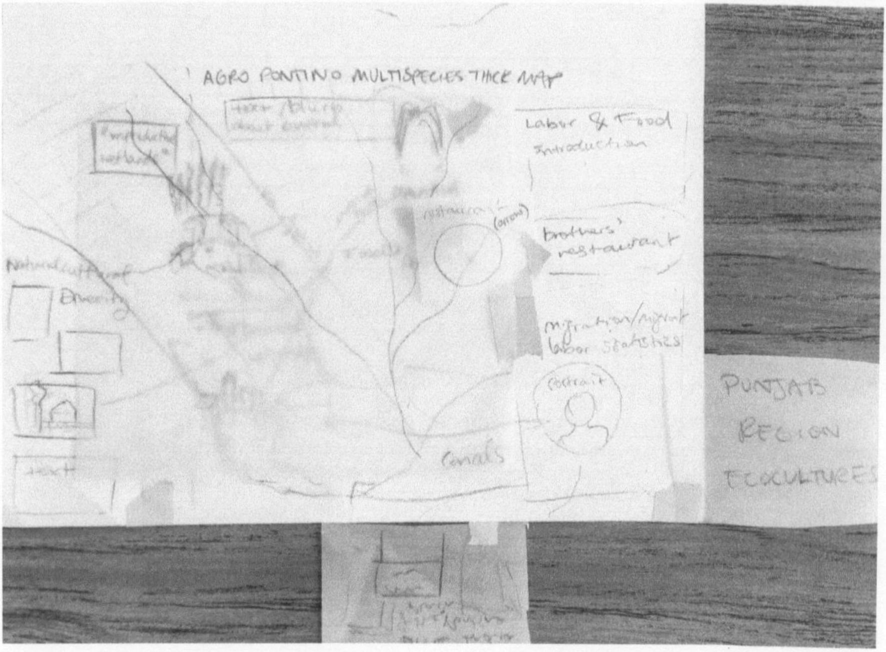

Figure 6.8 Adding layers to the "base" thick-map. We used translucent pieces of paper to imagine the layers visually. (Vetri Nathan)

version. This archive will also contain images and sound recordings from all "more or less animate" beings in different micro-ecosystems of the region.

In sum, multispecies thick-maps such as the Agro Pontino and *Scala dei Turchi* prototypes seek to strategically glitch thin colonial and purely anthropocentric narrations of places and spaces. Their contrasts with the colonial spectacle are evident:

COLONIAL THIN-ACCOUNTS	MULTISPECIES THICK-MAPS
Autopoetic—entry into the location is via a single perspective or cultural agenda.	Sympoetic—multiple points of entry are imagined and planned.
Oriented towards imagining isolated classified lifeforms via narrowly defined relationships based on resources. Derived from a hierarchical verticality of values, with humans at the top. Often represent the	Oriented towards imagining identities as products of shifting and intra-active multispecies relationships (oddkin). Many of the relations visualized are not only direct and linear, but often either a complex intra-action between cultural frameworks and natural

COLONIAL THIN-ACCOUNTS	MULTISPECIES THICK-MAPS
perspectives of certain dominant cultural identity groups. Limited naturalcultural interaction that is visualized, with often highly asymmetrical power differentials between the two entities that are interacting.	processes (such as how the Circeo National Park is a product of a specific historical concept of nature), or a result of indirect and cascading material consequences (for example, the success of the *Beccamoschino* being a result of ecocultural and labor activity by agricultural workers).
Limited anthropocentric time scales and de-situated knowledge that are often infused with strategic discursive capture-and-control-oriented narrations.	Thick-time accounts and different time scales (from split seconds to millions of years).
Carefully and narrowly curated to depict a controlled message about the space/place. Data collection is often done without local input.	Data collection includes input from local expertise, both human and non. Both the analog and digital versions of the map allow for spectator/consumer hallucinations. The digital version is especially well-designed for continued input from locals. For example, the digital thick-map can contain visual and narrative spaces for locals to contribute and represent their own place to visitors. It can also contain sound recordings of nonhuman contributions to the map.
Often uses de-situated quantitative and qualitative forms of knowledge citation as proof.	Engages in the broadest possible mix of quantitative, qualitative and multi-perspectival situated views. Situatedness means verifiable via embodied realities and relationships: attachment sites of lively agents.
Strategically promotes extractive and profit-oriented knowledges.	Strategically glitches extractive and profit-oriented knowledges and brings back value to sustainable naturalcultural relationships of kinship and care.

In Chapters 3 and 4, we explored how digital technology has changed our relationships with ourselves and our multispęcies kin (human and non). This section has showed how multispecies thick-maps can glitch the normative colonial mediatic mechanisms of capture-and-control that occur through the fetish-mapping machine of desire. However, these new kinds of maps are just one method by which the narrow focus of thin epistemologies can be glitched via naturalcultural discourse analysis. Many non-academic methods also disrupt colonial thin-accounts, such as artistic and creative ways of (re) presenting multispecies co-becoming, and even simply *being in relationships* of care and kinship. Naturalcultural discourse analysis can highlight these ways of being and of practicing multispecies co-becoming to celebrate and learn from them. The next two case studies explore two very different successful implementations of multispecies ethical glitches.

Transcorporeal Kinship: Digital Creative Nostalgia and Global Somali Cuisine

All three subsections of the Agro Pontino multispecies thick-map demonstrate how discourses, affects and practices of food production can become important vectors for pragmatic multispecies choices. Indeed, via a multispecies paradigm, food can be understood and valued as nonhuman bodies and life that we consume to survive and thrive, making food both a very intimate and a planetary-scale material manifestation of naturalcultural paradigms. One of the most evident examples of changed relationships in the Cybercene is our relationship with food, both personally and collectively. The next case study, of the Somali diaspora's multi-mediatic relationship with food, explores how opportunities do exist for cyberscreens to function as tools to aid in situating and embodying knowledges of care and kinship.[29] This kind of re-tooling of cyberscreen panopticons requires careful and strategic use of digital storytelling, always as a means towards fostering embodied sympoesis. Such use requires relational co-constitution of knowledges and affects within embodied multispecies realities, thus allowing them to be liberated from their obsessively autopoetic capture-and-control tendencies.

In her cookbook *In Bibi's Kitchen: The Recipes and Stories of Grandmothers from the Eight African Countries that Touch the Indian Ocean*, chef and author Hawa Hassan describes the distinctive blend of traditional foodways and Cybercene technology that led to its creation:

> In many ways, this is an old-fashioned cookbook that has nothing to do with trends or newness. It's filled with recipes from grandmothers and their rich histories, which have been handed down through generations. It reflects on times that have passed. Yet in so many ways this is an incredibly modern book that has relied on a web of technology and social connections to come together (including, but not limited to, multiple iPhones, many WhatsApp group chats, dozens of Dropbox folders and countless text messages).[30]

In this section, I explore a complex example of multispecies and multi-kind sympoesis: that of the (re)creation, (re)inscription and dissemination of Somali nationhood vis-à-vis diverse culinary foodways utilized by its global diasporic communities.[31] This story of diasporic belonging presents us with an important case study of an ethical glitch—the glitch being successful due to the emmeshing of virtual Cybercene communication with *embodied multispecies sympoesis*: the labor of imagining, procuring, cooking and eating (and *making-with*) many different species of plants and animals, as well as the sympoesis of intra-human activity that dining, rather than feeding, encapsulates. A moving account slowly emerged as I researched this story; one of the existence and persistence of a resilient global community through both traditional and new culinary—and therefore, multispecies—practices. To use Chef Hassan's words, the culinary pathways of the global Somali diaspora present remarkable examples of "multigenerational connection, cross-cultural connection, wireless connection [...] and ultimately, human connection" that are creating vital cultures and identities of belonging despite the complex geopolitical situation within their country of origin.[32]

Through an examination of multiple voices and media platforms, this case study explores the tension between the marvelously hybrid culinary profile of Somalia and its need to imagine a unified identity, which will serve to etch the contours of a recognizable modern nation state for both its own diasporic citizenry and for new food spectator/consumers not of Somali origin situated in the various international sites of community and consumption. To create this strong and empowered national/global culinary identity, I surmise that there must be a collective act of creative cultural imagination to bind together the dazzling hybridity of Somali cuisine, which consists of flavor principles and techniques drawn from local, Arabic, Persian, Indian, and Italian traditions, and which is further hybridized by the influence of the culinary profiles of the various host countries of the Somali diaspora.

Let's discuss just one example of the global multispecies web of life that Somali cuisine deploys via the *Xawaash* spice mixture, one of the ingredient assemblages that forms a key pillar in the flavor principle of the cuisine. This spice blend (pronounced Hawash), comes from the Arabic *Hawa'ij*, meaning "requirements" or "essential."[33] The mixture contains nothing less than seven species of plants:

- **Cumin:** *Cuminum cyminum*
- **Coriander:** *Coriandrum sativum*
- **Cardamom (green):** *Elettaria cardamomum*
- **Turmeric:** *Curcuma longa*
- **Black Peppercorns:** *Piper nigrum*
- **Cinnamon (Ceylon):** *Cinnamomum verum* or *Cinnamomum cassia*
- **Cloves:** *Syzygium aromaticum*

The proportions and variations of which ingredients are used and how can change from family to family, and also mirror regional preferences, sometimes featuring other plant species:

- **Fenugreek:** *Trigonella foenum-graecum*
- **Ginger:** *Zingiber officinale*
- **Nutmeg:** *Myristica fragrans*
- **Sage:** *Salvia officinalis*

A multispecies paradigm allows food studies scholars to explore food practices as essential multispecies naturalcultural practices, because cultural choices and preferences, honed via communal formations of tradition and innovation, result in direct and indirect cascading effects upon a whole range of agricultural ecosystems. Culinary choices also have profound multispecies implications, as illustrated by Stacy Alaimo's notion of *transcorporeality*. Alaimo configures bodies as essentially porous rather than self-sufficient entities. Her concept takes down perceived dualities of self/other, inside/outside, and us/the world.[34] What can be more indicative of the multispecies transcorporeality of our bodies than eating—a material and energetic process that allows us to imbibe the biopower of other species into our own bodies, to become part of our tissues?

With a multispecies culinary paradigm as a frame, I explore diasporic Somali cuisine as a distinctive result of multispecies identity formation and sympoesis. I call this sympoetic glitching process "creative nostalgia"—a somewhat contradictory process through which sympoetic food-related content and practices address a collective material and psychic yearning for a strong and stable Somali homeland (via citations of geographically, nationally, and regionally specific histories, traditions, tastes, bodies and locations), and also fulfills the antithetical need for adaptability, in order to transform these fixed tastes, bodies and places into far more mobile and mutable signifiers that can provide optimism, visibility and belonging within a globally dispersed diasporic community, and between these communities and their local "host" cultures and culinary contexts. The ethical glitch is the unlearning of naturalcultural norms of culinary and cultural knowledge situated within a physical nation state and the relearning of different naturalcultural meanings, affects and practices in very different situated contexts within new host cultures. The hybridity of Somali cuisine, compounded by this imaginative process of creative nostalgia, is mirrored by the hybridity of the cybermedia used to create and disseminate images and practices of this cuisine, and in the hybridity of the ecosystem of plant and animal species involved in its creation via ethical relationality. The examples analyzed in this case study will also demonstrate how, by deploying creative nostalgia, content creators of the Somali diaspora find personalized pathways to steer away from the damaging othering effects of digital culinary production and dissemination, such as extreme commodification, cultural appropriation, and disregard for the value of labor. Rather than promoting aestheticized and aspirational "food porn"—variously defined as "prose and recipes so removed from real life that they cannot be used except as vicarious experience"[35] and "food that is increasingly out of reach to the average cook or consumer"[36]—or utilizing national or "native" food traditions within a populist political framework that directly or indirectly excludes othered cultural groups, these diasporic culinary representations pull us towards cuisine's embodied, creative, sustainable, empowering, inclusive, nurturing and ecoculturally rich capabilities.

Analyses of both traditional print and digital/social media give rise to precise methodological and ethical considerations, as there must be some kind of selection due to the sheer amount of data. Akin to the methodology that grounds this entire book project, the primary sources used in this case study are deliberately contained yet diffuse in authorship and diverse in media, to properly reflect the contemporary hybrid multi-mediatic communication approach taken by global diasporic communities in the Cybercene. I analyze representations of Somali food/cyber practices by established chefs, and also blogs, social media posts and other digital entries by semi-professional home cooks, hobbyists, and everyday self-described food afficionados. My definition of the representation of "food practices" is similarly broad-based, and includes published recipes, presentations in cooking demos and videos of tastings on diverse platforms such as Facebook and YouTube. I also include cookbooks, a format that has always been considered by food historians to be gold standard for canonization of cuisines. While I include one print cookbook in my analysis (the aforementioned title by Chef Hassan), I would like to establish that the contemporary story of Somali diasporic cuisine, due to its global audience and reach and its diffusion in the Cybercene, goes far beyond culinary canonization through print publication and opens up pathways to such processes of identity creation and replication by far more heterogenous, equitable, democratic and, one could say, decolonial means.

The Somali language was principally transmitted orally until literacy campaigns in the 1970s adopted Somali characters written with the Latin alphabet.[37] This could be one possible factor contributing to the uniquely hybrid nature of Somali cuisine content creation, which privileges transmission more through audiovisual/cyber than canonical written/print media. Another is the more immediate practical factor of the resources needed to get a cookbook published. These reasons aside, even cookbooks are hybrid products in the Cybercene, as mentioned above in the quote by Chef Hassan, and it is important to consider how digital foodways have changed the very nature of the print cookbook. The aim of this case study is not to create an exhaustive list of references to fully map the diversity of global Somali cuisine and culinary practices, thus creating a new canon, but rather to use a select yet diverse collection of voices to understand how strength of national identity can persist or thrive, even given the dual hybrid circumstances of Somalia's richly hybrid cuisine, and hybridization driven by the experiences of the Somali diaspora. Such an approach will hopefully help subvert the "top-down" process of culinary and cultural canonization and empower the various protagonists of lived Somali culture who are creating a fluid, hybrid, continually mutating yet recognizable national identity through food. Some of the questions that I explore through my analysis of the culinary sources described above are: How do culinary authors manage colonial and postcolonial influences to create a unified national narrative about Somali cuisine? What are some of the key naturalcultural themes and culinary flavor principles that govern Somali cuisine as imagined by its diaspora? How are diasporic stories and experiences and culinary flavors connected via screens? How is multispecies sympoetic creative nostalgia deployed to negotiate and manage Somalia's hybrid culinary and cultural profile, rather than to divide and isolate?

In the Introduction to his volume *Food: The Key Concepts*, Warren Belasco considers an old axiom in food studies—dining is inherently different from feeding. While the latter signifies the physical act of providing sustenance to a biological entity, dining is a social and cultural act, defining what it is to be human. Drawing from Roland Barthes, Belasco goes on to define cuisine as a "shared set of protocols, usages, communications, behaviors, etc."[38] Benedict Anderson has famously defined the nation as an "imagined community"—meaning that culture forms the basis of a shared identity needed to strongly connect a vast number of strangers together.[39] If dining is an implicitly cultural activity and cultural imagination is key to forging national identity, then it follows that Somalia's cuisine must reflect directly on the particularities of its identity creation as a nation state. Such considerations of cuisine and nation become even more intriguing and complex in a study where the cuisine of interest is being represented by its global diasporic community, rather than those located physically within the geographical bounds of a culinary *terroir*. The Cybercene mediatic landscape, with its hyperconverged realities, becomes a key player due to this diffuse diasporic spectatorship seeking to create spaces of belonging.

To examine how these complex and sometimes contradictory forces of belonging are negotiated and managed, I will begin with an analysis of a re-cited televised cooking demonstration by noted Chef Jamal Hashi. As with many digital citations in the Cybercene, re-citation on a different digital platform gives content a far longer and broader reach and afterlife than traditional television. This is certainly the case with this demo, broadcast on the local culture section "Twin Cities Live" of ABC's 5 Eyewitness News, then reposted to Chef Hashi's Facebook page for his followers. One of the presenters, Ben Leber, frames the demo by publicizing "Somali Week" in Minneapolis and inscribing it within other national American celebrations: "This weekend the Twin Cities is going to be full of celebrations—not only is it the Fourth of July, but Somali Week is starting tomorrow!"[40] By placing the cooking demonstration of a typical favorite Somali snack, *sambusas*, within recognizable summer traditions of street fairs and cookouts featured in Independence Day celebrations in the United States, the show's writers create an unusually inclusive rather than exoticizing cultural framework. Chef Hashi begins his demo by further publicizing the upcoming Somali Week festivities to be held in the city center, mentioning "forty to fifty thousand visitors coming to town, big Somali artists are coming from Somalia to perform at the Target Center […] so people are coming from global (sic) to party here. […] People are travelling from across all the states, from Canada, to be here."[41] As Chef Hashi announces numbers, local street names and logistics about the event, the hosts provide constant interjections of wonder and appreciation of the scale of the event. The framing of the actual cooking demo with such impressive information about local community-building, the presentation of the event as an opportunity to provide outsized global attention to a relatively regional city and state and the casually mentioned yet detailed descriptions of city landmarks and street names are all key to establishing a strong presence of belonging by the chef, both for himself and for the general diasporic community he represents.

This creative interplay of two extremes to negotiate belonging—the hyper-local and the hyper-global—continues in the lead up to the actual cooking demo and is also inscribed visually in the bodies on screen. Both presenters of the show can be physically categorized by audiences as "quintessentially white" Midwesterners, in visual contrast to the chef's blackness. When Chef Hashi asks the presenters whether they have ever tried sambusas, they both reply in the negative, potentially creating a sense of anticipation in the spectator/consumer that they are about to experience a spectacle of captured-and-controlled exotic otherness through the dish (and its corresponding body). However, Chef Hashi masterfully ethically glitches any othering of the dish, and I would argue, of himself as a racialized body. He de-exoticizes and reinscribes sambusas within the local culture by saying "this is quickly becoming a local favorite. You can find this everywhere in Minneapolis right now."[42] Besides utilizing mimicry to break apart the stereotype of exotic African-ness, Chef Hashi utilizes four other key methods to make sambusas familiar, comforting, healthy and delicious rather than exotic and "niche": he involves the two hosts in the cooking, he adapts the recipe by using Mexican tortillas and Chinese eggroll wraps for the shells of the sambusas, rather than making these from scratch using flour, he uses a vegan lentil filling in addition to a beef version to "include our vegetarian friends" and finally, he uses an air-fryer, rather than deep-frying in oil as tradition would dictate, to promote ease of cooking in the contemporary middle-class American kitchen context and wide interest in this technique.

The result of this epistemic and pragmatic glitching of a dish relatively new to most viewers is summed up by the presenter Mr Leber, who exclaims that the open sambusa shell that needs filling "looks just like a snow cone!"[43] Chef Hassan's cultural insertion of Somali cuisine within the center of American, Minnesotan, and Minneapolitan/Saint Paulite spaces and bodies is one approach to creating both inclusivity and visibility. Rather than forming a "third space" through cuisine that is culturally hybrid—an in-between that is neither "us" nor "them," neither Somali nor American, but both and neither,[44] he describes and re-inscribes sambusas within local summer holiday traditions. While this move could be characterized as catering to the psychic demand of the hegemonic culture to normalize and neutralize perceived difference, I believe that Chef Hassan's choice, certainly made with careful consideration, finds the right balance between celebrating the uniqueness of Somali culture and also domesticating it within a Minnesotan context. This approach of "running away" from capture-and-control locates a delicate balance between affirming a unique history and identity and yet remaining inclusive and porous to possibilities of new belongings, glitching the potential othering narratives that the visible Black body on screen could instigate and providing a new and alternative path of belonging and kinship.

Due to the fact that gender-roles and norms are almost always deeply enmeshed within multispecies foodways, it was unsurprising that Cybercene strategies that glitch established sex and gender roles would find their way into diasporic representations of Somali cuisine. One of the most interesting websites in this regard was Bristol, UK-based Fozia Ismail's "Arawelo Eats." Created to publicize the author's supper club, the website frames the culinary practices on the landing page with the title "Somali food love Arawelo style." It goes on to explain:

> Arawelo was a semi mythical Somali Queen, who fought for female liberation and defied gender roles. After becoming destitute following the death of her husband, Arawelo formed an army of misfit women. They lived in wilderness in independence until their matriarchal community was threatened by male bandits. She led a series of successful raids and eventually bought peace by becoming the first Queen to rule Somalia over 2000 years ago.
>
> This is one version of the tale, there are many others.
>
> May all misfit women continue to forge their own paths.[45]

The historical origins of Queen Arawelo's legend are murky, but due to it being a story passed on orally through generations, it presents a fascinating number of alternative versions, some more transgressive than others: for example, some tell a story of a queen who castrated all males in order to subvert the patriarchy.[46] Fozia Ismail taps into this image of a strong Somali woman to accomplish several cultural negotiations and glitch established cyberdiscourses of gender roles: she proposes a woman-centered culture in opposition to the male-dominated culture of her own diasporic Somali community, she empowers female culinary entrepreneurs of all backgrounds in a professional field that is notoriously dominated by men while women are forced into the equally difficult yet financially non-remunerative task of domestic cooking, and finally, she seeks to democratize the elitist roots of all modern digitally enhanced foodways. As she says in a newspaper article:

> If I go into a fancy restaurant, I don't have to like it. Not because I grew up poor, Somali or as a refugee, or because I don't understand western flavours. I just might not like it. But if you're outside that world you can feel uncomfortable trusting your own judgment. It sometimes feels like the food world is for a particular class, a particular place, a particular kind of person who is a "foodie". Everyone eats food, so I don't know what defines a foodie.[47]

Keenly aware of the underlying inequitable power dynamics in server/customer, African/European, Black/White, woman/man, food/foodie relations,

Ms Ismail's supper club utilizes food as a vector to discuss and re-evaluate a multitude of identities in a post-Brexit world, whether Somali, East African, or British, rather than simply offering it as a commercial venture. As she succinctly puts it, "I didn't want to just pander to this idea that white people can just gaze and expect some kind of authentic African performance at the supper club [...]. I didn't also want to justify my humanity through the food."[48] Again, fundamental to the ultimate success of these negotiations and their ethical glitching of cultural capture-and-control is the *embodied activity* of a supper club where people meet in person to commune and get an experiential, rather than de-situated, understanding of these complicated gender, racial and national power dynamics.

While questions of gender do feature in Somali diasporic culinary content, they usually occur in broader biocultural contexts where transnational identities are also explored. A good example of the representation of the complex intermingling of gender, race and nationality can be seen in an episode of the cooking show "Tasty," hosted by actor and comedian Mindy Kaling with Somali guest chef Hawa Hassan and re-cited on YouTube. An ethical glitch to colonial authority is made visible by two women of color hosting the show, and this embodied presence extends its impact from a "simply" culinary context to its geopolitical reflections on Africa, refugee status, race, gender and national borders.[49] Due to this double theme of cuisine and identity, the demonstration of the famous Somali dish *Suugo Suqaar* becomes a reflection on transnational cultural solidarity, peppered with "infotainment" subsections where Somalia is geographically tethered and located visually on a map for American audiences. The cooking is explicitly connected to Indian diasporic experiences through Kaling's reflections on childhood and through the corporeal experience of smelling the *Xawaash* spice mixture we explored above, made of ingredients brought to Somalia via Indian Ocean spice trade routes. This cooking demonstration exemplifies how contemporary culinary authorities like Chef Hassan are harnessing creative nostalgia by utilizing Somali cuisine's transnational adaptability to create diverse communities of belonging. As *Suugo Suqaar* is a pasta dish, it is curious that Hassan does not dwell upon the tragic and difficult period of colonization by Italy. This is in line with other instances of Somali content creators, both professional and non, not characterizing Italian culinary influences as necessarily Italian, but rather as an integral part of the "core foods" that form the culinary profile of Somali cuisine. In many ways, this cultural absorption, the Somali-zation of pasta without any or only cursory colonial references, is a method of selective recollection to decolonize the national cuisine. Within such a strategic glitching, pasta is not by default the exclusive domain of Italian cuisine, but also an integral ingredient in Somali cuisine, part of just one of many successive waves of influence. Several short videos on platforms such as TikTok, YouTube and others show Somali eaters using their hands rather than forks to eat spaghetti—an act and representation of an act that glitches and nativizes the colonial way of eating pasta, and especially spaghetti, "properly" with a fork.

In most such online videos of Somali content creators, pasta is contextualized as being as "normal" and native to Somali cuisine as the ubiquitous peeled banana that accompanies many dishes (bananas, too, feature extensively in Somali cuisine due to the Italians cultivating the crop extensively in Somalia).[50] Indigenization of colonially imported plant life does not mean Italian colonization is glossed over completely in discussions of Somali cuisine. However, mentions of this period tend to go hand in hand with descriptions of other culinary influences, both benign and not, for example the connections to the Middle East and India via trade routes. This is an original and effective decolonial glitching strategy.

I will end this case study with a discussion of the cookbook I quoted from at the beginning: *In Bibi's Kitchen*. Featuring chapters on cuisines from Eritrea, Somalia, Kenya, Tanzania, Mozambique, South Africa, Madagascar and Comoros, *In Bibi's Kitchen* is a triumph of transnational creative nostalgia. It is able to effectively harness the power of tradition and yet re-present the stories and cuisines with a fresh urgency, specificity, and capability that situates these locations, bodies and tastes in the culinary center rather than the periphery. The authors note in their Introduction how "there are barely any cookbooks published by American publishing houses that feature African food, let alone food from one of the many parts of the continent (disturbingly, Africa is often mistaken as a country, not a continent)."[51] Recipes are divided into eight sections, moving geographically from the northernmost (Eritrea) to South Africa. The innovative structure of the book comes from two aspects: the connections it makes between seemingly disparate countries through their links to the historical Indian Ocean trade routes and interpenetrating histories, and the interviews and portrait-quality photos of the grandmothers (or *bibis* in Swahili) that generate an intimate relationship of comfort, care and kinship. These personal biographies and portraits give authorship and primary agency to the grandmothers. Interview questions and responses discuss not only the recipes, but all the other aspects of cuisine that are often equally important to the uniqueness of dishes—the societal norms, both generational and gender based, the placement or displacement of lives often lived in several places, even when rooted to the same spot. For example Ma Sahra, one of the grandmothers, "lives in the same place she was born, but because the border has since shifted, her birthplace, once part of Somalia, is now recognized as Kenya."[52] The book also elicits responses from the *bibis* about their different understandings of what home and community mean to them in the context of cooking and living in their communities. The deliberate co-mingling yet distinction between the eight countries' cuisines, and the interpenetration of recipes with geopolitical facts, personal interviews and micro-histories provides a fascinating (and ethically glitched) thick-account of the intimate hybrid (embodied and cyber) connection between cultural and culinary hybridity.

As this section's analysis of a diverse range of culinary representations demonstrates, the process of sympoetic creative nostalgia that leads to the intermingling of (hi)stories and foods mirrors the complex multispecies mix of ingredients and flavor principles of Somali cuisine itself, which, in turn, reflects the hybridity of media utilized to create and share recipes, techniques and etiquette. Whether through a "grammable" image, a re-cited YouTube video of a television show, a Facebook post or a cookbook, these diasporic culinary content creators, both professional and non, are actively imagining a past, present, and future for a dispersed yet deeply connected community. Colonial history—often elicited as the epistemic and discursive frame for a "proper" understanding of African culture and history—is diminished yet not forgotten in these culinary narratives: colonial culinary influence becomes just one of many other historical influences in a richly hybrid nation. Debilitating mediatic stereotypes of the Somali imagined community focused on hunger, misery, disease, and violence, which restrict its image to a very limited range that benefits continued neocolonial relationships with the more economically prosperous countries of the Global North, are glitched into a more complex and vibrant image. This reframing, retelling and reimagination of cultural history via a myriad of foodways makes it more inclusive of a large selection of ingredients, core foods, flavor principles and etiquette. It decenters the colonial experience without forgetting it and affirms the centrality and transcorporeality of creative bodies (often female), tastes, smells, textures, voices, and multispecies relationships. This loose conglomeration of culinary representations contributes to a stubborn persistence of belonging in many different permutations and combinations, despite dramatic disruptions and upheaval in its recent history. The sympoetic and inclusive creative nostalgia of Somali diasporic cuisine, with its multispecies protagonists connecting past, present and future practices of eating and cooking, is a naturalcultural practice of ethical glitching. It therefore presents an interesting case of how natural-cultural agency can sometimes be usurped by bodies, lives and diasporic ecocultural systems to generate deeply tethered alternative circles of care and kinship in a globalized world.

Our Buffalo Cousins: The Wind River Tribal Buffalo Initiative

> That the twentieth century America might exist the buffalo and the Indian had to go.
>
> Merrill Burlingame[53]

TallBear and Willey discuss how dominant settler-colonial discourses of gender and family roles that constitute and delimit what *home* should be have restricted and damaged other important forms of affinity and relationality. Dr Davis' performances unsettle these sclerotic forms of belonging, glitching their exclusionary discursive flows to diversify and expand the extent of love, care and nurturing of life. The dazzling diversity of Cybercene media used to communicate culinary knowledge mirrors Somali cuisine's hybrid profile, revealing the adaptability and openness between first, second and even third generation immigrants, and between diasporic communities and their neighbors, friends, and multispecies collaborators. I will now turn to the final and perhaps most exemplary case study of the incorporation of the ethical glitch within a community. The *Wind River Tribal Buffalo Initiative* story explores a profound entanglement between two species to explore how multispecies sympoesis is able to glitch restrictive Cybercene discourses of commodification and othering.[54] Indeed, few other examples can demonstrate the profound entanglement of the destinies of two communities than that of the Native American Indian tribes of the Great Plains and the North American Bison, also known as the American Buffalo.

The story of the almost-extinction of both communities by the 1880s and the healing from that brink back to some kind of restored habitability represents a profound story of multispecies care and kinship. I contend that the Indigenous/bison case also demonstrates some of the fundamental ways in which the right combination of scientific thought, cultural paradigm shift and multispecies inclusivity can be successful in turning the tide away from Cybercene othering and classification/commodification of lives. In this section I will connect the specific case of Indigenous/bison coexistence (as examined within the Eastern Shoshone and Northern Arapaho communities in the Wind River Tribal Reservation in Wyoming and the Greater Yellowstone National Park region) to wider questions of the role of global indigeneity and epistemic change in generating pathways towards healing and habitability in the Cybercene.[55]

It is important to not flatten the incredible diversity of Indigenous ecocultural systems and their very different colonial and postcolonial histories. It is nevertheless useful to understand that some commonalities exist, particularly in those communities that still have some form of situatedness within biodiverse lands and ecosystems. This proximity is complicated by the many different precolonial, colonial and postcolonial cultural paradigms that may often clash or coexist in unexpected collaborations and combinations within these communities. Vicente Diaz illustrates the political and analytic utility of investigating Indigenous identity and cultural innovation within the concept of *transindigeneity*, which can "help us imagine new ways of being indigenous but that also describes deep aboriginal cultural belongings to specific places while also permitting wide lateral reach across time and space, albeit in ways that do not lose familiar and signature indigenous belongings and accountabilities to place, to site, and cultural specificity."[56] I would like to take this "transindigenous approach" one step further and include my specific case study of buffalo/human revitalization in the Wind River community to reach across time and space to other anti-colonial glitching practices of all kinds. Weaving this case within a broader sympoetic coalition of situated knowledges connects the specific experience and perspectives of Wind River ecocultural leaders to concerns in other places and spaces.

Erika Zavaleta's 1999 investigation of Yup'ik waterfowl hunting presents an example of a nuanced analysis of Indigenous conservation that is attentive to the subtleties and complexities of discussing Indigenous ecocultural practices. She examines historical data on Yup'ik hunting and understanding of waterfowl population dynamics and governmental regulation, but she also provides unique data on Yup'ik motivation for conservation—she finds that while some hunters simply followed federal regulatory law most others were clearly positively motivated to conserve for the future.[57] Zavaleta's holistic approach to the problem can be a powerful model for future such studies. Raymond Hames (who also discusses Zavaleta's study) provides a frank and detailed assessment of the political discourse of the role of Indigenous communities in conservation efforts, and also produces a complex and nuanced conclusion, one that guides this study's humanistic approach to understanding the specific case of human/bison rejuvenation without stereotyping Indigenous communities within the extremes of ecologically noble or ecologically destructive:

> The ecologically noble savage debate has entered a new phase. […] The issue will move to a more detailed consideration of how people manage […] or engineer […] their environment and how management or engineering affects ecosystem stability and biodiversity.[58]

Such explorations may lead to more interest in understanding Indigenous ecocultural and multispecies paradigms. Native paradigms in the Greater Yellowstone ecosystem are deeply embedded within material practices of land use, food acquisition and consumption, tribal law and spiritual practice, and yet they are not equally and universally part of every member's worldview, as is the case with any community. For example, Richard Baldes strove for the tribal game code to be overhauled so that the two tribes would have a better say over the destinies of the wildlife in the reservation. It was immensely successful in making the reservation a haven for game and yet it was an action that many individual tribal members bitterly opposed at first. Complexity is key, and yet, via a colonial discursive imagination, Native Americans have been represented in popular culture as a one-dimensional stereotype. Ward Churchill describes how cinematic representations of Native Americans were fantasies of white settler-colonizers: "These depictions of Native Americans enable the perception that Natives are of another time and extinct in a manner that rids the United States of the 'moral or psychic obligation to confront the fact of Native America, as either an historical or topical reality'."[59]

A vast variety of mediatic images (from sports mascots to comic books to mainstream television) played and still play a role in cultivating highly restricted images of Indigenous communities globally, accomplishing the task of epistemic, discursive and psychic capture-and-control with the goal of gaining access to the vast and rich land and ecological resources that these communities held. Philip Deloria argues that the ambivalence of the newly formed American nation was easily located in a split and ambivalent emotion of both desire and repulsion:

> Euro-Americans had imprisoned themselves in the logical mind and the social order. Indians represented instinct and freedom. They spoke for the "spirit of the continent." Whites desperately desired that spirit, yet they invariably failed to become aboriginal and thus "finished." Savage Indians served Americans as oppositional figures against whom one might imagine a civilized national Self. Coded as freedom, however, wild Indianness proved equally attractive, setting up a "have-your-cake-and-eat-it-too" dialectic of simultaneous desire and repulsion.[60]

This double logic of desire and repulsion found within the constructed image of the "wild" Native Americans is not limited to cultural discourse in North America—it represents a global phenomenon of cultural ventriloquism. It is not hard for a multispecies studies scholar to immediately notice how the control discourses deployed to manage Indianness, as described by Deloria, resemble discourses also commonly used to cage-and-stage many animal lives. The doubled logic here also parallels with the desire/repulsion dynamic of the stereotype-as-fetish discussed earlier in the book. These Native American control-metaphorics are also a globalized strategy: Indian images have also been used a surprising number of times to make nativist arguments against immigration and multiculturalism and in support of other such expressions of racialized biopolitical conservatism in Europe, playing on the fascination and repulsion that the noble savage stereotype still incites in continental biopolitical discourses.

The malleable image of the almost-extinct Indigenous peoples has now been harnessed within collective psychic mechanisms of capture-and-control via cyber-citation. Analyzing these discursive patterns can foster an understanding of how global indigeneity can be liberated into more nuanced, situated and complex ecocultural contextualizations. The role of individual and place-specific Indigenous communities in ecocultural conservation is complex, diverse and contradictory, made doubly so due to the colonial ventriloquism they are still subject to. It is however far easier to ascertain that while they constitute only 5% of the world's population, they host vast amounts of Earth's remaining biodiversity.[61] Therefore, any strategy that involves protection of multispecies diversity on the planet in the face of the immense destruction taking place in the Cybercene will be far more successful if Indigenous visions, aspirations and motivations are taken into consideration. I would like to posit that by including Indigenous epistemologies in the conversation about and the practice of conservation, dated colonial notions of conservation that exclude human presence from conservation lands (what Hames describes above as "a strict definition of conservation") can be glitched, allowing for a more decolonial, complex and nuanced multispecies partnership within natural ecosystems to emerge.

It was ethically important for me in this case study to glitch the researcher/researched duality and center both Indigenous ecocultural leaders' voices and the viewpoints of ecosystem science experts without ventriloquizing. To do so, I had to actively reflect on my positionality and capacity to discuss Indigenous epistemologies despite not being a member of any of these communities. While I am not of Indigenous ancestry, I experienced, to my joy, that I was able to gain the trust of the Indigenous experts I consulted for this project partly due to my own positionality as a brown, postcolonial individual, with my own community's historical experience of colonialism, even if it was of a very different nature—that of British colonial imperialism rather than Anglo-European settler-colonial violence. Despite these differences in our personal histories, I sensed that our common experience of coloniality and my visible brownness, along with the spiritual animism that also informs part of the multifaith traditions of India, assisted me in respectfully "standing with" Indigenous experts and their communities with humility and affective recognition. While none of my work for this section was ethnographic or human-subject related, since my conversations with all experts focused on their professional expertise as community leaders and ecologists, I believe that my hybrid positionality did help to create a productive foundation for deep conversations about the interconnections between culture, identity and ecology.

In order to listen, learn and explore the wider context of multispecies relationalities within the Wind River Tribal Buffalo Initiative, I traveled widely within the Greater Yellowstone National Park ecosystem, visiting the prairie and intermountain states of Idaho, Montana, Wyoming, Colorado and South Dakota. I was fortunate to meet experts willing to share their perspectives—experts both Indigenous and non, human and non. I am incredibly grateful for the expert collaborators and knowledge sharers who were so generous with their time for this project.[62] Working on this section of the book was truly an act of sympoesis (*making-with*), where the thinkers and shapers of their communities (human and non) sometimes challenged and sometimes confirmed my abstract ideas of what research in the humanities (and the sciences) can and should be, allowing me to truly understand and experience some of the Indigenous ways of approaching *research through ethical relationality*. I was lucky to gain perspectives and access the complexity of the embodied realities of multispecies kinship via the forging of new relationships with the land, the water, the air, the community of experts (human and non), and the spiritual, cultural and epistemological paradigms of the places and spaces that welcomed me during my visit.

The regeneration of the historically decimated populations of American Buffalo or Bison (scientific name *Bison bison*) in the Greater Yellowstone ecosystem allows us to tie together the many strands of Cybercene ecocultural entanglements and consider solutions to combat its most corrosive trends. Diverse native names reflect the multiplicity of ways in which bison are differently configured within each specific tribal ecocultural system: for instance, the Lakota name for male bison is *tatanka*, for female *pte*, the

Blackfoot call them *iinniiwa*, the Salish *qweyaway*. Bison cultural expert Ken Zontek notes: "It's difficult to think of another group of humans who have become so intertwined with a wild animal species that it pervaded their culture. Bison permeate virtually everything material for Native Americans, as well as spiritual [...They] never lost their spiritual relationship with their kindred animal."[63] It is also well documented that bison were decimated at the same time and along the same lines as the extermination and uprooting of Indigenous tribes, tragically demonstrating the very strategic multispecies exploitation and extermination that forms the foundation of colonial thought and practice. The bison were reduced in the space of a few decades from an estimate that ranges from "the low millions to as high as 200 million"[64] *to only a few hundred individual animals* by the late 1800s. It was a truly staggering ecocultural tragedy for the whole continent, but especially for the Indigenous and multispecies communities that were deeply entangled with its naturalcultural activities. William Hornaday's now-famous map of the incredible shrinking range of the bison (Fig. 6.9) vividly represents the bison holocaust.

The crucial analytic naturalcultural "gap" in the map is the exploration how the decimation of the bison was both an explicit and an implicit settler-colonial naturalcultural policy led by individuals, groups and the United States government to weaken Indigenous power over their lands by taking away their cultural, spiritual and nutritional source of energy. Fur trade scholar Merrill Burlingame chillingly crystallizes this policy of Euro-American colonial capture-and-control with his observation: "That the twentieth century America might exist the buffalo and the Indian had to go."[65] The collapsing boundaries of the Hornaday map, therefore, also hauntingly represent the rapid and relentless decimation of Indigenous lives and communities across the continent. Ken Zontek cites the work of Immanuel Wallerstein and Fernand Braudel, suggesting that co-extermination of the Native American and bison communities happened via the general colonial process where "regions and countries around the globe, including their indigenous biota, became sucked into a world system that exploited them. [...] For these dependency theorists, North America's indigenous people simply failed to overcome outside pressures exerted upon them by Euro-American technology, materialism, biotic invasion and coercion."[66] In other words, this was a disastrous colonial naturalcultural capture-and-control desire machine, actualized with a multiplicity of biocultural weapons. Evidence seems to suggest that Native American communities were not only materially almost pushed to extinction, but that this physical extermination was concurrent with and intra-actively connected to the almost-extinction of any sustainable multispecies paradigm that may have existed before the arrival of Euro-American paradigms of unsustainable hunting and depletion of natural resources.

The renewal of bison herds from that brink to today brings together complex embodied considerations of multiple spaces, lives and desires: the local Native American tribes, cybernarrative-influenced local ranchers,

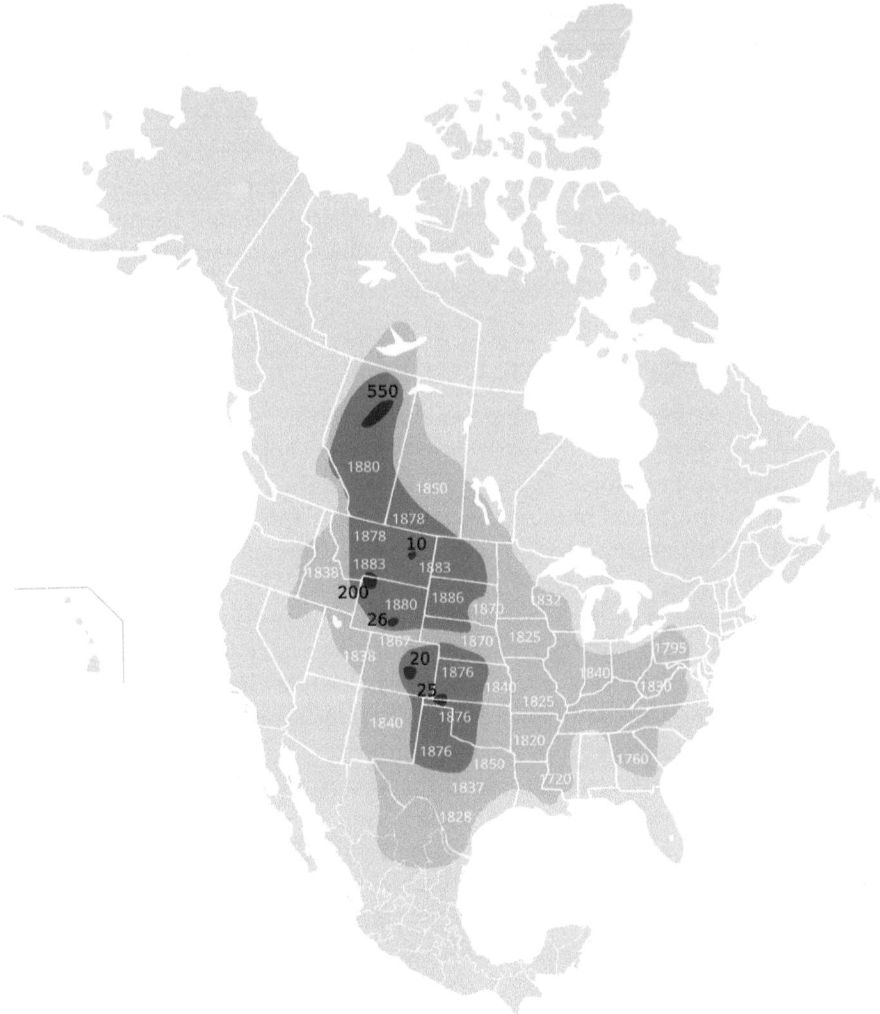

Figure 6.9 Map of the extermination of the American Bison to 1889. Adapted from a drawing of William Temple Hornaday in *Geographisches Handbuch zu Andrees Handatlas, vierte Auflage*, Bielefeld und Leipzig, Velhagen und Klasing, 1902. (Wikimedia Commons)

cyberscene-influenced tourists, and the park biologists and staff who are beholden to both science and governmental bureaucratic regimes. Very importantly, we also have the bison themselves, with their lives often captured-and-controlled yet tending to "run away" from the different management protocols imposed upon them, with usually fatal consequences for their lives. In the Cybercene, these wild lives are dependent on those very protocols and narratives of physical and epistemic management for their survival through relative freedom (rewilding) or capture-and-control and commodification as livestock (domestication). Furthermore, the bison case also entangles these considerations within questions of ecosystemic health (bison herds are fundamental to the success of a web of other species and in turn are supported by a web of other species); political borders (those between designated "park" boundaries and not, between different states in the U.S. federal system, between native lands and state/federal property and also between the United States and Canada); expansionist, settler-colonial and neo-colonial ecocultural power relations, and the science, economics and politics of ecological conservation, tourism and food systems (primarily beef production, potentially affected by the pathogen *Brucella abortus* carried by bison).

A Tethered Multispecies Origin Story

I began this chapter by discussing how observation, embodied connection, ceremony and storytelling are often described as fundamental to Indigenous epistemologies and axiologies. To honor this paradigm, I will begin this journey within the Native/bison multispecies community with a powerful story of multispecies relationality—an Emergence Story. There are different versions of this story within the tribes of the Black Hills of South Dakota, but here below is the one told in the Cheyenne Creek community on the Pine Ridge Indian Reservation of the Oglala Lakota tribe. The story was told by Wilmer Mesteth—a tribal historian and spiritual leader—to Sina Bear Eagle, and reproduced on the website of Wind Cave National Park for public outreach. Fundamental to the power of the Emergence Story is a specific and material place—the Wind Cave complex, one of the longest cave systems in the world. Along a pathway leading away from the Wind Cave Visitor Center, I was able to visit and pay my respects at a place of great spiritual significance—the natural opening to the cave system. This unassuming hole in the ground is the physical location, the very place where this intricate origin story of both bison and human becomes enmeshed in the material ecosystem of the land.

> This story begins at a time when the plants and the animals were still being brought into existence, but there were no people or bison living on the earth. People at that time lived underground in the *Tunkan Tipi* — the spirit lodge — and were waiting as the earth was prepared for them to live upon it.[67]

The opening in the ground to the Wind Cave complex is the very opening through which both people and bison entered the earth above.

> The Creator instructed them to follow that bison. From the bison, they could get food, tools, clothes, and shelter. The bison would lead them to water. Everything they needed to survive on the earth could come from the bison.
>
> When they left the cave, the Creator shrunk the hole from the size of a man to the size it is now, too small for most people to enter,

> to serve as a reminder so the people would never forget from where they'd come.[68]

While this elaborate origin story powerfully opens up the complex threads of multispecies, multi-kind relationality that exist between the landscapes, ecosystems, animals, plants and humans, it is given even more power by the visual and haptic presence of the actual natural opening of the Wind Cave system. This opening exists to this day. It *actually breathes*, as if to serve as a constant reminder of its life-giving properties. Stepping close to the opening, which is indeed now too small for humans or bison to emerge, I could hear and feel a constant, loud and deep *whoosh* of air being expelled, coming surely from what we would now explain as the pressure differentials created by the interaction of outside air and that within the vast chambers inside. However, understanding this story and the experience of listening to the opening breathe as an embodied epistemological pathway rather than simply a geological spectacle allowed me to understand this pressure differential in a very different way. The wind emerging so incessantly from that sacred spot is a vivid acoustic and tactile reminder of how communities can forge multispecies bonds via restoring disavowed perspectives and re-evaluating dominant cultural discourses—it forms an *embodied citation*, different from the de-situated cyber-citations of paranoid knowledge. These epistemic norms, grounded within the material workings of landscapes and bodies rather than de-situated visual spectacle, can be the groundwork upon which restorative actions may take place against Cybercene de-situatedness. Viewed from such a perspective, we can begin to understand how tethered stories can convincingly glitch the othering and commodifying implications of the hierarchical Linnaean classification of life, thus allowing us to finally provide respect and equity to the other forms of life that sustain our societies.

The movement of air, the *breath of the earth* emanating from Wind Cave can also be understood as the ecological metaphor for the freedom of space and movement that is so critical for the ecosystemic health of both humans and large land mammals such as the bison. Indeed, bison, unlike cattle, have a positive effect on the plant and animal ecosystem by how they graze, how they move from place to place, and how they migrate seasonally.[69] Native American tribal movement, which also followed the bison paths, was critical to ecologically sound use of the land and its resources, allowing tribes to harvest without depletion.[70] It is not a surprise that this strategic seasonal movement was characterized by settler-colonialists as aimless wandering, a false nomadism argument used by invaders to deny natives rights of land ownership. Yellowstone National Park was created in 1872 by President Ulysses S. Grant. The report of the House Committee on Public Lands describes the bill in these words:

> The bill now before Congress has for its objective the withdrawal from settlement, occupancy, or sale, under the laws of the United

> States a tract of land fifty-five by sixty-five miles, about the sources of the Yellowstone and Missouri Rivers, and dedicates and sets apart as a great national park or pleasure-ground for the benefit and enjoyment of the people.[71]

Reading these words via a decolonial epistemic framework, one can begin to ask: the benefit and enjoyment of *which* people exactly? At which community were the words "the withdrawal from settlement, occupancy, or sale" aimed? Yellowstone was sparsely inhabited by Euro-Americans, but was an important summer habitation for Indigenous and multispecies nomadic communities. Viewed via the positionality of these communities, the creation of the park seems to follow a settler-colonial strategy of conservation via usurpation, just like Mussolini's integral reclamation of the Agro Pontino wetland system. The bill to create Yellowstone conveniently repressed and forgot the 11,000 years of Indigenous history within the park's ecosystem and reformatted it as a natural wonderland, an Eden without humans, a *terra nullius*.

And yet, Yellowstone National Park's creation was a milestone in world conservation history. How much more just would its formation have been if it had been envisioned as a caring place for a truly multispecies ecocultural system, for Indigenous communities and other nonhumans? Today, many selfie-obsessed tourists echo the historical capture-and-control epistemologies that created a no-natives land at Yellowstone in the first place, ignoring the incredible natural embodied spectacles of life such as the bison, and the geological might of the geysers, in order to get the "perfect image" for their social media posts. Increasingly, park officials are spending their limited energy and resources on rescuing tourists who are harmed due to not respecting the embodied realities of natural ecosystems, as these visitors try and domesticate wild and inanimate bodies through Cybercene image-capture. Figure 6.10

Figure 6.10 Tourists watch and capture images on their phones as bison are corralled for vaccinations at the Annual "Custer State Park Buffalo Roundup."
(Vetri Nathan, with permission of Custer State Park)

illustrates this dynamic beautifully: as the bison are "rounded up" for vaccinations at Custer State Park in South Dakota by park managers, tourists watch the spectacle of their "taming" and capture these primal images on their phones. The seen becomes the capture-and-control scene, again and again and again.

As tourists insert their capturing-and-controlling presence into the park and saturate both physical and virtual space with their bodies and screens, as well as (over)populating cyberdiscourse via the mechanisms of hyperconvergence, another far more restorative kind of repopulation is finally happening. In *Restoring a Presence: American Indians and Yellowstone National Park*, Peter Nabokov and Lawrene Loendorf quote Mark Spence to remind us of the often-negated continuous presence of several tribes in the park: "Yellowstone [Park] is a geographic square that was grafted over native understandings of the place and the region. Consequently, to expect native stories, memories and actions to precisely distinguish between things and places on either side of current park boundaries is absurd."[72] It is important also to note that the vital material, spiritual and ecocultural connections that native tribes had with the lands of what is now Yellowstone National Park parallel animal and plant practices within and with the ecosystem—habits and traditions that do not conform to the political boundaries of public/private, federal/state/private park ownership. Seasonal migration patterns of bison and other animals were well understood by Indigenous tribes, who would follow bison herds to their wintering grounds, now found outside park boundaries, in locations such as Paradise Valley. Yet the migration of Yellowstone's rejuvenated bison populations often leads to tragic consequences, as bordering state regulations force park officials to slaughter any bison crossing this artificial border.[73] Shana Drimal, Senior Conservation Associate at the Greater Yellowstone Conservation Coalition—a non-profit advocacy group based out of Bozeman, Montana—describes how the Greater Yellowstone ecosystem "is a 20 million acre extraordinary system, comprising one of the planet's largest intact ecosystems. It is made up of five national forests, two national parks, ten designated wilderness areas, three national wildlife refuges, the 2.2-million-acre Wind River Indian Tribal Reservation. It is considered the birthplace of modern ecosystem-management and large-landscape conservation, in addition to being the traditional homelands of deep cultural significance to 49 different tribes."[74]

Figure 6.11 shows a map of the Greater Yellowstone ecosystem, encompassing two national parks and its far wider interconnected ecological boundaries. In order to restore the multispecies interconnections that have been severed by the many boundaries that now overlay the area, I refer to this entire ecosystem via a multispecies and Indigenous perspective as the Greater Yellowstone *Ecocultural System* (GYES). Just as the cleavage of the Punjab and Bengal ecocultural systems into two separate nations (India and Pakistan) led to all kinds of cultural and ecological problems that persist to this day, viewing the GYES as more than a square-shaped park via a

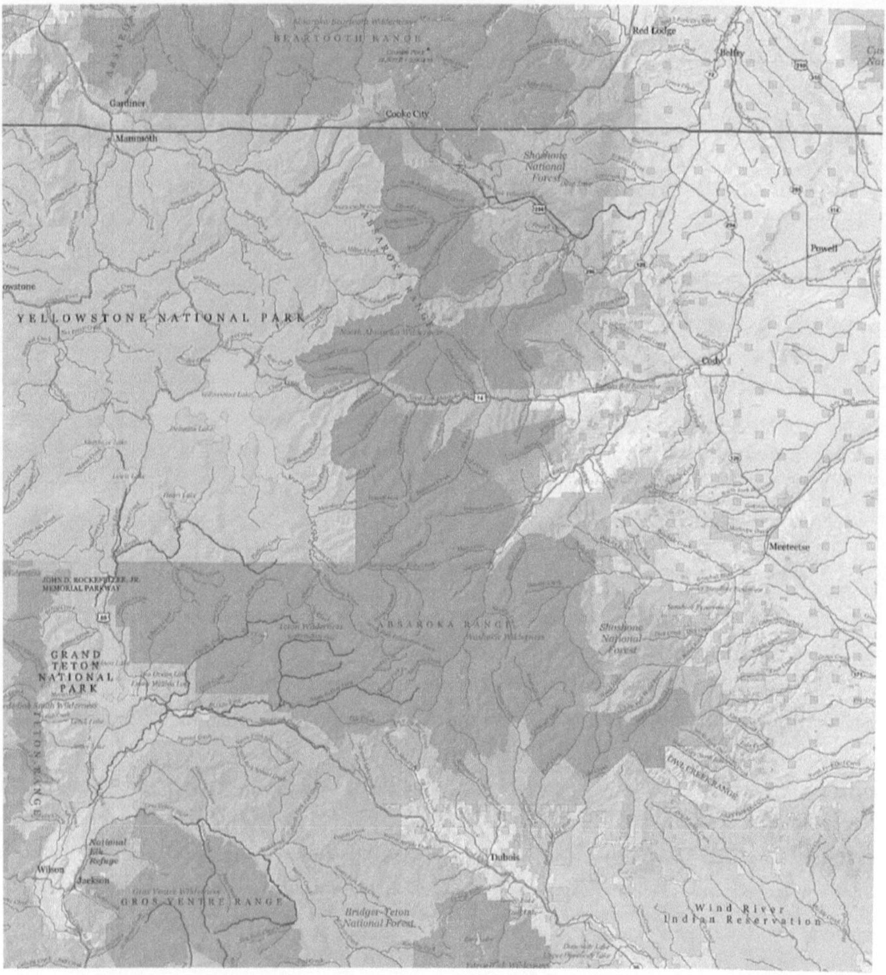

Figure 6.11 Detail of a map representing the Greater Yellowstone Ecocultural System, including most parts of Yellowstone National Park at top left, Grand Teton National Park at bottom left and Wind River Indian Reservation at bottom right. Map created by Charles Drimal and Shana Drimal, Greater Yellowstone Coalition. (https://greateryellowstone.org/)

multispecies and Indigenous paradigm allows the space to "run away" from its imperfectly envisioned conservation boundaries.

Jason Baldes (Fig. 6.12, pictured with his buffalo community), Executive Director and Eastern Shoshone Tribe Buffalo Manager of the Wind River Tribal Buffalo Initiative explained how all parts of the bison, when hunted, are used in both food and ceremony. Bringing back the bison, working diligently to allow it to thrive again as a wild animal, as well as hunting bison on these newly restored lands, has revived a keystone relationship within the community, providing a restored sense of place, space, identity and ecocultural

Figure 6.12 Jason Baldes with his Buffalo family at Wind River. (Vetri Nathan)

habitability. Wes Martel, Senior Wind River Conservation Associate at the Greater Yellowstone Coalition and Board Member of the Buffalo Initiative explained how community ceremony, which includes storytelling and hunting as well as spiritual ceremonies, and tribal youth education programs create and maintain a vital and vibrant embodied multispecies relationality. This relationality is, in turn, the strong foundation upon which tribal sovereignty can be restored. Seen in this way, ethical relationality, as practiced in both Indigenous ceremony and Indigenous research, becomes not only an epistemic attitude, but also a deeply integrated methodology of ecocultural restoration. As Sioux scholar Vine Deloria Jr. explains:

> We are all relatives when taken as a methodological tool for obtaining knowledge means that we observe the natural world by looking for relationships between various things in it. That is to say, everything in the natural world has relationships with every other thing.[75]

The Wind River Reservation buffalo were brought back to the land after a gap of more than 130 years through the Wind River Buffalo Initiative, thanks to a series of three preconditions: firstly, there was a strong Indigenous epistemic and axiological yearning in an important subset of the community for the severed physical connection with the buffalo and through it, a yearning for a restoration of ecocultural connection to the land. Secondly, the spiritual connection to the buffalo continued to be conserved even after the decimation of actual bison, forming a bridge of epistemic survivance via storytelling and ceremony to sovereignty over their lands. For example, on Indigenous Peoples Day in 2025, the Initiative hosted around 70 children from the Wyoming Indian Elementary School at a "Buffalo Bash" where "in teepees they made tiny clay Buffalo, learned about the importance of all the parts of the Buffalo, and were taught about important plant medicines."[76] The survivance and passing on of these axiologies to newer generations can only be called a triumph of previous generations that sought to conserve a connection to the land despite the disappearance of the bison and the violent imposition of contrarian dominant settler-colonial ideologies, forced upon the community via colonial ideas of land ownership and use as well as through hierarchical practices of multispecies relationships such as intensive farming and cattle ranching. Inspired by Foucault, Brendan Hokowhitu classifies both historical and ongoing colonialism into three types: the first is *sovereignty colonialism*, which would use colonial law to strip the imagined sub-human Indigenous peoples of "natural cultural claims to our epistemologies and lands." The second he calls *biopolitical colonialism* and the third *disciplinary colonialism*, which came later:

> [it] was also subjugatory, but through the institutionalized discipline of Indigenous bodies and the production of brown citizens through an array of discontinuous discourses, for instance, family structures [...] nutritional advice, unemployment subsidies, institutionalized pre-natal and natal care, war, statistics on childhood obesity, lifestyle magazines, educational curricula, mental health, daily and weekly schedules, prisons, workplaces, sport, time, the solar calendar; all of these seemingly discontinuous heterogenous "enunciations" productively disciplined the Indigenous body.[77]

Despite the imposition of these immense structural and material colonialisms, often brought about by different federal agencies, including the Bureau of Indian Affairs, via policies such as job placement of tribal youth outside the reservation in far flung places such as Seattle and Chicago, the germ of thought and spirit that provided the basis for Indigenous uses of land, with a sustainable and far lighter ecological footprint, continued to exist within the ceremonies and traditions of the community.[78] Sovereignty accorded to the Eastern Shoshone and Northern Arapahoe tribes at Wind River via treaty rights also allowed them the legal freedom to re-establish bison by protecting and restoring the land and water within the reservation.[79] Finally, a fundamental external precondition that served to pave the way for bison regeneration at Wind River was the establishment of the Inter-Tribal Buffalo Council (ITBC) in 1991:

> In February 1991, Tribes gathered in the Sacred Black Hills of South Dakota. Nineteen Tribes from all four directions attended. Lakota representatives from several of the Native Nations in South Dakota were there, as well as the Crow, Shoshone-Bannock, Gros Ventre/Assiniboine, and Blackfeet Nations of Montana. Various Pueblo representatives from New Mexico, and the Winnebago, traditionally called Ho Chunk, from both Nebraska and Wisconsin came.

> Choctaw Nation of Oklahoma, and some as far west as Round Valley of California arrived. Some of these Tribes were historically enemies, but they united for a common mission…
>
> *To restore buffalo to Tribal lands for cultural and spiritual enhancement and preservation.*[80]

Wind River Reservation Tribal Buffalo Initiative, established in 2022, worked with many entities including the ITBC, the Greater Yellowstone Coalition, as well as the biologists of Yellowstone National Park to establish its two herds of wild bison. Deriving its herd from a collection of diverse locations allowed the initiative to conserve diversity within the genetic makeup of the herd.

Keeping the bison wild has been a key tenet of this initiative, as it is the "wild" behavior of the bison such as family connections, grazing patterns, wallowing, birthing and migration that are bringing about a restoration of the many ecosystems that make up the Wind River Reservation within the GYES. When viewed through a multispecies paradigm shift, these "wild behaviors" are nothing other than bison "culture." The mission of the initiative is "land rematriation, community revitalization and youth education through restoration of conservation Buffalo"[81]—demonstrating the importance the experts leading this initiative place on multispecies ecocultural interconnections rather than simple "conservation" that does not include the Indigenous human element in the picture. The vision of the initiative is to ultimately have "thousands of buffalo on tens of thousands of acres, protected under tribal law as wildlife."[82] Several scholars have pointed to various ways in which Indigenous communities exercise sovereignty struggles within the limits imposed by state sovereignty.[83] The Wind River buffalo restoration initiative is a concrete example of a powerful project and vision that weaves together the intimate interconnections between Indigenous multispecies epistemology, axiology, affect, ceremony and spirituality as research, as well as the use of hybrid epistemic and cultural orientations to conserve and promote sovereignty over tribal lands.

Bison–Human Intra-action in the Greater Yellowstone Ecocultural System

Very importantly, we also have the bison of the GYES themselves. Very much like the Channel Island Fox in the Preface, naturalcultural discourse analysis allows us to ethically glitch anthropocentric epistemologies to revalue these magnificent beings as important ecocultural theorists and practitioners. They represent important lives that are often captured-and-controlled both literally and metaphorically to maintain politically established naturalcultural norms. Bison-centered agency consistently challenges these limits, which often result in deadly pushback on them by management protocols. Political borders exert intense pressures; those between designated "park" boundaries and not, between different states in the U.S. federal system, between native lands and state/federal property and also between the United States and Canada. In the tribal lands of Wind River, Director Jason Baldes' insistence on maintaining a *conservation herd* as opposed to a *commercial herd* allows the bison to better coexist in an ethical relationship with the tribal communities where their own agency and culture are taken into consideration, rather than simply being seen as a resource for commodification and consumption—a sort of blockchain bison farm. Keeping bison wild also means the conservation of the precious genetic diversity of bison, a diversity that was depleted as the few remaining herds were often interbred with cattle in the early twentieth century by ranchers seeking to create more "manageable" strains of life. In the Cybercene, bison have been captured-and-controlled for their lucrative meat in ranches, which do not allow them to become "ecologically alive"—that is to say be in numbers large enough to function as a keystone species. The livestock model, which treats bison as a woolly cow, is ultimately tethered to expansionist, settler-colonial and neo-colonial ecocultural power relations. The science, economics and politics of this certain type of ecological conservation, tourism and food production is tethered to these colonial forms of relationality. Tribal wild buffalo initiatives such as that of Wind River demonstrate how Indigenous embodied multispecies knowledge can glitch these dominant colonial epistemic systems of capture-and-control via rewilding that is strongly connected to both community *and* biodiversity revitalization. Jason and Richard Baldes have worked extensively with the Northern Arapahoe and Eastern Shoshone tribes in Wind River to reclassify bison as wildlife rather than livestock. While this reclassification may initially seem a relatively minor or formal detail, it not only provides a conceptual ethical glitch that changes the epistemic and affective way in which the bison community is understood, but also allows for the bison to be included in the game code. Jason Baldes explains: "Once they're in the game code, then we can protect that population, we can grow it, and designate more habitat."

> The resolution states support for the Wind River Tribal Buffalo Initiative and its own tribal buffalo restoration efforts and recognizes that bison "were and remain central to the culture, health and welfare of the Northern Arapaho Tribe since time immemorial." Further, it states that the large mammals shall be "managed in accordance with wildlife management principles to expand the herd and support the establishment of Buffalo within their traditional homelands."[84]

The Wind River Tribal Buffalo Initiative's insistence on a free-ranging and wildlife designation is not simply a fussy epistemic stance nor simply a question of territorial sovereignty. It is both of those but most importantly, it is an embodiment of a concrete, multispecies, multi-kind respect for the way in which free-ranging bison affect the entire ecocultural system. Designating bison as a wild animal will without doubt pose multiple challenges and opportunities in the future. The final section of this case study provides a multispecies thick-account that describes exactly how the bison becomes a keystone species not only for the ecosystem, but also for the cultural systems in the reservation.

Through my conversations and embodied presence among the experts (both human and non) in the GYES and at Wind River Tribal Reservation, it became clear that the Indigenous epistemological paradigm of bison *both as a related family member and a free-ranging wild animal* was and is fundamental to the resilience of the spiritual and cultural multispecies link between bison and Indigenous communities. This link survived via language and ceremony, despite this bond being materially severed due to both Indigenous and bison communities being exterminated and/or uprooted from their lands. Such a nurturing paradigm severely glitches the obsessive Linnaean classification and commodification of biological life. Furthermore, Indigenous axiology places value upon the *wildness* of bison, deeply different yet profoundly related to the tribe, rather than treating them simply as a domesticated resource for commodification and consumption. As Jason Baldes says, "The notion of the Wild, the notion of wilderness, 'untrammeled by man,' 'leave no trace'—those are foreign concepts to Native people."[85] This insistence on an Indigenous form of rewilding rather than colonial forms of rewilding as land usurpation and ranching is key to *both* ecosystemic revitalization *and* economic/nutritional benefits, rather than simply short-term economic gain. Insistence on wild bison/human multispecies communities is also intended to provide actual space for multispecies relationality rather than only paying lip-service to giving other species a right to exist and thrive on their own terms. One truly begins to understand spaces in terms of *our* multispecies land and *our* multispecies futures, when humans are imagined to be part of an intricate ecocultural system. Understanding wilderness differently becomes a way to care in a different, more ethical and ecologically sound way.

These threads of relationality between the wild bison and the Indigenous communities are representative of other such multispecies relationships within Indigenous communities. These strands extend out to a complex web of other relationships with mortal and "inanimate" objects and landscapes. Bringing back the bison as *free-ranging* rather than ranched communities is crucial to the many cascading naturalcultural effects they instigate. Jason Baldes mentioned how bison grazing is different from cattle grazing because bison have the cultural knowledge to keep on moving, allowing the plants to regenerate rather than completely denuding the landscape, a situation that occurs often with captured-and-controlled ranched animals such as cattle: "In areas with heavy bison grazing [...] plants were denser, shorter, and more nitrogen-rich. Not only did these plants grow as much as they would have if they weren't grazed [they were] 150 percent more nutritious than plants growing in areas with no bison."[86]

There are also immense benefits to what happens at the other end of the bison from its mouth—its surprisingly important dung. Bison dung "patties" and urine are now being understood by scientists to be key in bringing essential nutrients to support plants and insects, which in turn support the entire ecosystem of "higher" animals. The recent study by Chris Geremia et al. also shows how "Bison stabilized net aboveground production while accelerating nitrogen turnover, increasing aboveground nitrogen pools while carbon pools remained stable, which improved landscape nutritional quality."[87] The sympoetic story of bison dung unleashes a tethered story full of innumerable multispecies lively agents: nutrient cycling begins at digestion, where the rumen is broken down by a complex microbial community. Many of these microbes and undigested plant matter, including seeds, make up bison dung. Bison dung patty is also home for up to 300 species of insects and "1,000 individual insects will occupy that same patty from the time it's deposited to its removal. A single dung pat can produce up to 3,000 flies over two weeks. Luckily for the box turtle, bats and many birds, more dung beetles and flies in the area means more meals for them as well. Bison dung and urine also provide a rich food supply essential to microorganisms found within soil. An entire, nearly invisible micro-ecosystem is fueled by the deposition of bison poop and pee."[88]

Specialized dung beetles such as the spectacular Rainbow Scarab (*Phaneus vindex*) have co-become specifically with bison dung, and will simply not touch dung from other animals (Fig. 6.13). The extermination of bison led to direct effects on these beetle species. And why should these tiny creatures matter to us? Because "dung beetles have unique ways to decompose a bison patty. Tunneler and roller dung beetle species are especially important because they bury fresh manure below the soil surface, which places the nutrient-rich dung directly in the soil and activates microbial activity that begins to break down the dung."[89] Essentially, the dung beetle makes nitrogen, carbon and other nutrients available to the soil, allowing microorganisms to decompose them out of dung. This middle-man role then has innumerable cascading effects, allowing increased growth of nutritious plant life near dung beetle populations and from there, increased healthy elk and deer communities. Multispecies thick-mapping such vast, viscous and vulnerable connections of multispecies care and kinship, as I have tried to do here, brings back value to the many intra-actions that sustain us all. Such an intimate form of non-hierarchical, non-resource-based relationality provides a vital and important glitch in the obsessive Cybercene systems of hierarchical classification and commodification of bodies, lives and ecosystems: the dung beetle, the flies, the urine all connect equally in a web rather than a vertical chain of co-becomings. Bringing back balance to naturalcultural ecocultural systems gives back to the entire multispecies community a thousandfold.

These spiritual, cultural and epistemological relationships are not simply abstract or symbolic, but sympoetic. They are relationships via which native communities make and create with other bodies, with landscapes and with other interconnected lives. When hunted, bison are first asked permission to be killed, and forgiveness is asked of the individual bison that is culled.[90]

Figure 6.13 The spectacular Rainbow Scarab beetle. (Sebastian Eder)

Psychic mechanisms of stereotyping bison lives as "all the same" and killing without emotional distress are abandoned for a hybrid relationality where "every bison life taken is a cause of shared grief."[91] These kinds of multispecies embodied affective practices, underpinned by Indigenous axiologies of interconnected life, also have deep ecosystemic implications. When killed for the community, each and every part of the bison is used, the meat as a vital source of protein, the hide and bones used in ceremony to re-establish the important epistemic, psychic and embodied connections to the bison.[92]

The cascading effects of free-ranging bison beautifully demonstrate how Indigenous threads of multispecies relationality are forms of scientific storytelling of spatial and ecosystemic realities. These relationalities often extend along large swathes of unhindered ecosystem boundaries, such as the routes of seasonal bison migrations. This ecosystemic and multispecies conceptualization of land is very different from settler-colonialist concepts of land ownership and management via spaces that are often divided by walls and fences, such as the borders between national parks and private lands, between public and private lands, and between the private properties of two different owners. These dominant epistemic and economic paradigms of land use are the basis of material ecosystem fragmentation and lack of freedom of multispecies movement, essential for ecosystemic health and sustainability.

Tribes, settler-colonial forces and contemporary colonial political powers have always known that there is a convergence of interest between tribal interests and bison/ecosystem interests, and that the fates of both are deeply related. Treaty rights (for example, water rights) are ceaselessly being attacked or threatened to undermine these ecosystemic links and need to be constantly protected via legal channels.[93] The survivance of Indigenous as well as bison/ecosystem identity and culture in these ecosytemically understood spaces is therefore deeply linked to concrete questions of land sovereignty, that is to say rights that connect to political and legal questions of land and treaty declarations and agreements that determine who controls these ecosystems that these multispecies communities are dependent upon. The Wind River Reservation "was established in the 1860s and overseen by the Bureau of Indian Affairs with the goal of assimilating the Eastern Shoshone and Northern Arapaho into farmers and ranchers."[94] Enclaving these unwanted bodies spatially was therefore part of the settler-colonial masterplan to control-and-capture *multispecies* lives and ecosystems (both human and non) in order to exploit, extract and exterminate.

The Indigenous sovereignty paradigm provides a glitch within the dominant colonial notions of forced and controlled sovereignty via their ecosystemic understanding of their land. Indigenous sovereignty is intimately connected to the cultural and spiritual paradigms mentioned above, as well as with the uninterrupted movement of life within the ecosystem at large. Indigenous spirituality and sovereignty are also linked precisely because both spiritual and legal rights are sustained by a healthy and thriving multispecies ecosystem that is being actively restored and sustained on tribal lands. When these links between land sovereignty, spirituality and ecological resource commodification are made evident, it becomes very clear why both Native Americans and bison were caged-and-staged in shows and zoos. Such forced representations were essential precursors to actual capture-and-control of Indigenous and bison bodies, lands and ecosystems.

Despite these vast differences between colonial versus multispecies "borders," the very presence of State and National Parks (land imperfectly set aside for a multispecies and public common ecocultural good) as well as sovereign tribal lands (which in theory gives Indigenous people the agency and power to make land use decisions) are a glitch within the general ideological apparatus of Cybercene state capture-and-control of bodies, lives and ecosystems.

As I have noted, historical ecological paradigms played a role in removing Indigenous presence from public conservation lands. However, some of these historical injustices are slowly being remedied, with coalition-style presence of different multispecies links being nurtured to protect ecosystems. Discursive, legal and political pressures, coming directly out of settler-colonial paradigms, are still being applied, and are often couched within the performative political cybermediatic discourse of State-versus-Federal rights to try and diminish the power of such lands, in order to manage and control the glitch. For example, the State of Montana has pushed back against revitalization of wild bison populations beyond the artificial boundary of Yellowstone National Park, thus choosing to force park officials to "ship-to-slaughter" any bison that migrate outside park boundaries. Such capture-and-control perspectives pit two paradigms against each other: one that categorizes all life as domesticated and hence ultimately defined by as a resource for human communities via capture-and-control, and the other, Indigenous paradigm of life as a wild relation, and therefore worthy of an ethical relationship that should be carefully nurtured.

Which of these two paradigms will win out? The examination of this case within the multispecies studies theoretical framework of this book suggests that the most promising way for sustainable ecocultural models to be successful is to focus on transitioning the epistemologies and axiologies of more enrolled members of the tribes *and* the broader voting communities of the states of Wyoming and Montana from Cybercene-fuelled resource-based approaches to ethical embodied and multispecies relationship-based approaches. The restoration of a situated assemblage of knowledges, based on both Indigenous multispecies axiologies and cultural practices as well as the latest scientific know-how is a promising pathway. The story of the revitalization of Indigenous/bison lives in the American West provides a potent antidotal discourse to the de-situated follies of the Cybercene, and a reason to hope that more such successful efforts may be possible in other places and spaces. This is a story of deep multispecies collaboration between the bison and the GYES' multispecies communities, folks from the Wind River Tribal Buffalo Initiative, the Inter-Tribal Bison Council, Yellowstone National Park's scientists and organizations such as the Greater Yellowstone Coalition. Their passionate sympoetic efforts of biodiverse caring and kinship movingly demonstrate the power of "unexpected collaborations and combinations" fostered by a multispecies paradigm.

Conclusion

A Humanimal Guide to Staying with the Trouble and the Joy

The chapters of this book have explored parrot social networks, tragic partitions of land, charismatic mini-foxes, diligent dung beetles, ferocious biocultural violence, mosquito-hungry fish, campy drag queens, fragrant spice mixtures, cyberscreen-fed polarization and paranoia, curious dolphins, mysterious fungal networks, exclusive gated communities, trickster orchids and much more. All these examples of naturalcultural co-becomings worked sympoetically to shatter the anthropocentric perspectives of knowledge and culture that profoundly dominate human attentiveness. Such prevailing human-exclusive attentiveness in turn shapes our relentless strategic meaning-making and mapping via our mediatic, educational and other cultural systems, severely delimiting our multispecies futures. In *Kafka's Zoopoetics*, author Naama Harel cites Chabris and Simons to describe a famous psychological experiment—"The Invisible Gorilla"—to show how human meaning-making sustains the "systematic exclusion of the nonhuman":[1]

> In this experiment, subjects were shown a short video of two groups passing basketballs around and asked to count how many times the players in one of the groups pass the ball. Halfway through the video, someone in a gorilla suit strolls into the middle of the screen, pounds his chest, and then walks out of the frame. Focusing entirely on the given assignment, about half of the subjects missed the gorilla. When asked if they had noticed anything unusual, whether they had spotted anything walk through the scene, even more specifically—if they had seen a gorilla, the answer was no. This experiment demonstrates selective attention, also known as inattentional blindness, such that we fail to see things we are not prepared to see.[2]

This book calls for a multispecies paradigm shift to mitigate our chronic inattentional blindness towards our critter-kin. It asks us to change our viewing assignment drastically. Since we often fail to see relations we are not prepared to see or have consistently learned to ignore, it follows that we must transform our cultural seeing paradigms towards attentiveness to our multispecies past, present and futures. Such a paradigm shift also unsettles the biased ways in which we do create meaning and value when we pay attention to nonhuman lives. Given the importance of attentiveness, viewing and cultural interpretation, this book placed our highly curated media practices (both past and present) as central to meaning-making, identity creation, and therefore to our relations with those within our own species as well as our nonhuman critter-kin in the Anthropocene.

"Hello World!" is an iconic first phrase used by computer scientists as a traditional rite of passage to verify that new coding languages and systems are set up correctly.[3] This book also says "Hello World"—but to deploy an old/new code and system of exploring human cultures via a multispecies paradigm shift. By doing so, this research promotes a new humanities or posthumanities, where the "post-" does not vaguely signify "after" without proposing concrete answers to what that could mean. It is an invitation to an exciting and ethically glitching *multispecies* posthumanities. It is a rigorously collaborative and sympoetic endeavor that engages in transdisciplinary multispecies studies exploration in an age of accelerated digital transformation. It is an enhanced and biodiverse humanities in which *naturalculural studies* greatly expands (but also includes) the current scope and remit of cultural studies. This book makes the argument that, in the Greatest Acceleration, it would be myopic and detrimental to explore human–technology intra-action without also focusing on *Homo sapiens*' deep yet fraying ties to the natural world. Re-storying our ecocultural systems through such a paradigm allows us to understand circles of kinship in the Anthropocene and the Cybercene as complex matrices of intra-active material, ontological, ecological, affective and cultural attachments, all formed and sustained via direct and indirect cascading trophic connections. The environmental humanities have already engaged in explorations of ontological, material and affective turns in order to decenter the human and culture. However, a multispecies paradigm shift allows EH to interpret our places and spaces as rich and complex ecocultural systems in an even more deliberately multispecies way; to understand them to be replete with innumerable naturalcultural relationships of agency, co-becoming and intra-action. The case studies explored in this book demonstrate that relationships (family, kinship, community), deeply personal affects and psychic processes (loneliness, fear, hate, trust, paranoia) and social practices (meaning-making, storytelling, media habits, political strategy), normally relegated exclusively to the realm of human culture, actually have deep cause-and-effect connections with biodiverse systems. These are all ecocultural processes and phenomena. Conversely, natural systems have deeply constitutive intra-active relationships with human ontologies, affects and epistemologies of all kinds. Understanding these complex direct, indirect and cascading connections in what Donna Haraway calls *natureculture*, as consumed and interpreted via various media practices, has become even more urgent in the

Cybercene, because this is an ecocultural era that has unleashed even more powerful ways and means of inattentiveness towards our embodied naturalcultural realities. Relearning the arts of attentiveness via a multispecies paradigm shift can begin to heal the relentless violence caused by all our negligence.

Via multiscalar, transcultural and thick-time analyses, this book has sought to accomplish three main goals: firstly, to enhance the environmental humanities by decolonizing the field and decentering, yet not completely erasing, anthropocentric perspectives via a multispecies studies approach. Secondly, to propose new naturalcultural studies research methods. Thirdly, to gauge the health of ecocultural systems in the context of our digitally connected world. Situated knowledges, "standing with" marginalized knowledges (human and non) and complicating the researcher/researched duality allowed me to fix my positionality and intentionality in this exploration of the scope of multispecies studies as a young field of interdisciplinary research as well as a general cultural paradigm within the public sphere. The first part of this book explored how passionate immersion, cultivating the arts of attentiveness, radical transdisciplinarity and epistemic humility are other key presuppositions when embarking on multispecies studies projects. Via the two "island" case studies, Wi'ma and Sicily, I outlined the basic method of naturalcultural discourse analysis that is utilized throughout the book. This lab-incubated method ethically glitches the concept of *discourse*, which typically is understood to be exclusive to the realm of the cultural, and hence the human. By redefining discourse in naturalcultural contexts as "biocultural and biodiverse systems of shared material, ontological, ecocultural and cultural processes and relationships within any given multispecies ecocultural community," I enacted a multispecies paradigm shift within my research methodologies and suggested that the same could be attempted by others in the future.

Chapter 2 redefined *colonialisms* to include not only their human victims but equally how they implicate natural and multispecies lives, bodies and ecocultural systems. Such a new ecocultural framework opens up the study of colonial histories to far greater horizons than have been hitherto explored. Important work is being done in this direction in the field of environmental history, but these approaches still typically harbor anthropocentric perspectives and biases. Specifically, my second chapter examined the bio- and eco-political othering mechanisms of colonial spectacles such as those of the multispecies zoo. I suggested that this unique genre of colonial mediatic meaning-making provides us with a good case study to illustrate and analyze lingering historical legacies that haunt today's digital forms of multispecies epistemic capture-and-control. Further place- and space-specific research can help illuminate how speciesist epistemologies intra-act with other kinds of hegemonic identity constructions (gender, sexuality, race, nationality, class status, citizenship, etc.) to gain further power to capture-and-control lives, bodies and ecocultural systems. Other future studies can examine the intimate connections between our selective inattention towards nonhumans and crucial activities such as economic modeling, engineering, medicine, urban planning, architecture, law, tourism, social development programs and much more. Doing so could bring more integrated approaches to both human and nonhuman multispecies individuals and communities.

A detailed psychoanalytic and postcolonial analysis of the zoo spectacle in Chapter 2 also demonstrated how comparisons to the beast, the untamed and the wild were and still are some of the most fundamental ways in which the colonial double logic of desire and menace, lack and surplus is "made true." These mechanisms have only accelerated within de-situated Cybercene discourses through cybercitation and re-citation. Besides illustrating the foundational human/nonhuman duality in colonial discourses and demonstrating the revitalization of othering mechanisms within new "ecologies of self" in the Cybercene, another important conclusion that can be gleaned from these chapters is that the colonial moment is far from over. If colonialisms are understood as practices of distancing, domination and control upon Othered *multispecies* bodies, lives and ecosystems for profit, then the Cybercene's tech-mediatic-cultural practices have served to accelerate rather than diminish these kinds of othering and wasting relationships. Investigating and revealing the four *cultural-discursive-affective-psychic* mechanisms of colonial-cultural "capture-and-control" of multispecies lives, bodies and ecocultural systems described in Chapter 2 (multispecies othering, epistemic and affective distancing, colonial kinship and imagined surplus) allows us to qualitatively understand how colonial-cultural storymaking and storytelling need to constantly de-situate knowledge in order to make violent extractive actions palatable.

Colonialisms are therefore far from being done with our planet's multispecies lives. Any waning epistemologies and economies of coloniality since the major political decolonizations in the Global South in the twentieth century have now been reconstituted by Cybercene mediatic economies of desire where bizarre new versions of capture-and-control zoos circulate freely. They result in increased othering, epistemic polarization and embodied enclaving via reciprocal capture. The power centers of the colonial command economies that undertook multispecies othering and commodification have somewhat shifted away from the primary historical instigators (Western European colonial economies) to a multipolar range of cybermediatic-corporate-spectator-consumer hybrid command economies, directed by both former colonial powers and former colonies. Internal and internalized colonialisms of all kinds now abound within nation state frameworks. This book argues that the grand resurgence of colonialisms of many kinds has unsurprisingly coincided with the spread of widespread digital habits. The profound connection between ubiquitous and now-banal everyday digital media practices and the interconnected global ecocultural polycrises that we are facing has not been properly recognized, understood or analyzed. Unearthing this important link has been one of the primary findings of this book.

Given these transformations, the terms "postcolonial" and "postmodern," or even political categories such as "Left" and "Right" are no longer useful to map global cultural patterns and trends. Such diffusion of cultural power centers, but also resurgence of the illogical logic of nineteenth-century-style "Great Powers" resource-grabbing, can be witnessed at all scales of discourse—from global to national to regional to local down to individual practices of identity creation. If the colonial moment has been revitalized in the Cybercene, it is also important to understand how *everyone* is included in its gaping maw and drowned in its tsunami of digital slop—creating new and old forms of hegemonic and oppressed individuals and ecocultural communities. In the Cybercene, sentience-swapping, screen-life hyperconvergence and the quadruple role of the cyberscreen as map, mirror, mass communicator and procurer has brought about deep changes to the discursive-technological apparatuses. These gadgets form and reinforce our personal identities as well as who we exclude as others. Conversely, the cultural negotiations of our social selves deeply affect the material destinies of our bodies and lives. Since personal identities form and inform the collective, it logically follows that our personal negotiations via the Cybercene psychic-mediatic mechanisms of desire also constitute our collective bio- and ecopolitical attitudes and practices in this era. In short, the Cybercene has dramatically changed the stories by which we understand the world around us, via the lure of the four Cs—convenience, control, connection and consumption—and the quadruple role played by cyberscreens. My analysis of the digital capture-and-control of othered bodies illustrates the ways in which the Cybercene's epistemic, discursive and affective systems have brought about a rapid corrosion of the trust that underpins a sense of the commons. Captured bodies, such as those of Sarah Baartman, Cécile Kyenge, Satnam Singh and Emmanuel Nnamdi, and the millions of other named and unnamed human and nonhuman lives memorialized in this project, are representative of the vast number of multispecies bodies that have been othered by those cojoined twins of colonial othering—speciesism and biopolitical stereotyping—fueled by spectator intra-action and algorithmic co-becoming.

Chapters 2, 3 and 4 analyzed the complicated and often damaging play spectator/consumers engage in to control-and-capture life. These archives of mediatic ecocultural damage served as the foundation to then, in Part Three of the book, locate and analyze examples where communities and ecosystems have successfully glitched such corrosive othering mechanisms. I do not want to give a false impression of easy liberation from colonial forms of capture-and-control by illustrating examples of ethical glitching. On the contrary, Chapters 5 and 6 show the deeply complex and challenging nature of such paradigm shifts and the harm it can cause to those standing up and standing with communities of multispecies care and kinship. That said, it is also important to reject omnipresent epistemologies of crisis and instead embrace possible epistemologies of coordination as a response to crises.[4] Chapter 5 described the ethical glitch as a spontaneous or strategic error that, when harnessed within epistemic, discursive and affective cultural systems, can cause captured-and-controlled lives to run away in a more coordinated fashion. Some of the success stories of effective glitches examined Part Three also utilize the discursive-technological apparatuses of the Cybercene and turn them to restorative ends, thus proving my hypothesis that the problems we face are not technological per se, but epistemic, psychic, cultural and humanistic in nature. The cases of multispecies healing and habitability in the final chapter provide glimpses of possible solutions that are based on a proper understanding of cyber-discursive and cyber-affective mechanisms to move the planetary dial towards epistemologies and affects of multispecies attentiveness, care and kinship. The case of the new Somali diasporic relationality with food demonstrates adaptable and collaborative strategies of multispecies belonging via cyberscenes in the face of acute community spatial dispersal due to conflict. Ultimately, living and caring diasporic bodies and their digital and embodied intra-actions with other plant, animal and human communities via food practices make the glitch both strategic and material, causing colonial othering stereotypes within cybercontent to fall apart and become more complex, nuanced, hybrid and therefore not easily controllable. This example of multispecies diasporic nation-building through food demonstrates how mediatic national or racial stereotypes also contain the seeds of their own breakdown, which can be used strategically by subaltern content creators and cooks to glitch the system and provide transcorporeal pathways to belong and relate to the world. Instead of a spectacle optimized for narrow meaning and a frictionless interaction, we have an *intra-face* that exemplifies viscosity: the unavoidable stickiness, nonlinear and fluid inter-dependent nature of different naturalcultural agents that intra-act and co-become in any ecocultural system. The viscosity of things, and the trouble and joy this implies, allows for an understanding that glitches the colonial into a liberatory and open form of possible care and kinship: "ways of knowing and being with others involve careful attention *to what matters to them*."[5]

The glitching performances of Vaginal Davis and Fertile La Toyah Jackson as re-cited on YouTube show how disidentificatory citations, when combined with a reflective stance, can successfully complicate corrosive and reductive colonial definitions and discourses of family, femininity, masculinity, reproduction and love. Fertile's miraculous birthings and the caring friendship between Vaginal and Fertile open up the spectator/consumer to expansive and inclusive kinds of ethical relationality, care and kinship. By normalizing the liberatory sexual and reproductive body and its bio- and soulpower, their film accomplishes oddkin-making of queer sexuality, so often actively disowned and prosecuted by comparisons to the beast or the bestial. The final case study of the revitalization of the human-bison community in the Wind River Tribal Reservation celebrates an initiative that acknowledges a simple yet constantly repressed fact: we are all connected. The painstaking recovery of a free-ranging conservation herd rather than captured-and-controlled "ranched" bison is bringing a cascading series of naturalcultural effects to Wind River. Ethical glitches such as these constantly remind us that nothing exists in seclusion. Powerful colonial fantasies of unmitigated freedom to capture, commodify and consume continue to insist otherwise. The "ethical glitch" that all these case studies represent is so called because it still works within the system of colonial relations, but introduces agents who act with a thick-perspective rather than an obsessively "caged-and-staged" one. This is a sympoetic form of intra-action, rather than the colonial autopoetic mechanism of obsessively controlling the meaning attached to bodies, lives and entire ecocultural systems in order to other them for profit.

The overall question that all these cases explored was: how do we concretely move towards a thriving planet that is more capable of sustaining multispecies communities in a historical moment dominated by powerful Colonial-Cybercene "othering" mechanisms? The answer emerges from the generative situated perspectives of multispecies naturalcultural experts, such as those I have sought to celebrate in this book. Kyle Whyte argues:

> If the focus of knowing the world is on kinship relationships, then it is quite possible to wonder whether in today's time that it is possible to restore kin. For kinship relationships cannot be restored overnight. Many of the most important kinship relationships take time to develop. There is no guarantee that during the time it takes to develop those relationships, that certain issues within ecological systems will be simply repaired. The weight of kinship is very different from the sense of imminence in crisis epistemologies.[6]

Passionately revaluing and nurturing multispecies lives, bodies and ecocultural systems within sympoetic communities is the first step to releasing multi-being and multi-kind lives from the wasteful constraints of normative speciesist colonial forms of control and connectivity. Epistemic humility and adaptability could be the foundation for restorative ecocultural systems of belonging, care and ethical kinship for all multispecies bodies, lives and ecosystems. This book enacts such epistemic humility by ascribing nonhuman lives with the agency, power and role of co-equal producers of culture and theory.

Towards these restorative ends, the *Multispecies Futures Lab* continues to sustain ties to its older projects and pursue new ones. The study of food as a multispecies relationship has already become one thematic focus: the lab has incubated a *Global Food Studies: Ecocultural Diversity and Sustainability* course. At the time of writing, a new collaborative and long-term project that engages both undergraduate and graduate students is the "Multispecies Bruins Thick-Map." This initiative seeks to narrate the UCLA campus to its humanimal community via a multispecies paradigm, connecting quantitative and qualitative data via ever-expanding "layers" of plant, animal, fungi and micro-organism presence and intra-action. Creative writing, arts and performance studies students, as well as aspiring phycologists, microbiologists, zoologists, and mycologists, etc. will hopefully add future layers of multispecies meaning to this naturalcultural map. Other lab-based plans include building on the 2025 Channel Islands field trip to create an annual nomadic "Multispecies Futures Global Institute" that is open to graduate students and other scholars from around the world. Such a traveling compact-yet-generative experience would allow this book and its cojoined lab to become catalysts in fostering a sustainable and caring research global community around "all things multispecies."

A key conclusion in this project is that in the Colonial-Cybercene, *speciesism* continues to be foundational to dominant ways of thinking, feeling, acting, relating and belonging. It is the bedrock of the many kinds of otherings that have gained rather than lost momentum in the Cybercene's discursive-technological desire machine. The solution to steering away from wasted lives and towards equitable empowerment lies in providing Cybercene societies with a clear vision of how much all lives and ecosystems have in common, and the creation of such a common sense of belonging, rather than policies that isolate, apportion, divide and demarcate, making everything a zero-sum game between competing interests. Gandhi's nonviolent *satyagraha* was based on such an idea of a commons, even though it may not have been as explicitly multispecies as I am reformatting it to be: a community based upon "the selfless devotion (*bhakti*) to communicate with the 'Other' (ontologically conceived as the interdependence of all things), and the corresponding moral and political duty" to persistently engage in the search for the truth (*satyagraha*).[7] As several examples in this book have illustrated, speciesism and the alterity of animality are foundational to the othering of human lives and communities. Christopher Peterson argues that this foundational disavowal of animality must be addressed for meaningful change to occur:

> If negative stereotypes about nonhuman animals are the condition of possibility for negative stereotypes about social and political minorities, then avowing human animality would seem to hold much political promise.[8]

The avowal of human animality is key to arriving at a collective sense of the *multispecies commons*. To do this is to move the needle of discourse away from damaging multispecies identity politics and cultural wars towards the ecosystemic common good, without ignoring the realities of the different marginalized groups, human and non, that are being exploited. This broad public awareness of multispecies belonging can be most effectively accomplished by harnessing the powerful potential of the internet, creating a glitch within the othering system by using its own mechanisms of hyperconvergence, virality and algorithmic might. However, as my analysis of the Somali diaspora's reformatting of belonging via multispecies food connections demonstrates, the success of these discursive glitches can only be accomplished by also providing spectator/consumers with *trustworthy embodied tethers*, which can only come from sustained intra-active experiences of the multispecies commons. Therefore, while these solutions would be multifaceted as well as digital, it is important to recognize the fundamental need for embodied experiences of ecosystemic connection to truly empower persons and communities.

Both the proposed academic and community-based pathways that follow fundamentally emphasize the need for a multispecies commons where cooperation (caring and laboring with), sympoesis (making-with), and more generally being and becoming with selves, others, humans and nonhumans is a corrective to the arrogance, isolation and paranoia of the Cybercene. Although these conclusions have been discussed recently by psychologists, educators and many other researchers, the originality of what I propose in this study is its unwavering emphasis on multispeciesism. In a moment of rapid devaluation of human agency via new forms of automation, relearning and revaluing our deep humanimal connections to our vast multispecies family may just be that crucial strategic ethical glitch that throws open the doors to a more just, sustainable and joyful future.

Multispecies Studies: Further Avenues in Research and Teaching

Given the many directions our multispecies journey together has taken us, I'd like to summarize some thoughts on future directions that academic researchers and students could undertake within the young field of multispecies studies. Several scholars and institutions are already beginning to work in these and other directions, but far more needs to be done urgently to build and strengthen multispecies studies globally.

A meaningful new or post-humanities must be a multispecies one. The humanities must adapt and change in our transformed world to newer conceptualizations of culture as natureculture, thus glitching the inherent anthropocentrism that still thrives within its core tenets and approaches. Such a strategic renewal would lead to an ethical and methodological investment in our multispecies world, rather than abandoning it to the nihilistic propensities of the prefix "post-" (a post-apocalyptic post-human future) or to anthropocentric fantasies that do not corroborate with our deeply multispecies reality. Engaging in observation and analysis of natureculture via a multispecies paradigm would in turn call for a foundational *epistemic humility*. This attitude can help foster epistemic interdisciplinarity and diversity, thus building thicker accounts of any given research question in place- and space-specific studies. It moves beyond anthropocentric definitions of language, culture, communication, sentience and intelligence to allow nonhuman cultures to be possibly vastly and incomprehensibly *different* from ours, and yet (or perhaps, therefore) worthy of great respect. Building such strategic humility into naturalcultural research methodologies can lead to an openness that places new value on place- and space-specific embodied experiences of multispecies adaptation and living. Haraway's concept of situated knowledges allows the researcher to then form an assemblage of such expertise, where different perspectives form "unexpected collaborations and combinations." These methods allow for thicker decolonial rational knowledge claims that do not have to make the false choice between rigid autopoetic empiricism and easy relativism. For the creation of an ever-better scientific method, many cultural-colonial dichotomies need to be revisited via strategic epistemic humility such as, but not limited to: **human**/animal, **culture**/nature, **man**/woman, **heterosexual**/queer, **meaning**/affect, **epistemology**/ontology, **mind**/body, **meaning**/matter, **animate**/inanimate, **disciplinary expertise**/generalist knowledge and **researcher**/researched.

Transdisciplinarity is key to a robust multispecies posthumanities. This book has investigated how and why the creation of a rigorously multispecies posthumanities should also be a vital opportunity to urgently rethink siloed identity-based knowledge fields, units and departments separated into humanities, social science and science: these include "area" studies; national culture-based programs; gender, race, sexuality and other identity group studies, as well as detached units in biological, ecological and conservation studies that do not yet engage meaningfully with humanities units. By creating stronger links and bridges between these various important partial knowledge claims, all are able listen, learn and coordinate solutions for ecocultural sustainability in place- and space-specific contexts. Researchers in the biological and ecological sciences and beyond would benefit by opening up to more collaborative engagement with such transdisciplinary humanistic, philosophical and ethical pathways while humanities researchers could open up rich possibilities by engaging in the material and ontological realities of ecocultural systems.

Humanists would equally profit by engaging with a new multispecies posthumanities that connects the different experimental approaches within transdisciplinary configurations such as the environmental, digital, medical and urban humanities. Only by accomplishing such meaningful integration can the resulting transcultural and transdisciplinary assemblage of knowledge as the posthumanities more honestly propose itself to the public as central rather than peripheral to the mission of educational institutions. In this book, I suggested and then utilized a *trans-subaltern* naturalcultural approach that converges the attentiveness and interests of economically and naturalculturally marginalized sections of ecocultural systems. Akin to Vicente Diaz's transindigenous approach described in Chapter 6, trans-subaltern knowledges would continue to recognize and value the specific histories, identities, challenges and opportunities faced by unique place-specific groups. However, trans-subaltern conversations would then also liberate these separate studies and allow them be relevant to much broader ecocultural contexts and communities by fostering "lateral reach across space and time."[9] Via newer trans-subaltern approaches, the presence and knowledge of such communities and thinkers would become valued as lynchpins within a broad coalition of posthumanities multispecies knowledges, rather than being treated as marginalized voices, as the "stone guests" we encountered in Chapter 1,[10] or completely disregarded as inferior lives to be captured-and-controlled within dominant knowledge systems.

What kinds of *methods* would such a global multispecies, transcultural, transdisciplinary and trans-subaltern posthumanities use to better understand and revalue ecocultural systems? EH and ecocritical approaches have already provided many options for understanding the interconnections between the environment and human cultural texts. With full faith that scholars will continue to put forward many other techniques in the future, this book proposes *naturalcultural discourse analysis* as one basic valuable methodology to symptomatically read and analyze the various entanglements of culture and ontology in any given ecocultural system: the co-becoming of ideas, images, texts, values, but also of and with material, embodied, trophic relations. This kind of analysis enables the study of the effects and consequences of human cultural "meaning-making" on the natural world—meanings that consciously or unconsciously place relative values upon lives, bodies and ecosystems in those living natural systems, and vice versa. It also allows a further study of how epistemic, affective, psychic and cultural systems in place- and space-specific settings are constituted over time by the natural world, including by humanimal bodies. Finally, by glitching the definition of "discourse" to also include systems of material biological and ecological relations in ecocultural systems, naturalcultural discourse analysis allows multispecies studies to freely and humbly engage in and with biodiverse communities without ignoring or devaluing the importance of nonhuman epistemologies and multispecies ontological and affective agency.

In addition to deploying naturalcultural discourse analysis in a myriad of case studies, this book has also demonstrated the utility of an innovative visualization method for these analyses: *multispecies thick-mapping*. At its heart, this *Multispecies Futures Lab*-developed version of thick-mapping is a way to make visible the invisible or indirect connections between and within multispecies lives, bodies and ecocultural systems. Both prototypes of such maps in this book—the Agro Pontino and the *Scala dei Turchi*—focus on storytelling and meaning-making to recenter the researcher's and the spectator's attentiveness to direct, indirect and cascading ecocultural connections. Rather than obsessively capturing-and-controlling bodies, lives, places and spaces to create a highly narrowed-down focus on lives as resources, these new maps cultivate intra-species complexity and agency via layered investigations and storytelling. In this way, multispecies thick-mapping shows itself to be an important qualitative complement to more data-driven analyses of our digitally connected societies, places and spaces. Rather than frictionless lives, this visualization method honors viscosity—that inevitable stickiness of being—between the many attachment sites that characterize multispecies oddkin companionships in all ecocultural realities.

Both naturalcultural discourse analysis and multispecies thick-mapping bring attention back to the greatly ignored but fundamental importance of what I'd like to call *multispecies ecocultural knowledge* (MEK). Many examples throughout previous chapters have illustrated how plants, animals, fungi and microbes are experts of their own ecological and ecocultural niches.

Naturalcultural discourse analysis revalues these sympoetic situated knowledges, in opposition to colonial "thin" knowledges that are often presented with great arrogance and authority. These humbler yet powerful knowledges include the already well-known traditional ecological knowledge (TEK). This project makes the case for the inclusion of nonhuman ecological knowledge rather than limiting expertise to the realm of *Homo sapiens*. Adam J. Fix, Hugh Burnam and Ray Gutteriez propose that TEK itself should be understood as a portal towards valuable interspecies collaboration: "We propose that interspecies thinking as a collaborative concept should begin with the Indigenous cosmological foundations of TEK, which provide the basis for intercultural collaboration, and from this position choose to focus most intensively on the narratives and practices that comprise the human/nonhuman animal divide."[11] Valuing both MEK and TEK together is to recognize pathways of mutualistic intra-actions between the two. This approach can allow research to bring back value to what my colleague Professor Pasquale Verdicchio, called "artisanal intelligence," a kind of place- and space-specific and embodied intelligence that complements but also supersedes de-situated artificial intelligence. Is not earthworm engineering (class Oligochaeta) of our agricultural land and the beaver's (genus *Castor*) ecologically beneficial intra-action with water bodies as worthy of our attentiveness and respect as human works of ingenuity? Ethically integrating both MEK and TEK within command economies would mean decentralizing them and valuing multispecies innovation and ecocultural entrepreneurship—just as we accomplished in our naturalcultural discourse analysis of Chumash *tomol* in the Wi'ma Island case study in the Introduction. Nonhuman labor and innovation within their own ecocultural niches (for example, the bison community's crucial contributions to co-making a vibrant Greater Yellowstone ecocultural system) should be respected and recognized as important cultural knowledge. Valuing these kinds of situated MEK and TEK innovations require deep love of and curiosity towards one's naturalcultural and multispecies kin as well as a propensity to value both general as well as specialized knowledge. The MEK-TEK combination holds up far better in resolving pragmatic place-based problems than de-situated colonial knowledges flown in from distant and distancing command centers. Both MEK and TEK can also be verified by assessing the long-term equilibria of multispecies bodies, lives and ecosystems. Researchers can read and confirm what is being produced as "knowledge" within the assembled "accounts of the distinctive experiential worlds, modes of being, and biocultural attachments."[12] Finally, it is important that none of the newfound appreciation of MEK and TEK should ignore considerations of how to use this knowledge ethically, including providing fair compensation for authorship and labor.

The newfound attentiveness that naturalcultural discourse analysis and multispecies thick-mapping bring to these bountiful and deeply passionate naturalcultural co-becomings should be harnessed within the new posthumanities via *biodiverse teaching*. How can multispecies perspectives be creatively woven into teaching both within the classroom and beyond, and not only in subjects directly related to the environment, but within many other subjects, from mathematics to economics to business management to law? Students can develop their own research projects within courses that allow them to utilize the lens of their own specializations and majors, but use common posthumanistic tools such as thick-mapping, ecocritical analysis and naturalcultural discourse analysis. They may also begin to integrate their academic work with their affective and relational experiences, allowing for passionate immersion within their initiatives. These kinds of deeply situated and tethered courses can hopefully bring back value and resources to the multispecies posthumanities.

As Chapters 3 and 4 have demonstrated, a rigorous posthumanities multispecies studies curriculum cannot exclude an exploration of Cybercene media studies. Damage to or care for our natural and our cultural ecosystems are connected to the dramatically different and narrowed-down attentiveness that our cyberscreen-based lives provide, leading to a strong tendency of capture-and-control of attentions, affects and meanings. Future Cybercene media studies research and teaching within new units such as the digital humanities could analyze the many intra-actions between online digital epistemologies, affects, discourses and practices and naturalcultural contexts. Empowering spectator/consumers to recognize basic personal and collective Colonial-Cybercene strategies such as sentience-swapping would entail opening up access to Cybercene-relevant literacy, and from there, to more reflective media habits. It is essential that a sympoetic, decolonial and transdisciplinary posthumanities be fostered within academic communities globally and that these explorations be made more accessible to wider audiences. Critical digital humanities can further explore the lure of cyberscreen otherings and its psychic mechanisms of inattentional blindness and habitual capture-and-control. Students/spectators/consumers can be educated about the violent consequences of epistemic polarization and embodied enclaving.

Other theoretical frameworks can add much to the approach used in this book to exploring new media reception in the Cybercene. The role of the active spectator/consumer, made evident in Chapters 3 and 4, can be turned into useful knowledge to empower students via curricula that teach Cybercene media literacy to all generations. Such initiatives should begin to influence implementation of proper regulation measures and awareness of proper media usage in the Cybercene; for example a state bill implementing literacy curricula on media disinformation was signed into law in California in November 2023.[13] Australia's social media ban for under-16 year olds went into effect in December 2025.[14] Besides work on new media literacy, the digital humanities should honestly explore the many pros and cons of "A.I."-driven learning and "A.I."-driven command economies in addition to training students in A.I.-building and management. Such thoughtful and integrated approaches to innovation could possibly lead to ethical and pragmatic deployment of new tech systems that foster rather than degrade our multispecies commons.

Community Pathways Towards a Multispecies Paradigm

The previous section proposed some approaches by which a new multispecies posthumanities can revitalize the academic study of ecocultural systems on our battered and transformed, but still bountiful planet. This next and final part of the conclusion suggests some community-based pathways of fostering naturalcultural attentiveness based on the findings of previous chapters. This entire study demonstrates how none of the specific ecological mitigation measures proposed by the scientific community—for example, rewilding, conservation, re-engineering of urban spaces and industrial supply chains to consider multispecies lives and symbiosis—can be ultimately upscaled widely, quickly and efficiently enough unless the general public move away from their deeply embedded Anthropocene delusion—the master-narrative that humans are exceptional relative to other beings on the planet. For example, a study of Austin, Texas' incorporation of the agency of bat colonies in urban planning initiatives illustrate how concrete measures remain incomplete without a cultural multispecies paradigm shift. The research concluded that "while interspecies interactions shape urban environments, societal perceptions remain largely anthropocentric, underscoring the challenges in integrating non-human perspectives into urban planning and governance."[15]

The surest pathway towards giving the public a feeling a sense of care and kinship beyond species boundaries and zero-sum games begins with fostering nature-attentiveness and -connectedness—incrementally forging a differently tethered, deeply passionate and embodied kinship that unsettles unrelenting Cybercene mediatic reciprocal capture. The approaches discussed here should be regarded as gradual practices of mitigation and restoration rather than magical quick-fixes after centuries of unrelenting abuse and damage. Just as the deterioration of an ecocultural system can be rapid but its restoration a complex and slow process, so will multispecies ecocultural caring and restoration in the Cybercene take time and consistent effort from a broad coalition of multispecies actors. Despite this caveat, it is also true that paradigm shifts can occur surprisingly quickly, if they are done effectively and within a cultural context that is already ready to embrace such change. Readers will notice the interdependence and connections between the different pathways I discuss below. Indeed, the approaches explored here intersect and aid each other in an intra-active web-like fashion, rather than a linear structure, none being able to sustain the push towards care and kinship without the existence of the other pathways.

In a digitally connected world where de-situated, decontextualized citations are given new value, a key ethical glitch to promote rational knowledge claims would be to *foster assemblages of situated knowledges in communities.* As the world creeps towards ever more de-situated and de-situating "post truth" cyber-regimes that engage in othering to promote exploitation, extraction and extermination of multispecies lives, situating knowledge once again *within the contexts of specific bodies, communities and ecocultural systems* is the bedrock of any renewal of trust and value. Donna Haraway's concept of situated knowledges moves us away from the false duality of objectivity vs relativity of facts and expertise. Situated knowledges allow for the role of experts to be rehabilitated and respected, while simultaneously insisting on strong diversity within the ecosystem of knowledge creation to make sure expertise remains an inclusive, collaborative and collective effort, rather than a colonial or monocultural one. Perhaps as importantly, Kim TallBear's principle of "standing with" an assemblage of various situated knowledges finally allows us to glitch rigid categorizations of identities, thus going beyond the trap of identity politics and towards situated inclusive knowledge-producing groups.

No effective solutions can ignore the fact that the Cybercene is a product and the newest iteration of the power of colonial-minded command economies. In the Cybercene, tech-hype and sentience-swapping have greatly expanded profit-making via the lure of unlimited convenience, control, connectivity and consumption and to cyberscreen spectator/consumers around the world. Rather than throwing out the baby with the bathwater or reinventing the wheel, a meaningful curb on the excesses of unequal economic power via democratic checks can help create a more just and sustainable economic and cultural ecosystem while still encouraging ethical free enterprise—a fairer ecocultural system that mitigates and rebalances excessive accumulation of power and capital within a tiny minority of local, regional, national and international governmental institutions. These checks, in turn, will allow for a far more equitable, sustainable and thriving planetary economic system that reduces wasted lives and landscapes in the single-minded pursuit of short-term profit for shareholders. Corporatespeak and tech-washing—the Colonial-Cybercene's two major discursive strategies—need to be recognized and understood as cultural techniques that encourage as well as hide diverse forms of exploitation in plain sight while corroding any remaining democratic checks. Such de-situated knowledge could be identified and countered by an ecocultural system of multiple discourses all founded upon situated rational knowledge claims and multispecies interest-converging groups. Only thick-contexts generated within a polycentric democratic system of checks and balances can glitch such thin and violent accounts, while still supporting ethical innovation and transformation.

Chapter 3 discussed how sentience-swapping forms the fertile cultural terrain that allows for de-situated knowledges to thrive by building trust in automation. This foundational misrecognition of lively and ethical agency allows for a successful redistribution of the value placed on hierarchically classified lives, bodies and ecosystems, both human and non. Sentience-swapping attributes scarce and valuable biopower and soulpower to cyberscreen images and other tech simulacra while simultaneously devaluing multispecies and multi-kind ecocultural systems. This fundamental swap is a crucial ecocultural precondition to the exploitation and extraction of bio- and soulpower for profit within colonial systems of thought and practice. Some global movements have begun to ethically glitch these devaluations by going the other way: *legally redefining personhood to include nonhuman lives* as well as natural "more or less animate" systems such as mountains and rivers within these legal classifications.[16] For example, the Māori worked their cosmology into legislation in the 2017 *Te Awa Tupua* (Whanganui River Claims Settlement), claiming it as an "indivisible and living whole, comprising of the Whanganui River from the mountains to the sea, incorporating its tributaries and all its physical and metaphysical elements."[17] This incorporation of Indigenous cosmologies of care into contemporary legal systems in turn guarantees that these ecocultural values will be enshrined in law, which result in beneficial protections to specific situated communities as well as an accrual of planetary-wide health benefits. Such epistemic and legal ethical glitches are strategic anti-colonial moves that can receive much pushback precisely due to the way in which they counter historical uses of legalese to usurp bio- and soulpower from these living and nurturing entities. Providing multispecies ecocultural systems with legal and political protection is probably one of the most effective forms of demonstrating multispecies care and kinship. Such new kinds of naturalcultural multispecies citizenship can revitalize trust in battered democratic norms and institutions by conceiving the entire structure of our multispecies communities as an overarching system of *naturalcultural democracy*.

Interest convergence describes how the material or cultural interests of large and sometimes even mutually antagonistic cross-sections of society can be utilized to forge coalitions. This has often been done in the past to maintain the hegemonic status quo (for example, the interests shared by upper-class and working-class men in marginalizing women, or the interests shared by a broad cross-section of human society in sidelining nonhumans). These interests, in turn, have provided direct material and/or psychic benefits to bioculturally dominant bodies through political alliances and material privileges while depriving minority communities of the same. Interest convergence can however also be used productively to assist in "the difficult challenge of forging political solidarities across differences,"[18] and to tackle multispecies commodification, waste and damage, nurturing collaborative power via a strong coalition of situated bodies and knowledges. The Intertribal Buffalo Council, examined in the Indigenous/bison case, is one such example of successful interest convergence. Providing a voice to marginalized communities should be complemented by a pragmatic and strategic interest convergence between the widest possible coalition of communities rather than everyone acting separately. Doing this would involve fostering relationship-based rather than resource-based knowledges and connection rather than only connectivity. Such coalition building has to grow out of an attitude of ethical relationality rather than through what has been done until now: excessively focusing on narrow identity politics without completing the second half of the sentence: how does uplifting _____ 's lives (put any bioculturally marginalized identity group in the blank: dalit, women, animal, migrant, plant, working class) eventually help *everyone* in society, including those within dominant identity groups that form a majority? Interest convergence in human cultural contexts should also be connected to multispecies ecological mutualisms, which are really another form of naturalcultural symbiosis. A multispecies paradigm shift incorporating interest convergence can connect the material interests of both human and nonhuman communities to foster strategically effective care.

Besides creating the cultural conditions for multispecies affinity, it is also important to let animals, plants, fungi and microbes thrive *on their own terms* to whatever extent possible. When human societies ask themselves the question, "what matters to *all* of us, together?" we initiate epistemologies of care and coordination, rather than those of crisis. These can lead to better forms of multispecies co-habitation with human communities via ethical relationality. Such multispecies co-becomings include epistemic, affective and psychic avowals of the animality within humans. They also entail re-evaluating colonial axiologies that prefer hierarchical classifications of life in order to circumvent ethical considerations and reformatting dated notions of conservation that do not account for the possibility of productive human sympoesis. Epistemic and psychic re-evaluations such as these can glitch metro- and techno-normative perspectives and practices, allowing broader sections of society to co-create their own new inclusive axiologies. If we are products of the stories we tell ourselves and each other, then we must imagine the transformational potential of the creation and deployment of new multispecies stories based on actual embodied rewilding and co-habitation. The case of bison *and* Indigenous regeneration in the Wind River Tribal Reservation is an example of how such work can be and must be done. As seen in the Agro Pontino case study, rather than implementing rewilding as a return to Edenic forms of nature or Luddite fantasies of reneging on technology completely, it could be better conceptualized as a return to multispecies ecocultural relationality, where both human and nonhuman naturecultures ethically co-become in places and spaces.

Our Multispecies Futures has essentially explored how placing *trust* in naturalcultural relationships—full of co-becomings of interests, attachments, meanings, affects, but also flesh, matter, genetics, trophic connections—forms the foundation of healthy naturalcultural democracies. As we explored in Chapter 6, the magnificent Rainbow Scarab's expertise and existence is based on trust in a steady supply of bison dung, which in turn relies on healthy bison naturalcultural practices of knowledge transmission, family dynamics, grazing, movement and migration. The beetles' naturalcultural practices of dutiful dung retrieval and burial, in turn, form the trustful naturalcultural discursive humus upon which entire prairie ecological systems—both human and non—have evolved and been sustained. The recognition of the value of ever-expanding circles of kinship (and the acts of care that follow) is built on attentiveness and trust in the reliability of such mutualistic intra-active connections. Viewed within this context, the Cybercene represents an ecocultural era defined by the degradation of valuable trust in formerly reliable naturalcultural relationships; unbalanced climate systems and damaged ecological relationships mirror highly fragmented, insulated and isolated epistemic, affective and cultural human communities. Overcoming colonial mechanisms of distanced meaning-making for profit will require the re-establishment of embodied relations of naturalcultural trust once again. Turning the tide away from the Colonial-Cybercene's pervasive and corrosive epistemic, discursive, affective and psychic practices will require a broad multispecies paradigm shift. A widespread awareness of the mechanisms of the multispecies othering and commodification practices in the Greatest Acceleration may be able to prefigure reflection and pragmatic transformation, including wise and strategic use of new digital tools and capabilities.

With each passing day, even these partial knowledge claims and pathways towards a broad multispecies paradigm shift seem increasingly distant and untenable. In the Colonial-Cybercene, communities around the planet seem to be very quickly turning away from the joys of multispecies ethical relationality and careening towards narrowly focused hyper-efficient command economies of capture-and-control for consumption. Frictionless and optimized de-situated spectacles have become ubiquitous and preferred to "messier" tethered experiences of embodied naturalcultural connections. Image-lures on cyberscreens are rapidly gaining efficiency via exponential improvements in technology to create highly naturalistic deepfakes. The psychic, discursive and affective violence perpetuated via the active citation/participation of spectator/consumers in mediatic-cultural content continues to have profoundly damaging material effects on vibrant and vulnerable bodies, lives and ecocultural systems. In the space of barely two decades, Colonial-Cybercene paranoid de-situated knowledges have come to reign globally. This cultural context allows the increased and relentless decimation for profit of our planet's incredible biosphere and of the connective tissue that holds our multispecies communities (human and non) together. In such a rapidly degrading ecocultural environment, we further de-situate our epistemologies and affects away from multispecies kinship and nature-connectedness by tethering ourselves ever more intimately to the glow of our precious cyberscreens. I belong to the last generation that experienced childhood and youth without intensive digital mediatic intra-action. Will future generations even know what they are missing?

Fragile hope lies in the case studies analyzed in Chapters 5 and 6 that explore very different yet connected glitching possibilities to the damaging effects of Cybercene mechanisms of capture-and-control. Future specific studies on successful glitching ideas and practices could grow into a diverse and popular global compendium of such knowledges and practices—those magical instances when multispecies lives run away from the colonial violence of distancing, control, othering and waste and begin to thrive again within thick-matrices of lively belonging, care and kinship. Such a collective archive of place- and space-specific case studies can function as a resource and guide for future collective efforts to foster multispecies planetary rejuvenation—a humanimal manual, sympoetically written by a multitude of multispecies oddkin and gesturing towards vibrant spaces and places of biodiverse care and kinship.

The transformative returned gaze of the Channel Island Fox that we experienced in the Preface is a persistent invitation to keep on strategically unlearning anthropocentrism and relearning with humility and curiosity. We are constantly given such reminders of our relationships by an immense variety of multispecies and "more or less animate" beings, yet we are increasingly distracted or dismissive of these beautiful evocations. When will we realize that what we are dismissing as trivial has more power than we think?

Despite our accelerated retreat into the virtual that further encourages our Anthropocene delusions, our multispecies companions stay by our side, passionately and patiently "doing natureculture" in their own ways, sustaining our bodies and quietly enriching our daily existence.

Occasionally, they sometimes manage to break past our practiced inattentiveness towards them, our disregard for their labor, care and kinship. These moments strike us like a fleeting hallucination or flash of recognition of our deeply interconnected reality: simple things like the smell of rain on a parched sidewalk, a bellyache, the loving lick of a dog, the incessant chirping of a sparrow, a sip from a cup of sweet tea or an animal cartoon character on our screen can lead us down a cascading breadcrumb trail of interconnected affects, meaning-making and desires. Right after these moments of recognition, if we use simple "back of the envelope" informal naturalcultural discourse analysis, we could follow the evidence trail to care about these tiny embodied ethical glitches within a multiscalar and thick-time context: what prevailing global wind patterns brought the rain? How are my microbial gut-buddies doing today? How did millennia of co-becoming forge our canine companions' deeply affective bonds with us? How did sparrows become such innovative urban dwellers? What is the long global history of refined sugar that made this sweet substance so desirable? How do animated human-looking animal characters contribute to shaping human-nonhuman relationalities in young spectators? These considerations lead us into thick-stories of our complex ecocultural systems. With such passionate curiosity, we open ourselves up to enjoying marvelous matrices of interconnected and cascading meanings, desires, matter, affects, fears, bodies and lives. These are embodied, deeply tethered, profoundly situated and exceptionally nurturing kin- and oddkin-relationships that we can trust, and therefore truly care for. These kinds of casual multispecies meditative moments restore us into a fascinating multispecies paradigm and away from our habituated isolation. We sense and comprehend "complex 'ecologies of selves,' dynamic milieus that are continually shaped and reshaped, actively—even if not always knowingly—crafted through the sharing of 'meanings, interests and affects,' as well as flesh, minerals, fluids, genetic materials, and much more."[19]

As companions that have persisted with me to the end of this long journey, you too now belong within this vast, intricate and trustworthy multispecies assemblage. You're now part of a sprawling oddkin family, full of situated storymaking, storytelling and story-listening, all of it tethered deeply within the material and affective matrix of naturalcultural bodies, lives and ecocultural systems, always brimming with surprising connections and collaborations. Welcome—please make yourself at home in our multispecies futures.

Notes

Introduction: We Need a Multispecies Paradigm Shift

1. Donna Haraway, *Staying with the Trouble: Making Kin in the Chthulucene* (Durham, NC: Duke University Press, 2016), 4.
2. Gregory Cajete, *Native Science: Natural Laws of Interdependence* (Sante Fe: Clear Light Publishers, 2000), 57.
3. The coining of the term "humanimal," connecting humans to their animal kin in different ways, can be attributed to various people: Adam Rutherford in *Humanimal: How Homo sapiens Became Nature's Most Paradoxical Creature* (New York: The Experiment, 2019), Christopher Lloyd, who wrote the fabulous children's book *Humanimal: Incredible Ways Animals are Just like Us!* (London: What on Earth Books, 2019) and artist Kou Kitamura's 2022 solo exhibit "HUMANiMAL-KITAMURA," at art space co-jin, https://co-jin.jp/exhibition_en/humanimal-kitamura-kou-solo-exhibition/
4. Haraway, *Staying with the Trouble*, 4.
5. The Anthropocene is a term used to define a planetary era where human activity has caused a deep and lasting impact on Earth's ecosystems. I discuss this term in more detail in Chapter 3.
6. The Great Acceleration indicates the dramatic increase in human economic activity and consumption of natural resources since 1945.
7. A recent study by Miles Richardson demonstrated that when it comes to nature-connectedness, an *extinction of experience* has steadily occurred in human societies since the 1800s. By using a cultural data modelling methodology, he discovered "a clear decline of around 60% is revealed. Particularly from 1850, a time when industrialisation and urbanisation grew rapidly." See Richardson, "How we Lost Touch with Nature," *Finding Nature*, August 11, 2025, https://findingnature.org.uk/2025/08/11/model/ (accessed February 11, 2026).
8. A mass extinction is a short period of geological time in which a high percentage of biodiversity, or distinct species—bacteria, fungi, plants, mammals, birds, reptiles, amphibians, fish, invertebrates—dies out. In this definition, it's important to note that, in geological time, a "short" period can span thousands or even millions of years. The planet has experienced five previous mass extinction events, the last one occurring 65.5 million years ago and wiping out the dinosaurs from existence. Experts now believe we're in the midst of a sixth mass extinction, the first that is human-caused. For more information, see the World Wildlife Fund's resource "What is the sixth mass extinction and what can we do about it?"

   https://www.worldwildlife.org/stories/what-is-the-sixth-mass-extinction-and-what-can-we-do-about-it#:~:text=Unlike%20previous%20extinction%20events%20 caused,been%20converted%20for%20food%20production (accessed February 11, 2026).

9. *Metro-normative* is term coined by Jack Halberstam to describe the prejudicial attitude of urbanized humans towards their rural counterparts. See Halberstam, *In a Queer Time and Place: Transgender Bodies, Subcultural Lives* (New York: New York University Press, 2005).

10. Affect is a term that describes bodily and visceral experiences that precede meaning-making via conscious thought. In multispecies studies, the "affective turn" is an important development because it signifies paying closer attention to nonverbal and pre-"thinking" impacts of relations and interactions. Melissa Gregg and Gregory Seigworth name affect as "those visceral forces beneath, alongside, or generally other than conscious knowing that can serve to drive us toward movement, thought, and ever-changing forms of relation." Gregg and Seigworth (eds), *The Affect Theory Reader*, 1st edn (Durham, NC: Duke University Press, 2010), 1.

11. Oxford Bibliographies usefully explains the concept of "indirect effects": "Much ecology focuses on simple pairwise interactions, such as competition and predation; yet, species naturally exist in much more complex systems in which their abundances are determined by webs of species interactions. An important step for ecologists has been to understand how interactions may occur through loops and webs of connected species: it is these interactions that are now loosely collected together into what we call 'indirect effects.' Some types of indirect effects are thought not only to widely occur but also to be particularly important for determining both the abundances of individual species and community properties such as diversity and stability. These include trophic cascades, where predators enhance producer growth by feeding on consumer species, and keystone predation, where predators consume dominant competitors, thus allowing inferior competitors to persist. Identifying and quantifying indirect effects has become a major issue in ecology." Thomas E. Miller and Casey P. Terhost, "Indirect Effects in Communities and Ecosystems," *Oxford Bibliographies*, May 23, 2012, https://www.oxfordbibliographies.com/display/document/obo-9780199830060/obo-9780199830060-0020.xml#:~:text=obo/9780199830060%2D0020-,Introduction,a%20major%20issue%20in%20ecology (accessed February 11, 2026).

12. The lab was originally ideated in early 2022, launched in May 2023 as *The Cybercene Lab* during my tenure at Rutgers University, and then acquired its current name in May 2025 at UCLA to more accurately communicate its focus and priorities. For more information, see www.multispeciesfutureslab.org (accessed February 11, 2026).

13. The word *subaltern* denotes any marginalized group or community, typically those excluded from dominant power structures. The word was extensively theorized by Antonio Gramsci and taken up again famously by Gayatri Spivak in her essay "Can the Subaltern Speak?" (1985).

14. Vincent Slogan, quoted in Sweeney Windchief and Timothy San Pedro, *Applying Indigenous Research Methods* (New York: Routledge, 2019), ix.

15. Haraway, *Staying with the Trouble*, 23.

16. Kim TallBear, "Standing With and Speaking as Faith: A Feminist-Indigenous Approach to Inquiry," *Journal of Research Practice* 10.2 (2014): 2.

17. "Ecoculture" is a relatively new term that emphasizes the profound interconnections between human lived experiences and the larger nonhuman biological and

environmental systems that sustain and promote them. Rather than consider the ecological as an exclusive realm of "nature" and the cultural an exclusive concern of humans, "natureculture" and "ecoculture" both serve to impress on us the intimate connections between multispecies (human and non) natural and cultural lives.

18. Postcolonial (or post-colonial) studies is an interdisciplinary field that emerged primarily out of literature and cultural studies departments. It studies the political and cultural causes and effects of colonialism and imperialism, especially focusing on freedom movements and devalued perspectives of colonized communities globally. The closely related and complementary field of decolonial studies similarly seeks to locate alternative epistemologies and practices to those formed within colonial contexts.
19. "What is Ecology?" *The Ecological Society of America*, https://esa.org/about/what-does-ecology-have-to-do-with-me/ (accessed February 11, 2026).
20. See, for example, Rob Nixon, "Environmentalism and Postcolonialism," in Ania Loomba et al. (eds), *Postcolonial Studies and Beyond* (Durham, NC: Duke University Press, 2005), 233–251, Elizabeth M. DeLoughrey, *Allegories of the Anthropocene* (Durham, NC: Duke University Press, 2019), and DeLoughrey, Jill Didur and Anthony Carrigan (eds), *Global Ecologies and the Environmental Humanities: Postcolonial Approaches* (New York: Routledge, 2015).
21. See Haraway, *The Companion Species Manifesto: Dogs, People, and Significant Others* (Chicago: Prickly Paradigm Press, 2003).
22. Co-becoming is a concept best defined for the purposes of the environmental and multispecies humanities by Haraway in *The Companion Species Manifesto*. Haraway uses this to describe the coming into being and being of entities in existence via ongoing multiple human and/or nonhuman *relationships*. It is the recognition that all entities are in the process of constantly constituting and being co-constituted, complicating traditional identities that distinguish between the Self and its contrary, the Other.
23. Epistemic, meaning "about knowledge," refers to the internal logic that is contained within any given knowledge system. Epistemology, a central question in this book, relates to the theory of knowledge production, both in individual and collective spheres, i.e., how do we know what we know? How do systems of knowledge creation accept innovation and or resist it? How do disciplinary and discursive boundaries "police" ways of thinking about a given subject-matter?
24. For example, the use of DDT to eradicate malaria.
25. I had previously been calling this a "multispecies turn." I am grateful to Professor Claudio Fogu at the University of California, Santa Barbara, for suggesting that I find a bolder term to better reflect the systemic implications of such a shift in perspective.
26. Ontology, simply put, is the study of the reality of "being" in this world. Multispecies studies brings back value to simply "being" as opposed to "thinking" (epistemology) because it is the fulcrum upon which the concept of human exceptionality vis-à-vis other multispecies life rests: the normative idea is that other multispecies lives just exist, while humans live and *think*. Even the scientific name Carl Linnaeus gave our species, *Homo sapiens*, distinguishes us from others via exclusion—*sapiens* derives from the Latin *sapio*—to discern or to know. This connects our primary conception of ourselves as a species within the hierarchical chain of life to our capacity to generate epistemic frameworks.

27. "humanities," *Encyclopedia Britannica*, October 12, 2025, https://www.britannica.com/topic/humanities (accessed February 11, 2026).
28. Amitav Ghosh, *The Great Derangement: Climate Change and the Unthinkable* (Chicago: University of Chicago Press, 2016), 9.
29. Culture is accorded a higher value due to the fact that it involves "thinking" (epistemology) rather than "simply" living or existing (ontology). The scientific name of our species (*Homo sapiens*) automatically labels us as exceptional exactly because we are a thinking species. Logic dictates that this means that other species are automatically unthinking beings. Further assumptions stem from this fundamental epistemic norm, such as "culture" being exclusive to us "thinking humans" and that "thinking" is essentially a characteristic superior to simply "being." Another corollary assumption is that human life is exceptional and of higher value because thinking also probably means we feel meaning-related emotions that other species do not, including basic ones such as fear, love, jealousy, anger, etc.
30. Take for example, our own bodies. Science is finally giving more attention to the fact that our guts are multispecies entities, with great consequences for understanding how our digestive, immune, psychological, emotional and other systems continuously co-become.
31. Leroy Little Bear, "Jagged Worldviews Colliding," in Marie Battiste (ed), *Reclaiming Indigenous Voice and Vision* (Vancouver: University of British Columbia Press, 2007), 78.
32. Thom van Dooren, Eben Kirksey and Ursula Münster, "Multispecies Studies: Cultivating Arts of Attentiveness," *Environmental Humanities* 8.1 (2016): 4, quoting Dominique Lestel, Florence Brunois, and Florence Gaunet, "Etho-ethnology and Ethno-ethology," *Social Science Information* 45 (2006): 155.
33. Brendan Hokowhitu, "The Emperor's 'New Materialisms': Indigenous Materialisms and Disciplinary Colonialism," in Hokowhitu et al. (eds), *Routledge Handbook of Critical Indigenous Studies*, 1st edn (London: Routledge, 2020), 131–146, 135.
34. Ibid., 143.
35. Ibid., 143–44.
36. Cajete, *Native Science*, 152.
37. Haraway, *Staying with the Trouble*, 4.
38. Ibid.
39. van Dooren et al., "Multispecies Studies," 1
40. Anna Lowenhaupt Tsing, "Arts of Inclusion, or, How to Love a Mushroom," *Australian Humanities Review* 50 (2011): 19.
41. van Dooren et al., "Multispecies Studies," 6, my emphases.
42. The term "intra-action," coined by Karen Barad, describes how two entities (in this case, Wi'ma Island and its human visitors) co-constitute each other rather than simply interacting as separate entities. Understood in this way, the human visitor *partly becomes who they are* due to their profound entanglement, or intra-activity, associated with their passionate immersion in the multispecies ecocultural systems, just as the island is shaped by its visitors. See Barad, *Meeting the Universe Halfway: Quantum Physics and the Entanglement of Matter and Meaning* (Durham, NC: Duke University Press, 2007). I call an "ecocultural system" any place- and space-specific system of *intra-action* (input, output, feedback loops, both direct and indirect) between the natural and cultural agents.

43. Vetri Nathan and Teresa Fiore, "Ecocultures of the Scala dei Turchi: A Multispecies Thick-Map," 2025, https://storymaps.arcgis.com/stories/659a395c91c547f7949ba2b9a9a1deb5 (accessed February 12, 2026).
44. This project was created to be deployed within the framework of Agrigento as the 2025 *Italian Capital of Culture*. In addition to Professor Teresa Fiore, I am grateful to my two graduate student research assistants, Giacomo Frazzetta (who worked with Professor Fiore on the "cultural" pins), and Doğa Tekin (for her technical expertise in building the map).
45. Contributors to the stories in the *Scala dei Turchi* thick-map include faculty from the University of Palermo (Professors Valeria Scavone, Agostino Tomasello and Antonio Caruso), recent graduates from the University of Palermo (Fabiana Fiorica), historical and naturalistic guides (Marcello Mira, Antonino Dinolfo and Marco Falzone), photographers (Nuccio Zicari), and local chefs (Pino Cuttaia and Biagio Di Caro).
46. See Ron Broglio, *Surface Encounters: Thinking with Animals and Art* (Minneapolis, MN: University of Minnesota Press, 2011).
47. Sutirtho Roy, "Can the Non-Human Subaltern Speak? Addressing Injustice Through Parakeets, Penguins and Blue Macaws," *Studia Ubb Philologia* LXVII.2 (2022): 230. https://doi.org/10.24193/subbphilo.2022.2.13.
48. Édouard Glissant. *Poetics of Relation*. Translated by Betsy Wing (The University of Michigan Press, 1998)
49. Serenella Iovino and Serpil Oppermann, *Material Ecocriticism* (Bloomington, IN: Indiana University Press, 2014), 3.
50. Nathan and Fiore, "Ecocultures of the Scala dei Turchi."
51. Thick time conceives time as layered and multifaceted, an accumulation of past, present, and future, rather than a single, linear progression. It also accounts for subjective experiences (affective, emotional, generational) of the passage and accumulation of time as well as embodied understandings of time. In a multispecies studies context, thick time gains additional resonance by considering perspectives of time that are uniquely experienced by nonhumans. How does one of the world's oldest organisms, the Bristlecone Pine (average lifespan of around 3,000 years), experience time, compared to a mayfly (lifespan of 5 minutes to 24 hours)?
52. The "Golden Spike" in this pin refers to the presence of an official physical marker—a Global Boundary Stratotype Section and Points (GSSPs)—in the rock near the cliffs. This particular spike marks the boundary between the Piacenzian Stage (Middle Pliocene) and the Zanclean Stage (Lower Pliocene).
53. Cajete, *Native Science*, 52.
54. Studies like those undertaken by Jeremy T. Claisse et al. show that "oil and gas platforms off the coast of California have the highest secondary fish production per unit area of seafloor of any marine habitat that has been studied, about an order of magnitude higher than fish communities from other marine ecosystems. [...] Understanding the biological implications of these structures will inform policy related to the decommissioning of existing (e.g., oil and gas platforms) and implementation of emerging (e.g., wind, marine hydrokinetic) energy technologies." See Claisse et al., "Oil platforms off California are among the most productive marine fish habitats globally," *PNAS* 111.43 (2014): 15462. https://doi.org/10.1073/pnas.1411477111

55. See the National Oceanic and Atmospheric Administration's article on "The Ekman Spiral," https://oceanservice.noaa.gov/education/tutorial_currents/04currents4.html#:~:text=The%20Ekman%20spiral%2C%20named%20after,of%20water%20molecules%20below%20them (accessed February 16, 2026).
56. See the National Park Service's article on "Currents and Upswellings," https://www.nps.gov/chis/learn/nature/currents-upwelling.htm (accessed February 16, 2026).
57. See the California Department of Fish and Wildlife's article on "Scorpion State Marine Reserve/Federal Marine Reserve," https://wildlife.ca.gov/Conservation/Marine/MPAs/Scorpion/lc/5788311/lcv/s/Forest%20of%20giant%20kelp%20and%20algae%20understory%20at%20Scorpion%20SMR.
58. A big thanks to Professor Pasquale Verdicchio for giving me permission to use this term, which he mentioned during a conversation over coffee.
59. Jan Timbrook and Chris Chapman, *Chumash Ethnobotany: Plant Knowledge Among the Chumash People of Southern California*, 2nd ed. (Santa Barbara: Santa Barbara Museum of Natural History, 2007), 208.
60. "Chumash Painted Cave," *CyArk*, 2025, https://www.cyark.org/projects/chumash-painted-cave/in-depth (accessed February 16, 2026).
61. See Jonas Siehoff, "Species that never interact can still profoundly influence each other's evolution," *Phys.org*, August 27, 2025, https://phys.org/news/2025-08-species-interact-profoundly-evolution.html (accessed February 16, 2026).
62. "Santa Rosa Island Cloud Forest Restoration," *National Park Service*, https://www.nps.gov/chis/learn/nature/cloud-forest.htm (accessed February 16, 2026).
63. John R. Johnson, "The Arlington Spring Man," *West of the West: Tales from California's Channel Islands,* https://www.teachchannelislands.org/tales/arlington-springs-man (accessed February 16, 2026).
64. Gary Polakovic, "Channel Island Woman's Bones May Rewrite History," *Los Angeles Times*, April 11, 1999, https://www.latimes.com/archives/la-xpm-1999-apr-11-mn-26401-story.html (accessed February 16, 2026).
65. The term "symptomatic reading," coined by Louis Althusser, describes a widely accepted method of undertaking cultural analysis of literary and critical texts. In his formulation of "ideology," Althusser introduces the concept of the "problematic." A problematic is the underlying series of questions that command the answers given in the text; it is the ideological or theoretical structure that frames the text. The aim of the critical reader or theorist, then, is to deconstruct the underlying problematic of the text by inferring both what is said, *and what is not said*, within the text. This kind of reading is symptomatic reading. Althusser describes Marx's reading of Adam Smith as symptomatic, because it divulges the undivulged event in the text it reads, and in the same movement relates it to a different text, present as a necessary absence in the first. Like his first reading, Marx's second reading presupposes the existence of two texts, and the measurement of the first against the second. But what distinguishes this new reading from the old is the fact that in the new one the second text is articulated with the lapses in the first text. See more from Althusser and Étienne Balibar in *Reading Capital: The Complete Edition*, trans. Ben Brewster and David Fernbach (New York: Verso, 2016).
66. Steve Chawkins, "Ancient Bones Belonged to a Man—Probably," *Los Angeles Times*, September 11, 2006, https://www.latimes.com/archives/la-xpm-2006-sep-11-me-bonesside11-story.html (accessed February 16, 2026).

67. Ibid.
68. Deep time describes the vast and linear span of geological and cosmic history, while thick time is a non-linear concept where past, present, and future become entangled and superimposed. Deep time can be generally understood as objective (even though the theory of relativity complicates this stance), while thick time is subjective and experiential. It focuses more on how different temporalities co-become and coexist in a single moment or space.
69. Kathryn Yusoff, "*A Billion Black Anthropocenes or None*: A New Book About Race and Geology," interview by Leah Kaplan, *Black Perspectives*, April 12, 2019, https://www.aaihs.org/a-billion-black-anthropocenes-or-none-a-new-book-about-race-and-geology/ (accessed February 16, 2026).

Chapter 1: Staying With the Joy: Framing Multispecies Studies Approaches

1. van Dooren, "Multispecies Storytelling in Catastrophic Times," *Thom van Dooren*, June 24, 2025, https://www.thomvandooren.org/2025/06/24/multispecies-storytelling-in-catastrophic-times/ (accessed February 18, 2026).
2. Ibid.
3. It is crucial for multispecies studies to confront the damaging ecocultural consequences of speciesist categorizations and attitudes. Since Classical European times, hierarchical taxonomies such as the medieval Great Chain of Being have imagined and arranged species on vertical scales of inferior-to-superior. These greater "natural" orderings of life were adapted during the Age of Exploration, on the basis of the Linnaean classification system, to imagine human races in hierarchical terms. These imaginations are not exclusive to racialized thought in normative European traditions—for example, the caste system in India utilizes a classification and categorization system for human groupings that became strictly hierarchical and exclusionary in order to concentrate cultural power among an exclusive elite group of "superior" beings.
4. Scholars have also suggested alternatives such as "multi-being" and "trans-species" studies. Despite very valid reasons to choose one of these terms, I finally chose multispecies simply because it is the most readily understood within and beyond academia.
5. Linnaean classification was developed by the Swedish naturalist Carl Linnaeus (1707–1778) to organize living organisms into groups. The system utilizes seven taxonomic levels, each denoting increasing levels of kinship: kingdom, phylum, class, order, family, genus, and species. This system provides the Latin binomial nomenclature used to name organisms (Genus species).
6. Christopher Peterson, *Bestial Traces: Race, Sexuality, Animality* (New York: Fordham University Press, 2012), 2.
7. Ibid.
8. Ibid.
9. Haraway, *Staying with the Trouble*, 4.
10. Aileen Moreton-Robinson, "Relationality: A Key Presupposition of an Indigenous Social Research Paradigm," in Chris Andersen and Jean M. O'Brien (eds), *Sources and Methods in Indigenous Studies* (New York: Routledge, 2017), 232.
11. Haraway, *Staying with the Trouble*, 4.

12. For example, the share of net worth held by the top 1% of U.S. individuals rose from around 22% in 1990 to more than 30% by 2024. See Govind Bhutada, "Visualized: The 1%'s Share of U.S. Wealth Over Time (1989–2024), *Visual Capitalist*, February 8, 2025, https://www.visualcapitalist.com/visualized-the-1s-share-of-u-s-wealth-over-time-1989-2024/ (accessed February 18, 2026).
13. Haraway, *Staying with the Trouble*, 2.
14. Haraway, "Making Kin: An Interview with Donna Haraway," interview by Steve Paulson, *Los Angeles Review of Books*, December 6, 2019, https://lareviewofbooks.org/article/making-kin-an-interview-with-donna-haraway/ (accessed February 18, 2026).
15. van Dooren, "Care," *Environmental Humanities* 5.1 (2014): 291–294, 291–292. https://doi.org/10.1215/22011919-3615541
16. van Dooren et al., "Multispecies Studies," 4, quoting Lestel et al., "Etho-ethnology and Ethno-ethology," 155.
17. Edward Said, *Culture and Imperialism* (New York: Vintage Books, 1994), 9.
18. Haraway, "Feminist cyborg scholar Donna Haraway: 'The disorder of our era isn't necessary'," interview by Moira Weigel, *The Guardian*, June 20, 2019, https://theguardian.com/world/2019/jun/20/donna-haraway-interview-cyborg-manifesto-post-truth (accessed February 18, 2026).
19. Marco Armiero describes contemporary planetary ecological crises as symptoms of wasted *relationships, lives and places* in *Wasteocene: Stories from the Global Dump* (Cambridge: Cambridge University Press, 2021), 2.
20. Haraway, *The Companion Species Manifesto*; Haraway, *Staying with the Trouble*; Bruno Latour, *Reassembling the Social: An Introduction to Actor-Network-Theory* (New York: Oxford University Press, 2005); Barad, *Meeting the Universe Halfway*; Jane Bennett, *Vibrant Matter: A Political Ecology of Things* (Durham, NC: Duke University Press, 2009); Rosi Braidotti, *The Posthuman* (Cambridge: Polity Press, 2013); Ursula Heise, *Imagining Extinction: The Cultural Meanings of Endangered Species* (Chicago: University of Chicago Press, 2016); van Dooren et al., "Multispecies Studies"; Armiero, *Wastocene.*
21. Sarah McFarland and Ryan Hediger, *Animals and Agency: An Interdisciplinary Exploration* (Leiden: Brill, 2009); Sue Donaldson and Will Kymlicka, *Zoopolis: A Political Theory of Animal Rights* (Oxford: Oxford University Press, 2011); Alasdair Cochrane, *Sentientist Politics: A Theory of Global Inter-species Justice* (Oxford: Oxford University Press, 2018).
22. Cajete, *Native Science*; Hokowhitu, "The Emperor's 'New Materialisms': Indigenous Materialisms and Disciplinary Colonialism," in Hokowhitu et al. (eds), *Routledge Handbook of Critical Indigenous Studies*; Deborah Bird Rose, *Wild Dog Dreaming: Love and Extinction* (Charlottesville: University of Virginia Press, 2011); Kim TallBear, "An Indigenous Reflection on Working beyond the Human/Not Human," *GLQ: A Journal of Lesbian and Gay Studies* 21.2–3 (2015): 230–235; Kyle Whyte, "Against Crisis Epistemology," in Hokowhitu et al. (eds), *Routledge Handbook of Critical Indigenous Studies*, 52–64.
23. Cary Wolfe, *Animal Rites: American Culture, the Discourse of Species, and Posthumanist Theory* (Chicago: University of Chicago Press, 2003); Lori Gruen, *Entangled Empathy: An Alternative Ethic for Our Relationships with Animals* (New York: Lantern Books, 2015); Natasha Myers, "Ungrid-able Ecologies: Decolonizing the Ecological Sensorium in a Ten-Thousand-Year-Old NaturalCultural Happening," *Catalyst: Feminism, Theory, Technoscience* 3.2 (2017): 1–24, https://doi.org/10.28968/cftt.

v3i2.28848; Laura Foster, "Critical Perspectives on Plants, Race, and Colonialism: An Introduction," *Catalyst: Feminism, Theory, Technoscience* 5.2 (2019): 1–6. https://doi.org/10.28968/cftt.v5i2.32309

24. For an excellent overview of current approaches in multispecies studies and ecojustice, see Sophie Chao and Danielle Celermajer, "Introduction: Multispecies Justice," *Cultural Politics* 19.1 (2023): 1–17. https://doi.org/10.1215/17432197-10232431
25. Cajete, *Native Science*, 69.
26. Ibid., 52.
27. Ibid., 53.
28. Luke E. Painter, Robert L. Beschta, and William J. Ripple, "Changing aspen stand structure following large carnivore restoration in Yellowstone," *Forest Ecology and Management* 594 (2025). https://doi.org/10.1016/j.foreco.2025.122941
29. Haraway, "Situated Knowledges: The Science Question in Feminism and the Privilege of Partial Perspective," *Feminist Studies* 14.3 (1988): 589.
30. van Dooren et al., "Multispecies Studies," 6.
31. Priya Satia, "Stanford scholar explains the history of India's partition, its ongoing effects today," interview by Alex Shashkevich, *Stanford Report*, March 8, 2019, https://news.stanford.edu/stories/2019/03/partition-1947-continues-haunt-india-pakistan-stanford-scholar-says (accessed February 19, 2026).
32. Shakeeb Asrar and Omran Abazid, "How were the India-Pakistan partition borders drawn?" *Al Jazeera*, August 17, 2017, https://www.aljazeera.com/news/2017/8/17/how-were-the-india-pakistan-partition-borders-drawn (accessed February 19, 2026).
33. Estimates stand at 14 million humans displaced and 2 million killed. See Tuba Tanveer, Nixon Mathur and Rakesh Sarwal, "Catastrophic impact of 1947 partition of India on people's health," *Journal of Family Medicine and Primary Care* 13.2 (February 2024): 401–408.
34. British Information Services, *India, 1944*, 1944, color map, 76 × 51 cm, University of Wisconsin, Milwaukee Library, American Geographical Society Library Digital Map Collection, https://collections.lib.uwm.edu/digital/collection/agdm/id/27867/ (accessed February 19, 2026).
35. Ibid.
36. Omair Ahmad, "Opinion: Deep scars to South Asia's ecosystems are a legacy of Partition," *Dialogue Earth*, August 14, 2022, https://dialogue.earth/en/nature/opinion-deep-scars-to-south-asia-ecosystems-legacy-of-partition/ (accessed February 19, 2026).
37. van Dooren, "Care," 291–292.
38. van Dooren et al., "Multispecies Studies," 6.
39. For example, as we have seen with the British Information Services map of colonial India, if in a given community it is commonly agreed as "fact" via discourse that colonial spaces are meant to be resources that can be freely utilized for imperial power, then ecocultural wealth will be seen as such, rather than as a respectful mutual relationship to be cultivated and tended.
40. Manfred Steger, "Searching for *Satya* through *Ahimsa*: Gandhi's Challenge to Western Discourses of Power," *Constellations* 13.3 (2006): 341.
41. Ibid., 343, quoting Gandhi, *Autobiography: The Story of My Experiments with Truth* (Washington, DC: Public Affairs Press, 1948).
42. For a more detailed analysis of the ecological implications of Gandhian concepts such as *satyagraha* and *ahimsa*, see Madhav Gadgil and Ramachandra Guha, *This*

Fissured Land: The Ecological History of India (Oxford, New Delhi, Oxford University Press, 1992).

43. Steger, "Searching for *Satya*," 343.
44. Ibid., 347.
45. Alfred W. Crosby Jr., *The Columbian Exchange: Biological and Cultural Consequences of 1492* (Santa Barbara: Greenwood Publishing Group, 1972).
46. Ramachandra Guha, "Radical American Environmentalism and Wilderness Preservation: A Third World Critique (1997)," in Libby Robin, Sverker Sörlin, Paul Warde (eds), *The Future of Nature: Documents of Global Change* (New Haven: Yale University Press, 2013), 409–432.
47. These connections can and should be made without essentializing, homogenizing or simplifying the immense diversity of existing Indigenous communities, especially those in the Global South. By "Indigenous," I mean the definition used by the United Nations:

    > Those who self-identify as Indigenous peoples at the individual level and are accepted by the community as its members; have a historical continuity with the pre-colonial and/or pre-settler societies; have a strong link to territories and surrounding natural resources; have distinct social, economic or political systems; have distinct language, culture and beliefs; form non-dominant groups in society, and resolve to maintain and reproduce their ancestral environments and systems as distinctive peoples and communities. (Moreton-Robinson, "Relationality," in Andersen and O'Brien (eds), *Sources and Methods in Indigenous Studies*, 69).

    As we can see, this definition already prefigures the important interconnections Indigenous communities possess between culture, identity and natural ecosystems. This strong relationality between culture/identity and land/space/nature is in itself an important reason to understand how global Indigenous thought and practice may revitalize embodied healing and multispecies habitability.
48. Bioculture represents the interplay of cultural frameworks and biological systems. Biopolitics refers to the use of "biologically" and "ecologically" themed images, ideas and considerations (species, race, gender, nationality, age, family, labor productivity, reproduction, sexuality, agency) to gain and maintain political, economic, cultural and/or ecocultural power.
49. Jorge Marcone, "The Stone Guests: Buen Vivir and popular environmentalisms in the Andes and Amazonia," in Ursula Heise, Jon Christensen, and Michelle Niemann (eds), *The Routledge Companion to the Environmental Humanities* (London: Routledge, 2017), 288.
50. Zoe Todd, "An Indigenous Feminist's Take On The Ontological Turn: 'Ontology' Is Just Another Word For Colonialism," *Journal of Historical Sociology* 29.1 (2016): 7. https://doi.org/10.1111/johs.12124
51. TallBear, "Standing With and Speaking as Faith," 2.
52. Survivance, as defined by Anishinaabe scholar Gerald Vizenor, is a portmanteau term that connects survival to resilience. It is not simply a passive survival in the face of ecocultural extermination, but an active, creative persistence of life, an agency and life force that acts despite the many challenges faced by oppressive forces. See Vizenor, *Survivance: Narratives of Native Presence* (Lincoln, NE: Nebraska Paperback, 2008).
53. TallBear, "Standing With and Speaking as Faith," 2.

54. Douglas Schuler has published some articles using this term, the earliest of which I can date to June 2023 ("Welcome to the Cybercene?" LIMITS, 9th Workshop on Computing Limits, June 2023, https://limits.pubpub.org/pub/j542e5lp/release/1 [accessed May 16, 2026]). My *Cybercene Lab* was envisioned in 2022 and founded in May 2023. This may be a case of a term being concurrently developed separately. Schuler's use of Cybercene seems to be primarily focused on the direct rather than indirect environmental effects of the digital revolution (issues such as e-waste) while also stating some of the indirect socio-cultural impacts. My conceptualization of the term is different, as it stems from the bringing together of a media studies understanding of this epoch (how the spectator/consumer intra-acts with online, onscreen realities) and an environmental and multispecies humanities methodology (how those online onscreen intra-actions affect actual bodies, lives and ecosystems).
55. Vetri Nathan, "Multispecies Futures Lab: Ecocultural Wellbeing and Kinship in the Cybercene," *Multispecies Futures Lab*, 2023, www.multispeciesfutureslab.org (accessed February 23, 2026).
56. Haraway calls for tentacular rather than linear forms of knowledge, arguing for sympoesis (making-with) rather than autopoesis. Citing a doctoral thesis by M. Beth Dempster, Haraway defines sympoesis as "collectively-producing systems that do not have self-defined spatial or temporal boundaries. Information and control are distributed among components. The systems are evolutionary and have the potential for surprising change." Dempster posits sympoetic thought systems as contrasts to autopoetic knowledges that have "self-defined spatial or temporal boundaries that tend to be centrally controlled, homeostatic and predictable." Haraway, *Staying with the Trouble*, 33, quoting Dempster, "A Self-Organizing Systems Perspective on Planning for Sustainability" (PhD thesis, Environmental Studies, University of Waterloo, Ontario, 1998), 27–32. My two terms for these two contrasting kinds of knowledge systems are thin-accounts (for colonial, autopoetic narrations of control) and thick-accounts (that come in mutating, sympoetic and tentacular forms).
57. Erich Hörl and James Burton, *General Ecology: The New Ecological Paradigm* (London: Bloomsbury Academic, 2017), 3.
58. A recent example of this recovery can be found in the Salmon and other fish communities of the Klamath River in California: "Its migratory fish were the primary food—and central to the culture—of the Karuk, Yurok, Klamath and other tribes. But, beginning in the 1920s, four hydropower dams were built on the river, blocking salmon from swimming upstream to spawn and limiting them to a reduced stretch of the river." A restoration agreement was signed in 2010, leading to a return of the first migrating salmon communities only months after the dams were removed. See Jeff Opperman, "How the Return of Salmon to the Klamath River Shows Us What's Possible in Wildlife Conservation," *Scientific American*, November 26, 2024, https://www.scientificamerican.com/article/how-the-return-of-salmon-to-the-klamath-river-shows-us-whats-possible-in-wildlife-conservation/ (accessed February 26, 2026).
59. My work is in conversation with the general approaches of what has been categorized as "posthuman" critical theory, especially in the way it seeks to challenge anthropocentrism within the broader field of the humanities. However, my personal preference is to avoid the term "posthuman" due to its ambiguous and heterogenous meanings even within the field of ecocriticism, but especially in the public sphere, where it often seems to connote envisioning a "science-fiction" post-apocalyptic world (i.e., a

present or future where humans have been replaced or hybridized with cyborgs, etc.). My book can be placed within the general category of "posthumanist" or "posthumanities" approaches if we consider the broader and less literal sense of the term, where the "post-" in the "posthuman" does not mean "after," but signifies "moving beyond" anthropocentric epistemological frameworks and accommodating multispecies ontologies and perspectives in order to collapse the dichotomies of nature/culture, human/nonhuman, mind/body, meaning/affect, etc. Due to these reasons, also described in the book, I prefer "multispecies studies": a term that includes and recognizes the animality of the human and yet goes beyond *Homo sapiens*' species-specific conceptual boundaries.

60. Lowenhaupt Tsing, "Arts of Inclusion," 21.

Chapter 2: Capture-and-Control

1. Daniel Vandersommers, *Entangled Encounters at the National Zoo: Stories from the Animal Archive* (Lawrence, KS: University of Kansas Press, 2023), 4.
2. Ibid., 2
3. See Said, *Orientalism* (New York: Vintage Books, 1979); Said, *Culture and Imperialism*; Homi K. Bhabha, *The Location of Culture* (New York: Routledge, 1994); and Robert J.C. Young, *Colonial Desire: Hybridity in Theory, Culture and Race*, 1st edn (New York: Routledge, 1995).
4. Benedict Anderson famously defined nations as socially constructed and therefore "imagined" communities: they are imagined since most citizens of a nation will never actually see each other, but this does not mean that they are false. On the contrary, they show the immense constitutive force of power/knowledge to create the material realities of actual physical nation states. See Anderson, *Imagined Communities: Reflections on the Origin and Spread of Nationalism*, revised ed. (London: Verso, 2006).
5. See for example Laura Foster, "Critical Perspectives on Plants, Race, and Colonialism: An Introduction," *Catalyst: Feminism, Theory, Technoscience* 5.2 (2019): 1–6, https://doi.org/10.28968/cftt.v5i2.32309; Mason Durie, *Ngā Tai Matatū: Tides of Māori Endurance* (Oxford: Oxford University Press, 2005); Zakiyyah Iman Jackson, "Outer Worlds: The Persistence of Race in Movement 'Beyond the Human'," *GLQ* 21.2–3 (2015): 215–18; Donald S. Moore, Jake Kosek, and Anand Pandian, "The Cultural Politics of Race and Nature: Terrains of Power and Practice," in Moore, Kosek, and Pandian (eds), *Race, Nature, and the Politics of Difference* (Durham, NC: Duke University Press, 2003), 1–70.
6. Young, *Colonial Desire*, 180–181, my emphasis.
7. Mary Norton, *The Tame and the Wild: People and Animals after 1492* (Cambridge, USA: Harvard University Press, 2024).
8. Pope Alexander VI, "Demarcation Bull Granting Spain Possession of Lands Discovered by Columbus," *Gilder Lehrman Institute of American History*, https://gilderlehrman.org/sites/default/files/inline-pdfs/T-04093.pdf (accessed February 26, 2026).
9. Nicole Winfield, "Vatican formally rejects 'Doctrine of Discovery' after Indigenous calls," *PBS News*, May 30, 2023, https://pbs.org/newshour/politics/vatican-formally-rejects-doctrine-of-discovery-after-Indigenous-calls#:~:text=Those%20decrees%20underpin%20the%20%E2%80%9CDoctrine,because%20they%20%E2%80%9Cdiscovered%E2%80%9D%20it (accessed February 26, 2026).

10. Sir Walter Raleigh, *The Discovery of Guiana*, Project Gutenberg Etext; No. 2272 (Champaign, IL: Project Gutenberg, 2006), my emphasis, https://www.gutenberg.org/files/2272/2272-h/2272-h.htm (accessed February 26, 2026).
11. Amerigo Vespucci, *Mundus Novus: Letter to Lorenzo Pietro di Medici*, trans. George Tyler Northup (Princeton, NJ: Princeton University Press, 1916), 1, https://encyclopediavirginia.org/primary-documents/mundus-novus-1503/ (accessed February 26, 2026).
12. Ibid., 3.
13. Ibid., 5–6.
14. Said, *Orientalism*, 72.
15. Columbus' famous letter to Spain can be read as one of the most important pieces of "fake news" in global history.
16. Gilles Boëtsch and Pascal Blanchard, "From Cabinets of Curiosity to the 'Hottentot Venus': A Long History of Human Zoos," in Nicolas Bancel, Thomas David, and Dominic Thomas (eds), *The Invention of Race: Scientific and Popular Representations* (New York: Routledge, 2014), 191.
17. See Blanchard, Boëtsch, and Nanette Jacomijn Snoep (eds), *Human Zoos: The Invention of the Savage* (Arles: Actes Sud Editions, 2011).
18. Haley N. Beer (et al.), "The Evolution of Zoos as Conservation Institutions: A Summary of the Transition from Menageries to Zoological Gardens and Parallel Improvement of Mammalian Welfare Management." *The Journal of Zoological and Botanical Gardens* 4.4 (2023): 648.
19. Ibid., 653
20. Chokri Ben Chikha and Karel Arnaut, "Staging/caging 'otherness' in the postcolony: spectres of the human zoo," *Critical Arts* 27.6 (2013): 667. https://doi.org/10.1080/02560046.2013.867589
21. Ibid.
22. Boëtsch and Blanchard, "From Cabinets of Curiosity to the 'Hottentot Venus'," 185.
23. Nigel Rothfels, *Savages and Beasts: The Birth of the Modern Zoo* (Baltimore: The Johns Hopkins University Press, 2002), 135.
24. Said, *Orientalism*, 72, my emphasis.
25. Rothfels, *Savages and Beasts*, 93.
26. Chika and Arnaut, "Staging/caging 'otherness' in the postcolony," 669–670.
27. Boëtsch and Blanchard, "From Cabinets of Curiosity to the 'Hottentot Venus'," 188.
28. The mechanisms of the stereotype-as-fetish will be discussed in more detail in Chapter 4.
29. Boëtsch and Blanchard, "From Cabinets of Curiosity to the 'Hottentot Venus'," 187.
30. Ibid., 190.
31. Rothfels, *Savages and Beasts*, 95.
32. Ibid., 137.
33. Bhabha, "Of Mimicry and Man: The Ambivalence of Colonial Discourse," *October* 28 (1984): 127.
34. Frantz Fanon, *Black Skin, White Masks*, trans. Charles Lam Markmann (New York: Grove Press, 1967).
35. The "regime of visibility" that Bhabha describes in *The Location of Culture* also leads him to connect the working of the stereotype to another important psychoanalytic mechanism that is useful to the study of power dynamics in film: Lacan's Imaginary.

The Imaginary is a phase of childhood described by Lacan in which questions of identity arise for the first time. Lacan calls the point of entry into this phase the "mirror stage." When the child looks at their own complete image, he experiences a split feeling towards this image: there is a feeling of great pleasure in his identification with his own complete image, and there is also a feeling of displeasure, arising from the awareness that he is not his image, that he is like his image, but somehow different. This split feeling gives rise to, respectively, the feelings of narcissism and of aggressivity. For Bhabha, the construction of colonial representation relies on the two tropes discussed above: the stereotype-as-fetish and the Imaginary. Fetishism masks difference by providing a compromised substitute—it works, in other words, as a *metaphor*. In the representation as stereotype-as-fetish, the object that substitutes is adjacent to, or associated with, the actual subject that is being constituted—in other words, it is a *metonym*. For Lacan too, the unconscious is structured like language, as it works through metaphor and metonymy, which are respectively linked to the twin dream processes described by Freud—condensation and displacement.

36. Boëtsch and Blanchard, "From Cabinets of Curiosity to the 'Hottentot Venus'," 191.
37. David Hearst, "African woman going home after 200 years," *The Guardian*, April 30, 2002, https://www.theguardian.com/world/2002/apr/30/education.arts#:~:text=In%20London%2C%20the%20act%20consisted,funeral%20in%20August%20or%20September (accessed March 2, 2026).
38. Ibid.
39. Bhabha, *The Location of Culture*, 80.
40. This, along with the importance of the "visible" as the entryway into the stereotype, and the need to anxiously repeat the stereotype, completes the explanation of the way in which such discourses work in the algorithmic qualities of hyperconverged Cybercene media. Once the stereotyped image is fixed, it needs to be anxiously re-cited, via different platforms, and forwarded to as many contacts as possible in order to maintain its currency. The fact that the images may be deepfakes, or decontextualized to such an extent that they may have only the most tenuous links to actual lives, bodies, actions, and ecosystems makes these cyberscenes all the more powerful as lures.
41. Said, *Orientalism*, 3.
42. Ibid.
43. Jean-Louis Comolli, *Cinema Against Spectacle: Technique and Ideology Revisited*, ed. Daniel Fairfax (Amsterdam: Amsterdam University Press, 2009), 53.
44. Ibid., 9.
45. The psychic processes of scopophilia, disavowal and misrecognition are key to the creation of fantasies. We will explore them in more detail in Chapter 4.
46. Octava Mannoni, *Clefs pour l'imaginaire; ou, L'autre scène* (Paris: Editions du Seuil, 1969).
47. van Dooren et al., "Multispecies Studies," 2.
48. Gilles Deleuze and Félix Guattari, *A Thousand Plateaus* (Minneapolis, MN: University of Minnesota Press, 1987), 11.
49. Deleuze and Claire Parnet, *Dialogues* (New York: Columbia University Press, 1977), 2.
50. Jonathan Lambert, "Photos: Mother Nature must be really annoyed at our fakery," *NPR*, August 30, 2025, https://www.npr.org/sections/goats-and-soda/2025/08/30/

g-s1-86289/photos-nature-anthropocene-polar-bear-elephants-illusion (accessed March 2, 2026).

51. Zed Nelson, quoted by Lambert, "Mother Nature must be really annoyed."
52. Ibid.
53. Nelson, "Polar bears in shopping malls, fiberglass mountains, indoor rainforests: how humans got addicted to faking the natural world," *The Guardian*, July 24, 2025, https://www.theguardian.com/environment/2025/jul/24/humans-addicted-faking-natural-world-anthropocene-illusion-zed-nelson-aoe (accessed March 2, 2026).
54. These de-situating processes are having ever-accumulating planetary consequences via *gradualism*. Gradualism is a term that I borrow from the biological and geological sciences. It was originally used in these fields to explain how evolution and other major geological changes (such as the drift of tectonic plates) occur via small, incremental genetic or physical changes that initially might not be noticeable until a moment arrives when these changes, added together, show substantial deviation from the original moment—which will not seem like a deviation since the change has occurred in incremental moves. The same can be said for many cultural and political transformations, where complex discursive systems mirror incremental normalization of extreme natural thresholds: for example the move towards authoritarian regimes may initially occur in stages too small to be considered a threat, but when added up over years ends up catastrophically mirroring the minute daily or monthly rise in global average sea level.

Chapter 3: Locating Cybercene Desires

1. Karen Hao, *Empire of AI: Dreams and Nightmares in Sam Altman's OpenAI Dreams and Nightmares in Sam Altman's OpenAI* (New York: Penguin, 2025), 10.
2. Sean Cubitt, *EcoMedia* (Amsterdam: Rodopi, 2005), 2.
3. "Digital Around the World," *Datareportal*, https://datareportal.com/global-digital-overview (accessed March 2, 2026).
4. The Capitalocene was another one of the important "-cenes" that I considered including in this discussion. However, since I perceive sufficient commonalities and continuations between extreme extractive capitalistic practices and historical colonial-corporate practices, I have avoided redundancy here by limiting my use of jargon. All these different extreme extractive practices dovetail well enough for our purposes within my ecocultural definition of colonialisms.
5. "Anthropocene," *Wikipedia: The Free Encyclopedia*, last modified November 3, 2025, https://en.wikipedia.org/wiki/Anthropocene (accessed March 3, 2026).
6. Ibid., 17.
7. Dipesh Chakrabarty, "The Climate of History: Four Theses," *Critical Inquiry* 35.2 (2009): 197–222.
8. Simon L. Lewis and Mark A. Maslin, "Defining the Anthropocene," *Nature* 519 (2015): 171–180. https://doi.org/10.1038/nature14258
9. Ruth Howells, "Epoch-defining study pinpoints when humans came to dominate planet Earth," *UCL News*, March 11, 2015, https://ucl.ac.uk/news/2015/mar/epoch-defining-study-pinpoints-when-humans-came-dominate-planet-earth#:~:text=The%20researchers%20have%20named%20the,disconnected%20peoples%20becoming%20globally%20linked (accessed March 3, 2026).

10. Amy C.F. King et al., "Reconciling ice core CO_2 and land-use change following New World-Old World contact," *Nature Communications* 15.1735 (2024): 1–9. https://doi.org/10.1038/s41467-024-45894-9
11. Lewis and Maslin, "Defining the Anthropocene," 177.
12. Armiero, *Wasteocene*, 2.
13. Zygmunt Bauman, *Wasted Lives: Modernity and Its Outcasts*, 1st edn (Cambridge: Polity Press, 2003).
14. See Rob Nixon, *Slow Violence and the Environmentalism of the Poor* (Cambridge, MA: Harvard University Press, 2013).
15. Armiero, *Wasteocene*, 11.
16. Bruno Venditti, "Visualizing the Critical Metals in a Smartphone," *Visual Capitalist*, August 25, 2021, https://www.visualcapitalist.com/visualizing-the-critical-metals-in-a-smartphone/ (accessed March 3, 2026).
17. "Global Atlas of Environmental Justice," *EJAtlas.org*, last modified August 6, 2023, https://ejatlas.org/conflict/bayan-obo-world-biggest-rare-earths-mine-baogang-group-baotou-inner-mongolia-china (accessed March 3, 2026).
18. Dharna Noor, "Climate advocates outraged at Trump administration plans to fast-track AI sector," *The Guardian*, July 23, 2025, https://www.theguardian.com/us-news/2025/jul/23/trump-ai-climate-change (accessed March 3, 2026).
19. For more on this topic, see Peter Dauvergne, *AI in the Wild: Sustainability in the Age of Artificial Intelligence* (Cambridge, MA: MIT Press, 2020), and Kate Crawford, *Atlas of AI: Power, Politics, and the Planetary Costs of Artificial Intelligence* (New Haven: Yale University Press, 2021).
20. Xiaowei Wang, *Blockchain Chicken Farm and Other Stories of Tech in China's Countryside* (New York: Farrar, Straus and Giroux, 2020), 52.
21. Ibid., 53.
22. See Halberstam, *In a Queer Time and Place.*
23. Brian X. Chen, "How Tech Created a 'Recipe for Loneliness'," *The New York Times*, November 10, 2024, https://www.nytimes.com/2024/11/10/technology/personaltech/technology-loneliness.html (accessed March 3, 2026).
24. In May 2023, then U.S. Surgeon-General Vivek Murthy published an official advisory on "Our Epidemic of Loneliness and Isolation." As of August 2025, it can still be found through the U.S. Department of Health and Human Services at https://hhs.gov/sites/default/files/surgeon-general-social-connection-advisory.pdf (accessed March 3, 2026).
25. Haraway, "Making Kin: An Interview with Donna Haraway."
26. Department for Environment, "Lobsters, octopus and crabs recognised as sentient beings," *GOV.UK*, November 19, 2021, https://www.gov.uk/government/news/lobsters-A-and-crabs-recognised-as-sentient-beings#:~:text=Amendment%20to%20Animal%20Welfare%20(Sentience,cephalopod%20molluscs%20as%20sentient%20beings (accessed March 3, 2026).
27. Ibid.
28. George Bernard Shaw's famous heroine in his 1912 play *Pygmalion*, cinematically adapted into the hit 1964 movie *My Fair Lady.*
29. The conclusion of the study states: "Among the possible explanations for the observed cetacean behavioural responses to dead conspecifics are strong attachment resulting in a difficulty of 'letting go'—possibly related to grieving—or perhaps individuals

failing to recognise or accept that an offspring or companion has died. Our current understanding is challenged by small sample size, incomplete descriptions, and lack of information on the physiology and neural processes underpinning the observed behaviour." Giovanni Bearzi et al., "Whale and dolphin behavioral responses to dead conspecifics," *Zoology* 128 (2018): 1. https://doi.org/10.1016/j.zool.2018.05.003

30. Ashley P. Taylor, "Why Do Parrots Talk?" *Audobon*, August 6, 2015, https:/audubon.org/news/why-do-parrots-talk#:~:text=Parrots%20are%20vocal%20learners%2C%20meaning,until%20his%20death%20in%202007 (accessed March 3, 2026).
31. Irene Pepperberg, quoted in Taylor, "Why Do Parrots Talk?"
32. Ibid.
33. Emily M. Bender et al., "On the Dangers of Stochastic Parrots: Can Language Models Be Too Big?" FACCT, Proceedings of the 2021 ACM Conference on Fairness, Accountability, and Transparency, March 3–10, 2021, 617. https://doi.org/10.1145/3442188.3445922
34. Dylan Freedman, "The Day ChatGPT Went Cold," *The New York Times*, August 19, 2025, https://www.nytimes.com/2025/08/19/business/chatgpt-gpt-5-backlash-openai.html (accessed March 3, 2026).
35. Andrew R. Chow and Angela Haupt, "A Psychiatrist Posed As a Teen With Therapy Chatbots. The Conversations Were Alarming," *TIME*, June 12, 2025, https://time.com/7291048/ai-chatbot-therapy-kids/ (accessed March 3, 2026).
36. For more information, see Adrian F. Ward et al., "Brain Drain: The Mere Presence of One's Own Smartphone Reduces Available Cognitive Capacity," *Journal of the Association for Consumer Research* 2.2 (2017): 140–154. https://www.journals.uchicago.edu/doi/10.1086/691462
37. A co-authored study by a group of MIT researchers found that "by 2026, the electricity consumption of data centers is expected to approach 1,050 terawatt-hours (which would bump data centers up to fifth place on the global list, between Japan and Russia)." See Adam Zewe, "Explained: Generative AI's environmental impact," *MIT News*, January 17, 2025, https://news.mit.edu/2025/explained-generative-ai-environmental-impact-0117 (accessed March 3, 2026). For more information about the study, see Norman Bashir et al., "The Climate and Sustainability Implications of Generative AI," *An MIT Exploration of Generative AI*, March 27, 2024, https://doi.org/10.21428/e4baedd9.9070dfe7 (accessed March 3, 2026).
38. A Stanford study shows that data centers "are reshaping local grids, plumbing, and politics. […] Masheika Allgood, an AI ethicist who consults with municipal governments, believes that so much water is evaporated for cooling that recycling can't replace what was lost. 'When you are cooling a data center – a significant amount of the water is evaporated during the cooling process … Even if you don't evaporate all the water, you have to treat it. You can't recycle that …'." Felicity Barringer, "Thirsty for power and water, AI-crunching data centers sprout across the West," *& the West*, April 8, 2025, https://andthewest.stanford.edu/2025/thirsty-for-power-and-water-ai-crunching-data-centers-sprout-across-the-west/ (accessed March 3, 2026).
39. For example, "content moderators working for a company contracted by Meta in Ghana's capital Accra told *The Guardian* that they have suffered from depression, anxiety and insomnia and substance abuse. They believe this is a direct result of their work as content moderators. They also claim that the psychological support offered to help process disturbing social media content was inadequate." Rachel Hall and

Claire Wilmot, "Meta faces Ghana lawsuits over impact of extreme content on moderators," *The Guardian*, April 27, 2025, https://www.theguardian.com/technology/2025/apr/27/meta-faces-ghana-lawsuits-over-impact-of-extreme-content-on-moderators (accessed March 3, 2026).

40. Emily M. Bender and Alex Hanna, "What's Really Artificial About AI is AI Itself," interview by Sarah Jaffe, *In These Times*, May 29, 2025, https://inthesetimes.com/article/big-tech-the-ai-con-book-silicon-valley (accessed March 3, 2026).
41. Ibid.
42. Kylie Robinson, "OpenAI Is Poised to Become the Most Valuable Startup Ever. Should It Be?" *Wired*, August 19, 2025, https://www.wired.com/story/openai-valuation-500-billion-skepticism/#:~:text=OpenAI%20Is%20Poised%20to%20Become,Italiano%20%E6%97%A5%E6%9C%AC%E8%AA%9E%20%E7%B9%81%E9%AB%94%E4%B8%AD%E6%96%87 (accessed March 3, 2026).
43. Bender and Hanna, "What's Really Artificial about AI."
44. "The intuition behind diffusion models is inspired by physics, treating pixels like the molecules of a drop of ink spreading out in a glass of water over time. Much like how the random movement of the ink molecules will eventually lead to their even dispersal in the glass, the random introduction of noise into an image will eventually result in what looks like TV static. By modeling that diffusion process, then somehow learning to *reverse* it, an artificial intelligence model can generate new images by simply 'denoising' samples of random noise.

    Diffusion models are most prominently associated with image generation and other image processing tasks such as inpainting and super-resolution, but their applications extend to other domains including audio generation, drug design and molecule generation." Dave Bergmann and Cole Stryker, "What are diffusion models?" *IBM*, https://www.ibm.com/think/topics/diffusion-models#:~:text=Diffusion%20models%20are%20generative%20models,to%20generate%20high%2Dquality%20images (accessed March 3, 2026).
45. Louisa Clarence-Smith, "Why the tech gods of Silicon Valley have turned to Christianity," *The Times*, August 26, 2025, https://www.thetimes.com/us/business/article/why-tech-gods-silicon-valley-turned-christianity-72gnb9cpg (accessed March 3, 2026).
46. Jack Duffe (@jackscarab.bsky.social), "What trillion-dollar problem is AI trying to solve?" *Bluesky*, December 27, 2024, https://bsky.app/profile/jackscarab.bsky.social/post/3lecjwbc5ts2o (accessed May 17, 2026).
47. For example, the extractive intentions described here stand somewhat in contrast to China's state policy of implementing A.I. strategically within existing technologies that can have immediate applications to the general public.
48. Chet Van Duzer, "Horror vacui," *Stanford University Libraries*, https://exhibits.stanford.edu/blrcc/feature/horror-vacui (accessed March 3, 2026).
49. Bender and Hanna, "What's Really Artificial about AI."
50. Haraway, "Deanimations: Maps and Portraits of Life Itself," in Caroline A. Jones and Peter Galison (eds), *Picturing Science, Producing Art* (New York: Routledge, 1998), 181.
51. Said, *Orientalism*, 3.
52. Ibid.
53. It is no wonder that tech folks have been developing competing A.I. models before "someone else does it" by using the analogy of the Manhattan Project.

54. Meher Ahmad, Jessica Grose, and Tressie McMillan Cottom, "What A.I. Really Means for Learning," *The New York Times*, August 12, 2025, https://www.nytimes.com/
2025/08/12/opinion/ai-college-classrooms-chatgpt.html (accessed March 3, 2026).
55. Curio Home Page, https://heycurio.com/ (accessed March 3, 2026).
56. Curio FAQ, https://heycurio.com/faq (accessed March 3, 2026).
57. Amanda Hess, "They're Stuffed Animals. They're Also A.I. Chatbots," *The New York Times*, August 15, 2025, https://www.nytimes.com/2025/08/15/arts/ai-toys-curio-grem.html (accessed March 3, 2026).
58. Ibid.
59. See Timothy Morton, *Hyperobjects: Philosophy and Ecology after the End of the World* (Minneapolis, MN: University of Minnesota Press, 2013).
60. Anne Friedberg, "Virilio's Screen: The Work of Metaphor in the Age of Technological Convergence," *Journal of Visual Culture* 3.2 (2004): 183.
61. Tiziana Terranova, "Free Labor: Producing Culture for the Digital Economy," *Social Text* 18.2 (2000): 34, https://web.mit.edu/schock/www/docs/18.2terranova.pdf (accessed March 4, 2026).

Chapter 4: Colonial-Cybercene Desire Machines

1. Citing de Angelis (2017) and Bollier and Helfrich (2012), Armiero says "communing relationships […] produce well-being through care and inclusion. These commoning bonds are contrasted against 'wasting relationships [that] produce profit from exploitation and othering." *Wasteocene*, 3.
2. van Dooren et al., "Multispecies Studies," 6.
3. Marcos Martínez, "Burned to death because of a rumour on WhatsApp," *BBC*, November 11, 2018, https://www.bbc.com/news/world-latin-america-46145986 (accessed March 4, 2026).
4. Chloe Xiang, "'He Would Still Be Here': Man Dies by Suicide After Talking with AI Chatbot, Widow Says," *Vice*, March 30, 2025, https://www.vice.com/en/article/man-dies-by-suicide-after-talking-with-ai-chatbot-widow-says/ (accessed March 4, 2026).
5. Associated Press, "Florida teenagers who filmed drowning death will not be charged over failure to help," *The Guardian*, July 22, 2017, https://www.theguardian.com/us-news/2017/jul/22/florida-teenagers-who-filmed-drowning-death-will-not-be-charged-over-failure-to-help (accessed March 4, 2026).
6. Said, *Orientalism*, 72, my emphasis.
7. "Paranoia and Delusional Disorders," *Mental Health America*, https://mhanational.org/conditions/paranoia-and-delusional-disorders/#:~:text=Some%20beliefs%20and%20behaviors%20of,advantage%20of%2C%20trouble%20relaxing%2C%20or (accessed March 4, 2026).
8. "sycophancy," *Oxford Learner's Dictionaries*, https://www.oxfordlearnersdictionaries.com/definition/english/sycophancy#google_vignette (accessed March 4, 2026).
9. Mark Wilson, "Smartphones Are Killing The Planet Faster Than Anyone Expected," *Fast Company*, March 27, 2018, https://www.fastcompany.com/90165365/smartphones-are-wrecking-the-planet-faster-than-anyone-expected (accessed March 4, 2026).

10. Fanon, *Black Skin, White Masks.*
11. Janet Staiger, *Interpreting Films: Studies in the Historical Reception of American Cinema* (Princeton, NJ: Princeton University Press, 1992), 9.
12. See André Bazin, "The Ontology of the Photographic Image," *What Is Cinema? Volume I*, trans. Hugh Gray (Berkeley, CA: University of California Press, 2004), 9–16.
13. Ibid., 9.
14. Meher Ahmad et al., "What A.I. Really Means for Learning."
15. Elizabeth J. Miller et al., "AI Hyperrealism: Why AI Faces Are Perceived as More Real Than Human Ones," *Psychological Science* 34.12 (2023): 1–14, 1.
16. Ibid., 11
17. Comolli, *Cinema Against Spectacle*, 3.
18. Reality TV is one such very obvious kind of play between believing and disbelieving.
19. Boëtsch and Blanchard, "From Cabinets of Curiosity to the 'Hottentot Venus'," 192.
20. "La questione Calderoli-Kyenge," *Il Post*, July 15, 2013, http://www.ilpost.it/2013/07/15/audio-calderoli-kyenge-orango/, translation my own (accessed March 4, 2026).
21. Bhabha, "Of Mimicry and Man," 132.
22. Ibid., 129.
23. Piers Douglas Lionel Howe et al., "Self-certification: A novel method for increasing sharing discernment on social media," *PLOS One* 19.6 (2024): 1–17. https://doi.org/10.1371/journal.pone.0303025
24. Sarah Becklake and Elisa Wynne-Hughes, "The touristic transformation of postcolonial states: human zoos, global tourism competition, and the emergence of zoo-managing states," *Tourism Geographies* 26.5 (2004): 790. https://doi.org/10.1080/14616688.2023.2231410
25. Ibid.
26. Arundhati Roy, "Approaching Gridlock: Free Speech and Failing Democracy," *The Wire*, March 24, 2023, https://thewire.in/politics/approaching-gridlock-free-speech-and-failing-democracy (accessed March 4, 2026).
27. Devra Waldman and Asher Ghertner, "The enclaved body: Crises of personhood and the embodied geographies of urban gating," *Progress in Human Geography* 47.2 (2023): 288. https://doi.org/10.1177/03091325221150792
28. Alessandro Leogrande, "Viaggio nella zona grigia che ha portato all'omicidio di Fermo," *Internazionale*, March 18, 2017, https://www.internazionale.it/reportage/alessandro-leogrande/2017/03/18/emmanuel-chidi-nnamdi-fermo-omicidio, translation my own (accessed March 4, 2026).
29. Russell Lee, "Negro drinking at 'Colored' water cooler in streetcar terminal, Oklahoma City, Oklahoma," July 1939, *Library of Congress*, http://hdl.loc.gov/loc.pnp/fsa.8a26761 (accessed March 4, 2026).

Chapter 5: Running Away

1. Patrick Greenfield, "The elephant whisperer: one Thai woman's lifelong quest to protect a rescued herd," *The Guardian*, August 8, 2025, https://www.theguardian.com/environment/2025/aug/08/animal-welfare-asian-elephants-thailand-conservation-aoe (accessed March 5, 2026).
2. Ibid.

3. Armiero, *Wasteocene*, 3.
4. van Dooren et al., "Multispecies Studies."
5. Armiero, *Wasteocene*, 3.
6. Joshua Dhivyan Gnanasekhar, "Mutualism (biology)," *EBSCO Research Starters*, 2024, https://www.ebsco.com/research-starters/biology/mutualism-biology (accessed March 5, 2026).
7. "The Microbiome," *Harvard: The Nutrition Source*, https://www.hsph.harvard.edu/nutritionsource/microbiome/ (accessed March 5, 2026).
8. Haraway, *Staying with the Trouble*, 101, 57.
9. Ibid., 33.
10. For more on ethical relationality, see Dwayne Donald's doctoral dissertation, "Forest, Curriculum, and Indigenous Métissage: Imagining Decolonization of Aboriginal-Canadian Relations in Educational Contexts," University of Alberta, 2009, https://galileo.org/pl/wp-content/uploads/004_Donald.pdf (accessed March 5, 2026).
11. Shawn Wilson, *Research Is Ceremony: Indigenous Research Methods* (Halifax: Fernwood Publishing, 2008), 38.
12. F. Richard Georgi, "Postcolonial and decolonial differences: Do prefixes matter?" *Virtual Encyclopaedia – Rewriting Peace and Conflict*, July 4, 2024, https://rewriting-peaceand
conflict.net/postcolonial-and-decolonial-differences/ (accessed March 5, 2026).
13. Haraway, *Staying with the Trouble*, 4.
14. Guattari, *Soft Subversions: Texts and Interviews 1977–1985*, ed. Sylvère Lotringer, trans. Chet Wiener and Emily Wittman (New York: Semiotext[e], 1996), 37, my emphasis.
15. "Orangutan," Karima 2G, *2G*, 2014.
16. Marjorie M. Spiegel, *The Dreaded Comparison: Human and Animal* (New York, NY: Mirror Books, 1996).
17. John Cannon, "Orangutan culture in focus in 'Person of the Forest': Q&A with researchers Cheryl Knot and Robert Rodriguez Suro," *Mongabay*, February 23, 2018, https://news.mongabay.com/2018/02/orangutan-culture-in-focus-in-person-of-the-forest-qa-with-researchers-cheryl-knott-and-robert-rodriguez-suro/#:~:text=Scientists%20argue%20that%20the%20existence%20of%20orangutan,creep%20into%20the%20habits%20of%20other%20orangutans (accessed March 5, 2026).
18. Michael Krützen et al., "Culture and Geographic Variation in Orangutan Behavior," *Current Biology* 21 (2011): 1808–12.
19. van Dooren et al., "Multispecies Studies," 6.
20. The uploaded video, accessible initially on YouTube, has since been "made private" by the anonymous citer, thus making it in accessible to casual visitors of the page. Its appearance and subsequent disappearance are yet another illustration of the ephemeral nature of Cybercene citations.
21. Cyrus Grace Dunham, "The 'Terrorist Drag' of Vaginal Davis," *The New Yorker*, December 12, 2015, https://www.newyorker.com/culture/culture-desk/terrorist-drag-vaginal-davis (accessed March 5, 2026).
22. Ibid.

23. "No One Leaves Delilah – A (W)rap on Riots," Vaginal Davis artist talk, ACUD Berlin, January 27, 2018, https://www.youtube.com/watch?v=BztgEUa5Y74&t=15s (accessed March 5, 2026).
24. José Esteban Muñoz, *Disidentifications: Queers of Color and the Performance of Politics* (Minneapolis, MN: University of Minnesota Press, 1999), 95.
25. Ibid., 97.
26. Ibid., 99.
27. Ibid., 100.
28. See Bhabha, *The Location of Culture*.
29. Original link last accessed by author in July 2025, since made private: https://www.youtube.com/watch?v=Pbks6cZyBgA.
30. Vaginal Davis and Fertile La Toyah Jackson, *That Fertile Feeling* part 1, YouTube, May 20, 2008, 2:34-2:52, https://www.youtube.com/watch?v=Pbks6cZyBgA.
31. Davis and Jackson, *That Fertile Feeling* part 2, YouTube, May 20, 2008, 2:59-3:03, https://www.youtube.com/watch?v=-6Ej9inqGg0.
32. Ibid., 3:07-3:09.
33. Ibid., 3:15-3:17.
34. Ibid., 3:28-3:40.
35. TallBear and Angela Willey, "Critical Relationality: Queer, Indigenous, and Multispecies Belonging Beyond Settler Sex & Nature," *Imaginations* 10.1 (2019): 8.
36. Jack Halberstam, *Wild Things: The Disorder of Desire* (Durham, NC: Duke University Press, 2020), x.
37. Nicole Seymour, *Strange Natures: Futurity, Empathy, and the Queer Ecological Imagination* (Champaign, IL: University of Illinois Press, 2013), 184.
38. Ibid.

Chapter 6: Our Multispecies Futures

1. Little Bear, "Jagged Worldviews Colliding," 77.
2. Manu Aluli Meyer, "Acultural Assumptions of Empiricism: A Native Hawaiian Critique," *Canadian Journal of Native Education* 25.2 (2001): 192.
3. Angela Giuffrida, "Satnam lost his arm and was allegedly left to die on the roadside. This is the horror of exploitation on Italian farms," *The Guardian*, July 20, 2024, https://www.theguardian.com/media/article/2024/jul/20/satnam-lost-his-arm-and-was-allegedly-left-to-die-on-the-roadside-this-is-the-horror-of-exploitation-on-italian-farms (accessed March 5, 2026).
4. Redazione Roma, "Satnam Singh, morto dopo essere stato scaricato ferito davanti casa," *La Repubblica*, July 20, 2024, https://roma.repubblica.it/cronaca/2024/06/20/news/morte_satnam_singh_latina_lovato-423262714/ (accessed March 5, 2026).
5. Haraway, "Deanimations: Maps and Portraits of Life Itself," 181.
6. Todd Presner, David Shepard, and Yoh Kawano, *Hypercities: Thick Mapping in the Digital Humanities* (Cambridge, MA: Harvard University Press, 2014), 7, my emphasis.
7. van Dooren et al., "Multispecies Studies," 6.
8. Paolo Gruppuso, "In-between Solidity and Fluidity: The Reclaimed Marshlands of Agro Pontino," *Theory, Culture & Society* 39.2 (2021): 53. https://doi.org/10.1177/02632764211038669

9. *Ecotone* (from the Greek "oikos" – home and "tonos" – tension) is a transitional zone between two ecosystems: "Two houses in tension: an ecotone is a border zone, where ecological systems meet and mingle, sometimes forming a new and different community. Ecotones occur at edges and physical boundaries, where fresh water meets salt water and water meets land, where tides roll up and down coasts, where woodlands become pastures and the fir trees of taiga forests give way to the lichen and grass of tundra." "Ecotone Explained," *Ecological Society of America*, https://esa.org/esablog/about/ecotone-explained/#:~:text=Credit%2C%20Jesse%20Allen%2C%20Robert%20Simmon,as%20a%20field%20of%20study (accessed March 5, 2026).
10. Gruppuso, "In-between Solidity and Fluidity," 68.
11. Mario De Mandato, *La primitivita' dell'abitare umano. Studi e ricerche* (Torino: Fratelli Bocca, 1933), 71–72.
12. Caleb Leeke, "Triturus carnifex, Triturus cristatus, Triturus dobrogicus, Triturus karelinii," *Caudata Culture*, December 2002, https://www.caudata.org/cc/species/Triturus/Triturus_sp.shtml (accessed March 5, 2026).
13. Ibid.
14. Indeed, funding for this capture-and-control project in the Agro Pontino was reduced when, after the invasion of Ethiopia in 1935–1937, it needed to be diverted to colonial projects external to Italian borders.
15. Gruppuso, "In-between Solidity and Fluidity," 60.
16. Halberstam, *Wild Things*, 15.
17. Armiero, Roberta Biasillo, and Wilko Graf von Hardenberg, *Mussolini's Nature: An Environmentalist History of Italian Fascism*, trans. James Sievert (Cambridge, MA: MIT Press, 2022), 24.
18. Ibid., 46. Mussolini also famously "tamed" a female lion named *Italia* in 1923, given to him by a circus owner. This story of multispecies domination configures a marvelous confabulation of multispecies, gender and national/patriotic discourse.
19. Margherita G. Sarfatti, *The Life of Benito Mussolini*, trans. Frederic White (New York: Frederick A. Stokes Company, 1927), 60.
20. For an excellent and detailed cultural analysis of Mussolini's environmental interventions, see Armiero et al., *Mussolini's Nature.*
21. Ibid., 63.
22. A "Battle against Malaria" public service information poster. *Museo della Terra Pontina, Latina*, photograph by Vetri Nathan.
23. Martin F. Gomon and Dianne J. Bray, "Gambusia holbrooki," *Fishes of Australia*, https://fishesofaustralia.net.au/home/species/3636 (accessed March 8, 2026).
24. "Punjab contributed 31% wheat, 21% rice to country's food reserves," *Times of India*, December 30, 2022, https://timesofindia.indiatimes.com/city/chandigarh/state-contributed31-wheat-21-rice-to-countrys-food-reserves/articleshow/96610974.cms (accessed March 8, 2026).
25. Sandeep Singh and Elena Cavallone, "India to Italy: a trail of broken dreams, loss and abuse," *Context*, November 1, 2024, https://www.context.news/money-power-people/india-to-italy-a-trail-of-broken-dreams-loss-and-abuse (accessed March 9, 2026).
26. Abusaleh Sharif and Ataulla Khan, "What's the 'Shadow Value' of an Illegal Indian Migrant to the US?" *The Wire*, October 24, 2024, https://thewire.in/world/indian-dunki-migration-us-shadow-value (accessed March 9, 2026).

27. Singh and Cavallone, "India to Italy."
28. "Zitting Cisticola," *eBird*, https://ebird.org/species/zitcis1 (accessed May 17, 2026).
29. For a more detailed analysis of this case from the perspective of African postcolonial and diaspora studies, see Vetri Nathan, "Edible Nation: Imagining Somalia Through Cuisine," in Rosetta G. Caponetto, Giusy Di Filippo and and Martina Di Florio (eds), *Longing for the Future: Mal D'Afrique and Afro-Optimism in Perspectives on Somalia* (New York: Routledge, 2023). https://doi.org/10.4324/9781003283102
30. Hawa Hassan and Julia Turchen, *In Bibi's Kitchen: The Recipes and Stories of Grandmothers from the Eight African Countries that Touch the Indian Ocean* (New York: Ten Speed Press, 2020), 1. This volume won a James Beard Foundation Prize in 2022.
31. I am grateful for all the content creators who are sharing their work digitally or otherwise and celebrating Somali cuisine and culture globally.
32. Hassan, *In Bibi's Kitchen*, 1.
33. Abdullahi Kassim and Leila Kassim, "Xawaash (Somali Spice Mix) Mélange d'épices Somali حوائج – بهارات صومالية," *Xawaash*, August 7, 2011, https://xawaash.com/?p=12#:~:text=www.xawaash.com,is%20disabled%20in%20your%20browser (accessed March 9, 2026).
34. See Stacy Alaimo, *Bodily Natures: Science, Environment, and the Material Self* (Bloomington, IN: Indiana University Press, 2010).
35. Molly O'Neil, "Food Porn," *Columbia Journalism Review*, September 1, 2003, https://www.cjr.org/from_the_archives/food-writing-cookbook.php (accessed March 9, 2026).
36. Anne McBride, "Food Porn," *Gastronomica: The Journal of Critical Food Studies* 10.1 (2010).
37. Jörg H.A. Janzen and Ioan M. Lewis, "Revolt in British Somaliland," *Encyclopedia Britannica*, November 11, 2025, https://www.britannica.com/place/Somalia/Revolt-in-British-Somaliland#ref419656 (accessed March 9, 2026).
38. Warren Belasco, *Food: The Key Concepts* (Oxford: Berg, 2008), 15.
39. See Anderson, *Imagined Communities*.
40. Chef Jamal Hashi, "Thank you for having me again Twin Cities Live. Elizabeth Ries & Ben Leber Page Always fun times sharing new foods with you," *Facebook*, July 3, 2022, 0:00-0:08, https://www.facebook.com/watch/?v=1493712504376887 (accessed March 9, 2026).
41. Ibid., 0:49-1:11.
42. Ibid., 1:34-1:39.
43. Ibid., 3:18-3:20.
44. For more on "third spaces," see Bhabha, *The Location of Culture*.
45. Fozia Ismail, "Somali food love, Arawelo style," *Arawelo Eats*, https://www.araweloeats.com/ (accessed March 9, 2026).
46. Mike Rosenwald, host, "Queen Arawelo," *Retropod*, April 10, 2019, https://www.washingtonpost.com/podcasts/retropod/queen-arawelo/?tid=aud_rsslink (accessed March 9, 2026).
47. Ismail, "The recipe that reminds me of home," interview by Killian Fox and Holly O'Neill, *The Guardian*, July 14, 2019, https://www.theguardian.com/food/2019/jul/14/recipe-that-reminds-me-of-home-souvlaki-somali-bean-stew (accessed March 9, 2026).

48. James Hansen, "Celebrated Somali Supper Club Makes London Debut," *Eater*, October 3, 2018, https://london.eater.com/2018/10/3/17931912/arawelo-eats-somali-supper-club-london-asma-khan-chefs-table-darjeeling-express (accessed March 9, 2026).
49. Tasty, "Mindy Kaling and Hawa Hassan Make Spicy Somali Pasta," *YouTube*, September 8, 2021, https://www.youtube.com/watch?v=PMqj2BSW0Yk (accessed March 9, 2026).
50. Isma'il Kushkush, "After Barren Years in Somalia, Signs of Growth by the Bunch," *The New York Times*, December 13, 2014, https://www.nytimes.com/2014/12/14/world/after-barren-years-in-somalia-signs-of-growth-in-bananas.html (accessed March 9, 2026).
51. Hassan, *In Bibi's Kitchen*, 2.
52. Ibid., 1.
53. Merrill G. Burlingame, "Buffalo in Trade and Commerce," *North Dakota Historical Quarterly* 3.4 (1929): 290.
54. For more information on this initiative, see https://windriverbuffalo.org/ (accessed March 10, 2026).
55. While the specific case studied in this book is of the Native American Indian tribal communities (even more specifically, the Eastern Shoshone and Northern Arapaho tribes at Wind River Reservation in Wyoming), I will use the word *Indigenous* here in order to create bridges of epistemic solidarity between the specific experience and perspectives of the Wind River community elders and ecocultural leaders and other such global cultures and communities. An important caveat to this kind of research is to not fall into essentializing all Indigenous individuals as ecologically noble savages. It is vital to underline how these communities are ultimately made up of individuals who present all the facets found in other groups or communities. Also, it is important to remember that all these individuals are no longer located in a timeless place outside history (or within deep time), beyond the realms of geopolitics and biopolitics, but inserted as spectator/consumers within the all-encompassing Colonial-Cybercene era.
56. Vicente Diaz, "Oceania in the Plains: The Politics and Analytics of Transindigenous Resurgence in Chuukese Voyaging of Dakota Lands, Waters, and Skies in Mní Sóta Makhóčhe," *Pacific Studies* 42.1/2 (2019): 2.
57. For more information, see Erika Zavaleta, "The Emergence of Waterfowl Conservation Among Yup'ik Hunters in the Yukon-Kuskokwim Delta, Alaska," *Human Ecology* 27 (1999): 231–266.
58. Raymond Hames, "The Ecologically Noble Savage Debate," *Annual Review of Anthropology* 36 (2007): 186. https://doi.org/10.1146/annurev.anthro.35.081705.123321
59. Ward Churchill, "Fantasies of the Master Race: Categories of Stereotyping of American Indians in Film," in Robert Stam and Toby Miller (eds), *Film and Theory: An Anthology* (Oxford: Blackwell, 2000), 703.
60. Philip J. Deloria, *Playing Indian* (New Haven: Yale University Press, 1999), 3.
61. Olivia Amitay, "How the UN Biodiversity Conference Impacts Indigenous Communities," *PBS*, April 5, 2023, https://www.pbs.org/wnet/peril-and-promise/2023/04/how-the-un-biodiversity-conference-impacts-Indigenous-communities/#:~:text=Indigenous%20peoples'%20lands%20host%2080,5%25%20of%20the%20world's%20population (accessed March 10, 2026).

62. My heartfelt thanks go to a wide range of dedicated buffalo experts from all over the American West for sharing their expertise with me for this study: Susan Ricci, Former Associate Intertribal Bison Council; Ken Zontek, Author and Faculty Member, Yakima Valley Community College; Chad Kremer, Custer State Park Bison Herd Manager; Richard Baldes; Jason Baldes, Executive Office, Board President and Eastern Shoshone Tribe Buffalo Manager; Wesley Martel, Senior Wind River Conservation Associate; Ina Tillman (Grandma Ina), Wind River community Elder; Shana Drimal, Senior Wildlife Conversation Associate, Greater Yellowstone Coalition; Charles Drimal, Deputy Director of Conservation, Greater Yellowstone Coalition; Shannon Dennison, Director, Denver Mountain Parks; and Ryan Philian, Park Ranger Supervisor, Denver Mountain Parks.
63. Zontek, *Buffalo Nation: American Indian Efforts to Restore the Bison* (Lincoln, NE: University of Nebraska Press, 2007), xiv.
64. Ibid., 17.
65. Burlingame, "Buffalo in Trade and Commerce," 290.
66. Zontek, *Buffalo Nation*, 18–19.
67. You can view the intricate and complex story in its entirety here: "The Lakota Emergence Story," *National Park Service*, September 11, 2024, https://www.nps.gov/wica/learn/historyculture/the-lakota-emergence-story.htm (accessed March 10, 2026).
68. Ibid.
69. Interview with Jason Baldes by Vetri Nathan.
70. Zontek, *Buffalo Nation*, 34
71. Elizabeth U. Mangan, "Yellowstone, the First National Park," *Library of Congress*, https://www.loc.gov/collections/national-parks-maps/articles-and-essays/yellowstone-the-first-national-park/ (accessed March 10, 2026).
72. Peter Nabokov and Lawrence Loendorf, *Restoring a Presence: American Indians and Yellowstone National Park* (Norman, OK: University of Oklahoma Press, 2004), 3.
73. Interview with Shana Drimal by Vetri Nathan.
74. Ibid.
75. Vine Deloria, *Spirit & Reason: The Vine Deloria, Jr., Reader*, ed. Barbara Deloria, Kristen Foehner, and Sam Scinta (Golden, CO: Fulcrum Publishing, 1999), 34.
76. Taylar Dawn Stagner, "WRTBI Marks Indigenous Peoples Day with Third Annual Buffalo Bash Celebration," October 20, 2025, *Wind River Tribal Buffalo Initiative*, https://windriverbuffalo.org/third-annual-buffalo-bash/ (accessed March 10, 2026).
77. Hokowhitu, "The Emperor's 'New Materialisms'," 140.
78. Interview with Wesley Martel by Vetri Nathan.
79. Interview with Richard Baldes by Vetri Nathan.
80. "History," *International Buffalo Council*, https://www.itbcbuffalonation.org/itbchistory (accessed March 10, 2026).
81. "About Wind River Tribal Buffalo Initiative," *Wind River Tribal Buffalo Initiative*, https://windriverbuffalo.org/explore/initiative/ (accessed March 10, 2026).
82. Ibid.
83. See Vine Deloria and Clifford M. Lytle, *The Nations Within: The Past and Future of American Indian Sovereignty* (New York: Pantheon Books, 1984); Andersen, *"Métis": Race, Recognition, and the Struggle for Indigenous Peoplehood* (Vancouver: University of British Columbia Press, 2014); Glen Sean Coulthard, *Red Skin, White Masks: Rejecting The Colonial Politics of Recognition* (Minneapolis, MN: University of Minnesota Press,

2014); J. Kēhaulani Kauanui, *Paradoxes of Hawaiian Sovereignty: Land, Sex and the Colonial Politics of State Nationalism* (Durham, NC: Duke University Press, 2018).

84. Mike Koshmrl, "Buffalo (almost) officially wildlife on some 2M new acres of Wyoming, a step toward roaming free," *WyoFile*, August 15, 2025, https://wyofile.com/buffalo-almost-officially-wildlife-on-some-2m-new-acres-of-wyoming-a-step-toward-roaming-free/ (accessed March 10, 2026).
85. Stephen Lezak, "Wild horses, buffalo and the politics of belonging," *High Country News*, December 1, 2021, https://www.hcn.org/issues/53.12/indigenous-affairs-wildlife-wild-horses-buffalo-and-the-politics-of-belonging (accessed March 10, 2026).
86. Madison Dapcevich, "The Key to Yellowstone's Ecological Health? Bison Poop," *Outside*, August 29, 2025, https://www.outsideonline.com/outdoor-adventure/environment/
bison-poop-yellowstone-park/ (accessed March 10, 2026).
87. Chris Geremia, E. William Hamilton, and Jerod A. Merkle, "Yellowstone's free-moving large bison herds provide a glimpse of their past ecosystem function," *Science* 389.6763 (2025): 904. https://www.science.org/doi/10.1126/science.adu0703
88. "Bison Bellows: A Healthy Prairie Relies on Bison Poop," *National Park Service*, https://www.nps.gov/articles/bison-bellows-10-6-16.htm (accessed March 10, 2026).
89. Ibid.
90. Interview with Jason Baldes by Vetri Nathan.
91. Ibid.
92. Interview with Ina Tillman by Vetri Nathan.
93. Interview with Richard Baldes by Vetri Nathan.
94. Lezak, "Wild horses, buffalo and the politics of belonging."

Conclusion: A Humanimal Guide to Staying with the Trouble and the Joy

1. Naama Harel, *Kafka's Zoopoetics: Beyond the Human–Animal Barrier* (Ann Arbor, MI: University of Michigan Press, 2020), 3.
2. Ibid., citing Christopher F. Chabris and Daniel J. Simons, *The Invisible Gorilla: And Other Ways Our Intuitions Deceive Us* (New York: Crown, 2010).
3. The first documented use of this phrase is by Brian Kernighan, a Canadian computer scientist who worked at Bell Labs and contributed to the creation of several programming languages. Interestingly for this study, Kernighan says that he was inspired to use this phrase when he saw a cartoon of a chick emerging out of an egg saying "Hello World."
4. Kyle Whyte, "Against Crisis Epistemology," in Hokowhitu et al. (eds), *Routledge Handbook of Critical Indigenous Studies*, 52–64.
5. van Dooren et al., "Multispecies Studies," 6.
6. Whyte, "Against Crisis Epistemology," 60.
7. Steger, "Searching for *Satya*," 343, quoting Gandhi, *Autobiography*, 6.
8. Peterson, *Bestial Traces*, 8.
9. Diaz, "Oceania in the Plains," 2.
10. Marcone, "The Stone Guests."

11. Adam J. Fix, Hugh Burnam, and Ray Gutteriez, "Toward Interspecies Thinking as a Collaborative Concept: Autoethnographies at the Intersection of Traditional Ecological Knowledge and Animal Studies," *Humanimalia* 10.2 (2019): 132.
12. van Dooren et al., "Multispecies Studies," 6.
13. Robin Buller, "'There's nothing more critical': California makes schools teach kids to spot fake news," *The Guardian*, December 5, 2023, https://www.theguardian.com/education/2023/dec/05/california-media-literacy-class-schools-misinformation (accessed March 11, 2026).
14. Social media ban," *UNICEF Australia*, https://www.unicef.org.au/unicef-youth/staying-safe-online/social-media-ban-explainer (accessed March 11, 2026).
15. Jacek Kotus and Qihang Qiu, "Multi-species placemaking or urban attraction? The human perception of Austin's bat colony," *Cities* 171 (2026).
16. Ibid.
17. Maria Bargh, "Diverse Indigenous environmental identities: Māori resource management innovations," in Hokowhitu et al. (eds), *Routledge Handbook of Critical Indigenous Studies*, 427
18. Miriam Tola, "Intersectionality and the Environmental Humanities: Notes on Elisions and Encounters," *AboutGender* 11.22 (2022): 4.
19. van Dooren et al., "Multispecies Studies," 3.

Bibliography

"About Wind River Tribal Buffalo Initiative," *Wind River Tribal Buffalo Initiative*, https://windriverbuffalo.org/explore/initiative/

"Anthropocene," *Wikipedia: The Free Encyclopedia*, last modified November 3, 2025, https://en.wikipedia.org/wiki/Anthropocene

"axiology," *Merriam Webster*, https://www.merriam-webster.com/dictionary/axiology

"Bison Bellows: A Healthy Prairie Relies on Bison Poop," *National Park Service*, https://www.nps.gov/articles/bison-bellows-10-6-16.htm

"Chumash Painted Cave," *CyArk*, 2025, https://www.cyark.org/projects/chumash-painted-cave/in-depth

"Digital Around the World," *Datareportal*, https://datareportal.com/global-digital-overview

"Global Atlas of Environmental Justice," *EJAtlas.org*, https://www.ejatlas.org/conflict/bayan-obo-world-biggest-rare-earths-mine-baogang-group-baotou-inner-mongolia-china

"History," *International Buffalo Council*, https://itbcbuffalonation.org/history-and-mission/

"humanities," *Encyclopedia Britannica*, October 12, 2025, https://www.britannica.com/topic/humanities

"La questione Calderoli-Kyenge," *Il Post*, July 15, 2013, http://www.ilpost.it/2013/07/15/audio-calderoli-kyenge-orango/

"Orangutan," Karima 2G, *2G*, 2014.

"Paranoia and Delusional Disorders," *Mental Health America*, https://mhanational.org/conditions/paranoia-and-delusional-disorders/#:~:text=Some%20beliefs%20and%20behaviors%20of,advantage%20of%2C%20trouble%20relaxing%2C%20or

"Punjab contributed 31% wheat, 21% rice to country's food reserves," *Times of India*, December 30, 2022, https://timesofindia.indiatimes.com/city/chandigarh/state-contributed31-wheat-21-rice-to-countrys-food-reserves/articleshow/96610974.cms

"Santa Rosa Island Cloud Forest Restoration," *National Park Service*, https://www.nps.gov/chis/learn/nature/cloud-forest.htm

"sycophancy," *Oxford Learner's Dictionaries*, https://www.oxfordlearnersdictionaries.com/definition/english/sycophancy#google_vignette

"The Lakota Emergence Story," *National Park Service*, September 11, 2024, https://www.nps.gov/wica/learn/historyculture/the-lakota-emergence-story.htm

"The Microbiome," *Harvard: The Nutrition Source*, https://www.hsph.harvard.edu/nutritionsource/microbiome/

"What is Ecology?" *The Ecological Society of America*, https://esa.org/about/what-does-ecology-have-to-do-with-me/

"Zitting Cisticola," *eBird*, https://ebird.org/species/zitcis1

Ahmad, Meher, Jessica Grose, and Tressie McMillan Cottom, "What A.I. Really Means for Learning," *The New York Times*, August 12, 2025, https://www.nytimes.com/2025/08/12/opinion/ai-college-classrooms-chatgpt.html

Ahmad, Omair, "Opinion: Deep scars to South Asia's ecosystems are a legacy of Partition," *Dialogue Earth*, August 14, 2022, https://dialogue.earth/en/nature/opinion-deep-scars-to-south-asia-ecosystems-legacy-of-partition/

Alaimo, Stacy, *Bodily Natures: Science, Environment, and the Material Self* (Bloomington, IN: Indiana University Press, 2010).

Amitay, Olivia, "How the UN Biodiversity Conference Impacts Indigenous Communities," *PBS*, April 5, 2023, https://www.pbs.org/wnet/peril-and-promise/2023/04/how-the-un-biodiversity-conference-impacts-Indigenous-communities/#:~:text=Indigenous%20peoples'%20lands%20host%2080,5%25%20of%20the%20world's%20population.

Andersen, Chris, *"Métis": Race, Recognition, and the Struggle for Indigenous Peoplehood* (Vancouver: University of British Columbia Press, 2014).

Anderson, Benedict, *Imagined Communities: Reflections on the Origin and Spread of Nationalism*, revised ed. (London: Verso, 2006).

Armiero, Marco, Roberta Biasillo, and Wilko Graf von Hardenberg, *Mussolini's Nature: An Environmentalist History of Italian Fascism*, trans. James Sievert (Cambridge, MA: MIT Press, 2022).

Armiero, Marco, *Wasteocene: Stories from the Global Dump* (Cambridge: Cambridge University Press, 2021).

Asrar, Shakeeb and Omran Abazid, "How were the India-Pakistan partition borders drawn?" *Al Jazeera*, August 17, 2017, https://www.aljazeera.com/news/2017/8/17/how-were-the-india-pakistan-partition-borders-drawn

Associated Press, "Florida teenagers who filmed drowning death will not be charged over failure to help," *The Guardian*, July 22, 2017, https://www.theguardian.com/us-news/2017/jul/22/florida-teenagers-who-filmed-drowning-death-will-not-be-charged-over-failure-to-help

Barad, Karen, *Meeting the Universe Halfway: Quantum Physics and the Entanglement of Matter and Meaning* (Durham, NC: Duke University Press, 2007).

Barringer, Felicity, "Thirsty for power and water, AI-crunching data centers sprout across the West," *& the West*, April 8, 2025, https://andthewest.stanford.edu/2025/thirsty-for-power-and-water-ai-crunching-data-centers-sprout-across-the-west/

Bauman, Zygmunt, *Wasted Lives: Modernity and Its Outcasts*, 1st edn (Cambridge: Polity Press, 2003).

Bazin, André, "The Ontology of the Photographic Image," *What Is Cinema? Volume I*, trans. Hugh Gray (Berkeley, CA: University of California Press, 2004), 9–16.

Bearzi, Giovanni et al., "Whale and dolphin behavioral responses to dead conspecifics," *Zoology* 128 (2018): 1–15. https://doi.org/10.1016/j.zool.2018.05.003

Becklake, Sarah and Elisa Wynne-Hughes, "The touristic transformation of postcolonial states: human zoos, global tourism competition, and the emergence of zoo-managing states," *Tourism Geographies* 26.5 (2004): 778–95. https://doi.org/10.1080/14616688.2023.2231410

Beer, Haley N. et al., "The Evolution of Zoos as Conservation Institutions: A Summary of the Transition from Menageries to Zoological Gardens and Parallel Improvement of Mammalian Welfare Management," *The Journal of Zoological and Botanical Gardens* 4.4 (2023): 648–664.

Belasco, Warren, *Food: The Key Concepts* (Oxford: Berg, 2008).

Ben Chikha, Chokri and Karel Arnaut, "Staging/caging 'otherness' in the postcolony: spectres of the human zoo," *Critical Arts* 27.6 (2013): 661–83. https://doi.org/10.1080/02560046.2013.867589

Bender, Emily M. and Alex Hanna, "What's Really Artificial About AI is AI Itself," interview by Sarah Jaffe, *In These Times*, May 29, 2025, https://inthesetimes.com/article/big-tech-the-ai-con-book-silicon-valley

Bender, Emily M. et al., "On the Dangers of Stochastic Parrots: Can Language Models Be Too Big?" FACCT, Proceedings of the 2021 ACM Conference on Fairness, Accountability, and Transparency, March 3–10, 2021, 610–23. https://doi.org/10.1145/3442188.3445922

Bennett, Jane, *Vibrant Matter: A Political Ecology of Things* (Durham, NC: Duke University Press, 2009).

Bergmann, Dave and Cole Stryker, "What are diffusion models?" *IBM*, https://www.ibm.com/think/topics/diffusion-models#:~:text=Diffusion%20models%20are%20generative%20models,to%20generate%20high%2Dquality%20images

Bhabha, Homi K., "Of Mimicry and Man: The Ambivalence of Colonial Discourse," *October* 28 (1984). https://doi.org/10.2307/778467

Bhabha, Homi K, *The Location of Culture* (New York: Routledge, 1994).

Bird Rose, Deborah, *Wild Dog Dreaming: Love and Extinction* (Charlottesville: University of Virginia Press, 2011).

Blanchard, Pascal, Gilles Boëtsch, and Nanette Jacomijn Snoep (eds), *Human Zoos: The Invention of the Savage* (Arles: Actes Sud Editions, 2011).

Boëtsch, Gilles and Pascal Blanchard, "From Cabinets of Curiosity to the 'Hottentot Venus': A Long History of Human Zoos," in Nicolas Bancel, Thomas David, and Dominic Thomas (eds), *The Invention of Race: Scientific and Popular Representations* (New York: Routledge, 2014).

Braidotti, Rosi, *The Posthuman* (Cambridge: Polity Press, 2013).

British Information Services, *India, 1944*, 1944, color map, 76 × 51 cm, University of Wisconsin, Milwaukee Library, American Geographical Society Library Digital Map Collection, https://collections.lib.uwm.edu/digital/collection/agdm/id/27867/

Broglio, Ron, *Surface Encounters: Thinking with Animals and Art* (Minneapolis, MN: University of Minnesota Press, 2011).

Buller, Robin, "'There's nothing more critical': California makes schools teach kids to spot fake news," *The Guardian*, December 5, 2023, https://www.theguardian.com/education/2023/dec/05/california-media-literacy-class-schools-misinformation

Burlingame, Merrill G., "Buffalo in Trade and Commerce," *North Dakota Historical Quarterly* 3.4 (1929).

Cajete, Gregory, *Native Science: Natural Laws of Interdependence* (Sante Fe: Clear Light Publishers, 2000).

Cannon, John, "Orangutan culture in focus in 'Person of the Forest': Q&A with researchers Cheryl Knot and Robert Rodriguez Suro," *Mongabay*, February 23, 2018, https://news.mongabay.com/2018/02/orangutan-culture-in-focus-in-person-of-the-forest-qa-with-researchers-cheryl-knott-and-robert-rodriguez-suro/#:~:text=Scientists%20argue%20that%20the%20existence%20of%20orangutan,creep%20into%20the%20habits%20of%20other%20orangutans

Chawkins, Steve, "Ancient Bones Belonged to a Man—Probably," *Los Angeles Times*, September 11, 2006, https://www.latimes.com/archives/la-xpm-2006-sep-11-me-bonesside11-story.html

Chef Jamal Hashi, "Thank you for having me again Twin Cities Live. Elizabeth Ries & Ben Leber Page Always fun times sharing new foods with you," *Facebook*, July 3, 2022, https://www.facebook.com/watch/?v=1493712504376887

Chen, Brian X., "How Tech Created a 'Recipe for Loneliness'," *The New York Times*, November 10, 2024, https://www.nytimes.com/2024/11/10/technology/personaltech/technology-loneliness.html

Chow, Andrew R. and Angela Haupt, "A Psychiatrist Posed As a Teen With Therapy Chatbots. The Conversations Were Alarming," *TIME*, June 12, 2025, https://time.com/7291048/ai-chatbot-therapy-kids/

Churchill, Ward, "Fantasies of the Master Race: Categories of Stereotyping of American Indians in Film," in Robert Stam and Toby Miller (eds), *Film and Theory: An Anthology* (Oxford: Blackwell, 2000).

Clarence-Smith, Louisa, "Why the tech gods of Silicon Valley have turned to Christianity," *The Times*, August 26, 2025, https://www.thetimes.com/us/business/article/why-tech-gods-silicon-valley-turned-christianity-72gnb9cpg

Cochrane, Alasdair, *Sentientist Politics: A Theory of Global Inter-species Justice* (Oxford: Oxford University Press, 2018).

Comolli, Jean-Louis, *Cinema against Spectacle: Technique and Ideology Revisited*, ed. Daniel Fairfax (Amsterdam: Amsterdam University Press, 2009).

Contois, Emily J.H. and Zenia Kish, *Food Instagram: Identity, Influence and Negotiation* (Champaign, IL: University of Illinois Press, 2022).

Coulthard, Glen Sean, *Red Skin, White Masks: Rejecting The Colonial Politics of Recognition* (Minneapolis, MN: University of Minnesota Press, 2014).

Crawford, Kate, *Atlas of AI : Power, Politics, and the Planetary Costs of Artificial Intelligence* (New Haven: Yale University Press, 2021).

Crosby Jr., Alfred W., *The Columbian Exchange: Biological and Cultural Consequences of 1492* (Santa Barbara: Greenwood Publishing Group, 1972).

Cubitt, Sean, *EcoMedia* (Amsterdam: Rodopi, 2005). https://doi.org/10.1163/9789004454910

Dapcevich, Madison, "The Key to Yellowstone's Ecological Health? Bison Poop," *Outside*, August 29, 2025, https://www.outsideonline.com/outdoor-adventure/environment/bison-poop-yellowstone-park/

Dauvergne, Peter, *AI in the Wild: Sustainability in the Age of Artificial Intelligence* (Cambridge, MA: MIT Press, 2020).

Davis, Vaginal and Fertile La Toyah Jackson, "*That Fertile Feeling* part 1," *YouTube*, May 20, 2008.

Davis, Vaginal and Fertile La Toyah Jackson, "*That Fertile Feeling* part 2," *YouTube*, May 20, 2008.

De Mandato, Mario, *La primitivita' dell'abitare umano. Studi e ricerche* (Torino: Fratelli Bocca, 1933), 71–72.

Deleuze, Gilles and Claire Parnet, *Dialogues* (New York: Columbia University Press, 1977).

Deleuze, Gilles and Félix Guattari, *A Thousand Plateaus* (Minneapolis, MN: University of Minnesota Press, 1987).

Deloria, Philip J., *Playing Indian* (New Haven: Yale University Press, 1999).

Deloria, Vine and Clifford M. Lytle, *The Nations Within: The Past and Future of American Indian Sovereignty* (New York: Pantheon Books, 1984).

Deloria, Vine, *Spirit & Reason : The Vine Deloria, Jr., Reader*, eds. Barbara Deloria, Kristen Foehner, and Sam Scinta (Golden, CO: Fulcrum Publishing, 1999).

Department for Environment, "Lobsters, octopus and crabs recognised as sentient beings," *GOV.UK*, November 19, 2021, https://www.gov.uk/government/news/lobsters-octopus-and-crabs-recognised-as-sentient-beings#:~:text=Amendment%20to%20Animal%20Welfare%20(Sentience,cephalopod%20molluscs%20as%20sentient%20beings.

Dhivyan Gnanasekhar, Joshua, "Mutualism (biology)," *EBSCO Research Starters*, 2024, https://www.ebsco.com/research-starters/biology/mutualism-biology

Diaz, Vicente, "Oceania in the Plains: The Politics and Analytics of Transindigenous Resurgence in Chuukese Voyaging of Dakota Lands, Waters, and Skies in Miní Sóta Makhóčhe," *Pacific Studies* 42.1/2 (2019).

Donaldson, Sue and Will Kymlicka, *Zoopolis: A Political Theory of Animal Rights* (Oxford: Oxford University Press, 2011).

Duffe, Jack (@jackscarab.bsky.social), "What trillion-dollar problem is AI trying to solve?" *Bluesky*, December 27, 2024, https://bsky.app/profile/jackscarab.bsky.social/post/3lecjwbc5ts2o

Dunham, Cyrus Grace, "The 'Terrorist Drag' of Vaginal Davis," *The New Yorker*, December 12, 2015, https://www.newyorker.com/culture/culture-desk/terrorist-drag-vaginal-davis

Durie, Mason, *Nga- Tai Matatu-: Tides of Ma-ori Endurance* (Oxford: Oxford University Press, 2005).

Fanon, Frantz, *Black Skin, White Masks*, trans. Charles Lam Markmann (New York: Grove Press, 1967).

Fix, Adam J., Hugh Burnam, and Ray Gutteriez, "Toward Interspecies Thinking as a Collaborative Concept: Autoethnographies at the Intersection of Traditional Ecological Knowledge and Animal Studies," *Humanimalia* 10.2 (2019): 128–149. https://doi.org/10.52537/humanimalia.9504

Foster, Laura, "Critical Perspectives on Plants, Race, and Colonialism: An Introduction," *Catalyst: Feminism, Theory, Technoscience* 5.2 (2019): 1–6. https://doi.org/10.28968/cftt.v5i2.32309

Freedman, Dylan, "The Day ChatGPT Went Cold," *The New York Times*, August 19, 2025, https://www.nytimes.com/2025/08/19/business/chatgpt-gpt-5-backlash-openai.html

Freeman, Jo, "The Tyranny of Structurelessness," *Jo Freeman.com*, https://www.jofreeman.com/joreen/tyranny.htm

Friedberg, Anne, "Virilio's Screen: The Work of Metaphor in the Age of Technological Convergence," *Journal of Visual Culture* 3.2 (2004): 183–93. https://journals.sagepub.com/doi/pdf/10.1177/1470412904045027

Gadgil, Madhav and Ramachandra Guha, *This Fissured Land: An Ecological History of India* (Berkeley: The University of California Press, 1993).

Georgi, F. Richard, "Postcolonial and decolonial differences: Do prefixes matter?" *Virtual Encyclopaedia – Rewriting Peace and Conflict*, July 4, 2024, https://rewritingpeaceandconflict.net/postcolonial-and-decolonial-differences/

Geremia, Chris, E. William Hamilton, and Jerod A. Merkle, "Yellowstone's free-moving large bison herds provide a glimpse of their past ecosystem function," *Science* 389.6763 (2025).

Ghosh, Amitav, *The Great Derangement: Climate Change and the Unthinkable* (Chicago: University of Chicago Press, 2016).

Giuffrida, Angela, "Satnam lost his arm and was allegedly left to die on the roadside. This is the horror of exploitation on Italian farms," *The Guardian*, July 20, 2024, https://www.theguardian.com/media/article/2024/jul/20/satnam-lost-his-arm-and-was-allegedly-left-to-die-on-the-roadside-this-is-the-horror-of-exploitation-on-italian-farms

Glissant. Édouard, *Poetics of Relation*. Translated by Betsy Wing (The University of Michigan Press, 1998).

Gomon, Martin F. and Dianne J. Bray, "Gambusia holbrooki," *Fishes of Australia*, https://fishesofaustralia.net.au/home/species/3636

Greenfield, Patrick, "The elephant whisperer: one Thai woman's lifelong quest to protect a rescued herd," *The Guardian*, August 8, 2025, https://www.theguardian.com/environment/2025/aug/08/animal-welfare-asian-elephants-thailand-conservation-aoe

Gregg, Melissa and Gregory Seigworth (eds), *The Affect Theory Reader*, 1st edn (Durham, NC: Duke University Press, 2010).

Gruen, Lori, *Entangled Empathy: An Alternative Ethic for Our Relationships with Animals* (New York: Lantern Books, 2015).

Gruppuso, Paolo, "In-between Solidity and Fluidity: The Reclaimed Marshlands of Agro Pontino," *Theory, Culture & Society* 39.2 (2021): 53–73. https://doi.org/10.1177/02632764211038669

Guattari, Félix, *Soft Subversions: Texts and Interviews 1977–1985*, ed. Sylvère Lotringer, trans. Chet Wiener and Emily Wittman (New York: Semiotext[e], 1996).

Guha, Ramachandra, "Radical American Environmentalism and Wilderness Preservation: A Third World Critique (1997)," in Libby Robin, Sverker Sörlin, and Paul Warde (eds), *The*

Future of Nature: Documents of Global Change (New Haven: Yale University Press, 2013), 409–432.

Gunther, Tim, "Marine Food Pyramid," *National Geographic*, https://education.nationalgeographic.org/resource/marine-food-pyramid-1/

Halberstam, Jack, *In a Queer Time and Place: Transgender Bodies, Subcultural Lives* (New York: New York University Press, 2005).

Halberstam, Jack, *Wild Things: The Disorder of Desire* (Durham, NC: Duke University Press, 2020).

Hames, Raymond, "The Ecologically Noble Savage Debate," *Annual Review of Anthropology* 36 (2007): 177–90. https://doi.org/10.1146/annurev.anthro.35.081705.123321

Hamilton, Clive, *Defiant Earth: The Fate of Humans in the Anthropocene* (Malden, MA: Polity Press, 2017).

Hansen, James, "Celebrated Somali Supper Club Makes London Debut," *Eater*, October 3, 2018, https://london.eater.com/2018/10/3/17931912/arawelo-eats-somali-supper-club-london-asma-khan-chefs-table-darjeeling-express

Hao, Karen, *Empire of AI: Dreams and Nightmares in Sam Altman's OpenAI Dreams and Nightmares in Sam Altman's OpenAI* (New York: Penguin, 2025).

Haraway, Donna, *Staying with the Trouble: Making Kin in the Chthulucene* (Durham, NC: Duke University Press, 2016).

Haraway, Donna, "Deanimations: Maps and Portraits of Life Itself," in Caroline A. Jones and Peter Galison (eds), *Picturing Science, Producing Art* (New York: Routledge, 1998).

Haraway, Donna, "Feminist cyborg scholar Donna Haraway: 'The disorder of our era isn't necessary'," interview by Moira Weigel, *The Guardian*, June 20, 2019, https://www.theguardian.com/world/2019/jun/20/donna-haraway-interview-cyborg-manifesto-post-truth

Haraway, Donna, "Making Kin: An Interview with Donna Haraway," interview by Steve Paulson, *Los Angeles Review of Books*, December 6, 2019, https://lareviewofbooks.org/article/making-kin-an-interview-with-donna-haraway/

Haraway, Donna, "Situated Knowledges: The Science Question in Feminism and the Privilege of Partial Perspective," *Feminist Studies* 14.3 (1988): 575–599. https://doi.org/10.2307/3178066

Haraway, Donna, *The Companion Species Manifesto: Dogs, People, and Significant Others* (Chicago: Prickly Paradigm Press, 2003).

Hassan, Hawa, *In Bibi's Kitchen: The Recipes and Stories of Grandmothers from the Eight African Countries that Touch the Indian Ocean* (New York: Ten Speed Press, 2020).

Hearst, David, "African woman going home after 200 years," *The Guardian*, April 30, 2002, https://www.theguardian.com/world/2002/apr/30/education.arts#:~:text=In%20London%2C%20the%20act%20consisted,funeral%20in%20August%20or%20September

Heise, Ursula, *Imagining Extinction: The Cultural Meanings of Endangered Species* (Chicago: University of Chicago Press, 2016). https://doi.org/10.1007/978-3-658-13213-2_92

Hess, Amanda, "They're Stuffed Animals. They're Also A.I. Chatbots," *The New York Times*, August 15, 2025, https://www.nytimes.com/2025/08/15/arts/ai-toys-curio-grem.html

Hiltzik, Michael, "Say farewell to the AI bubble, and get ready for the crash," *LA Times*, August 20, 2025, https://www.latimes.com/business/story/2025-08-20/say-farewell-to-the-ai-bubble-and-get-ready-for-the-crash

Hokowhitu, Brendan, "The Emperor's 'New Materialisms': Indigenous Materialisms and Disciplinary Colonialism," in Brendan Hokowhitu et al. (eds) *Routledge Handbook of Critical Indigenous Studies*, 1st edn (London: Routledge, 2020).

Hokowhitu, Brendan, "Indigenous Materialisms and *Disciplinary Colonialism*," *Somatechnics* 11.2 (2021): 157–73. https://doi.org/10.3366/soma.2021.0349

Hörl, Erich and James Burton, *General Ecology: The New Ecological Paradigm* (London: Bloomsbury Academic, 2017).

Howe, Piers Douglas Lionel et al., "Self-certification: A novel method for increasing sharing discernment on social media," *PLOS One* 19.6 (2024): 1–17. https://doi.org/10.1371/journal.pone.0303025

Howells, Ruth, "Epoch-defining study pinpoints when humans came to dominate planet Earth," *UCL News*, March 11, 2015, https://www.ucl.ac.uk/news/2015/mar/epoch-defining-study-pinpoints-when-humans-came-dominate-planet- earth#:~:text=The%20researchers%20have%20named%20the,disconnected%20peoples%20becoming%20globally%20linked

Iovino, Serenella and Serpil Oppermann, *Material Ecocriticism* (Bloomington, IN: Indiana University Press, 2014).

Ismail, Fozia, "Somali food love, Arawelo style," *Arawelo Eats*, https://www.araweloeats.com/

Ismail, Fozia, "The recipe that reminds me of home," interview by Killian Fox and Holly O'Neill, *The Guardian*, July 14, 2019, https://www.theguardian.com/food/2019/jul/14/recipe-that-reminds-me-of-home-souvlaki-somali-bean-stew

Jackson, Zakiyyah Iman, "Outer Worlds: The Persistence of Race in Movement 'Beyond the Human'," *GLQ* 21:2–3 (2015): 215–18.

Janzen, Jörg H.A. and Ioan M. Lewis, "Revolt in British Somaliland," *Encyclopedia Britannica*, November 11, 2025, https://www.britannica.com/place/Somalia/Revolt-in-British-Somaliland#ref419656

Johnson, John R, "The Arlington Spring Man," *West of the West: Tales from California's Channel Islands*, https://www.teachchannelislands.org/tales/arlington-springs-man

Kassim, Abdullahi and Leila Kassim, "Xawaash (Somali Spice Mix) Mélange d'épices Somali حوائج – بهارات صومالية," *Xawaash*, August 7, 2011, https://xawaash.com/?p=12

Kauanui, J. Kēhaulani, *Paradoxes of Hawaiian Sovereignty: Land, Sex and the Colonial Politics of State Nationalism* (Durham, NC: Duke University Press, 2018).

King, A.C.F., Bauska, T.K., Brook, E.J. et al., "Reconciling ice core CO_2 and land-use change following New World-Old World Contact," *Nat Commun* 15, 1735 (2024): 1–9. https://doi.org/10.1038/s41467-024-45894-9

Koshmrl, Mike, "Buffalo (almost) officially wildlife on some 2M new acres of Wyoming, a step toward roaming free," *WyoFile*, August 15, 2025, https://wyofile.com/buffalo-almost-officially-wildlife-on-some-2m-new-acres-of-wyoming-a-step-toward-roaming-free/

Kotus, Jacek and Qihang Qiu, "Multi-species Placemaking or Urban Attraction? The human perception of Austin's bat colony," *Cities* 171 (2026).

Krützen, Michael, et al., "Culture and Geographic Variation in Orangutan Behavior," *Current Biology* 21 (2011): 1808–12. https://doi.org/10.1016/j.cub.2011.09.017

Kushkush, Isma'il, "After Barren Years in Somalia, Signs of Growth by the Bunch," *The New York Times*, December 13, 2014, https://www.nytimes.com/2014/12/14/world/after-barren-years-in-somalia-signs-of-growth-in-bananas.html

Lambert, Jonathan, "Photos: Mother Nature must be really annoyed at our fakery," *NPR*, August 30, 2025, https://www.npr.org/sections/goats-and-soda/2025/08/30/g-s1-86289/photos-nature-anthropocene-polar-bear-elephants-illusion

Latour, Bruno, *Reassembling the Social: An Introduction to Actor-Network-Theory* (New York: Oxford University Press, 2005).

Lee, Russell, "Negro drinking at 'Colored' water cooler in streetcar terminal, Oklahoma City, Oklahoma," July 1939, *Library of Congress*, http://hdl.loc.gov/loc.pnp/fsa.8a26761

Leeke, Caleb, "Triturus carnifex, Triturus cristatus, Triturus dobrogicus, Triturus karelinii," *Caudata Culture*, December 2002, https://www.caudata.org/cc/species/Triturus/Triturus_sp.shtml

Leogrande, Alessandro, "Viaggio nella zona grigia che ha portato all'omicidio di Fermo," *Internazionale*, March 18, 2017, https://www.internazionale.it/reportage/alessandro-leogrande/2017/03/18/emmanuel-chidi-nnamdi-fermo-omicidio

Lestel, Dominique, Florence Brunois, and Florence Gaunet, "Etho-ethnology and Ethno-ethology," *Social Science Information* 45 (2006): 155–77. https://doi.org/10.1177/0539018406063633

Lewis, Simon and Mark Maslin, "Defining the Anthropocene," *Nature* 519 (2015): 171–80. https://doi.org/10.1038/nature14258

Lezak, Stephen, "Wild horses, buffalo and the politics of belonging," *High Country News*, December 1, 2021, https://www.hcn.org/issues/53.12/indigenous-affairs-wildlife-wild-horses-buffalo-and-the-politics-of-belonging

Little Bear, Leroy, "Jagged Worldviews Colliding," in Marie Battiste (ed.), *Reclaiming Indigenous Voice and Vision* (Vancouver: University of British Columbia Press, 2007), https://doi.org/10.59962/9780774853170-009

Lowenhaupt Tsing, Anna, "Arts of Inclusion, or, How to Love a Mushroom," *Australian Humanities Review* 50 (2011): 5–21. https://doi.org/10.1007/BF03341595

Mangan, Elizabeth U., "Yellowstone, the First National Park," *Library of Congress*, https://www.loc.gov/collections/national-parks-maps/articles-and-essays/yellowstone-the-first-national-park/

Mannoni, Octave. *Clefs pour l'imaginaire; ou, L'autre scène* (Paris: Editions du Seuil, 1969).

Marcone, Jorge, "The Stone Guests: Buen Vivir and popular environmentalisms in the Andes and Amazonia," in Ursula Heise, Jon Christensen, and Michelle Niemann (eds), *The Routledge Companion to the Environmental Humanities* (London: Routledge, 2017).

Martín, Bruno, "The Hidden Environmental Toll of Smartphones," *BBVA*, February 24, 2020, https://www.bbvaopenmind.com/en/science/environment/the-hidden-environmental-toll-of-smartphones/

Martínez, Marcos, "Burned to death because of a rumour on WhatsApp," *BBC*, November 11, 2018, https://www.bbc.com/news/world-latin-america-46145986

McBride, Anne, "Food Porn," *Gastronomica: The Journal of Critical Food Studies* 10.1 (2010). https://doi.org/10.1525/gfc.2010.10.1.38

McFarland, Sarah and Ryan Hediger, *Animals and Agency: An Interdisciplinary Exploration* (Leiden: Brill, 2009).

Meyer, Manu Aluli, "Acultural Assumptions of Empiricism: A Native Hawaiian Critique," *Canadian Journal of Native Education* 25.2 (2001). https://doi.org/10.3368/npj.2.1.25

Miller, Elizabeth J. et al., "AI Hyperrealism: Why AI Faces Are Perceived as More Real Than Human Ones," *Psychological Science* 34.12 (2023). https://doi.org/10.1177/09567976231201853

Miller, Thomas E. and Casey P. Terhost, "Indirect Effects in Communities and Ecosystems," *Oxford Bibliographies*, May 23, 2012, https://www.oxfordbibliographies.com/display/document/obo-9780199830060/obo-9780199830060-0020.xml#:~:text=obo/9780199830060%2D0020-,Introduction,a%20major%20issue%20in%20ecology

Moore, Donald S., Jake Kosek, and Anand Pandian, "The Cultural Politics of Race and Nature: Terrains of Power and Practice," in Donald S. Moore, Jake Kosek, and Anand Pandian (eds), *Race, Nature, and the Politics of Difference* (Durham, NC: Duke University Press, 2003), 1–70.

Moreton-Robinson, Aileen, "Relationality: A Key Presupposition of an Indigenous Social Research Paradigm" in Chris Andersen and Jean M. O'Brien (eds), *Sources and Methods in Indigenous Studies* (New York: Routledge, 2017), 69–77.

Morton, Timothy, *Hyperobjects: Philosophy and Ecology after the End of the World* (Minneapolis, MN: University of Minnesota Press, 2013).

Mueller, Pam A. and Daniel M. Oppenheimer, "The Pen Is Mightier Than the Keyboard: Advantages of Longhand Over Laptop Note Taking," *Psychological Science* 25.6 (2014): 1159–68. https://www.jstor.org/stable/24543504

Muñoz, José Esteban, *Disidentifications: Queers of Color and the Performance of Politics* (Minneapolis, MN: University of Minnesota Press, 1999).

Myers, Natasha, "Ungrid-able Ecologies: Decolonizing the Ecological Sensorium in a Ten-Thousand-Year-Old NaturalCultural Happening," *Catalyst: Feminism, Theory, Technoscience* 3.2 (2017): 1–24. https://doi.org/10.28968/cftt.v3i2.28848

Nabokov, Peter and Lawrence Loendorf, *Restoring a Presence: American Indians and Yellowstone National Park* (Norman, OK: University of Oklahoma Press, 2004).

Nathan, Vetri, "Multispecies Futures Lab: Ecocultural Wellbeing and Kinship in the Cybercene," *Multispecies Futures Lab*, 2023, www.multispeciesfutureslab.org

Nathan, Vetri and Teresa Fiore, "The Ecocultures of the *Scala dei Turchi*," 2025, https://storymaps.arcgis.com/stories/659a395c91c547f7949ba2b9a9a1deb5

Nathan, Vetri, "Edible Nation: Imagining Somalia Through Cuisine," in Rosetta G. Caponetto, Giusy Di Filippo and Martina Di Florio (eds), *Longing for the Future: Mal D'Afrique and Afro-Optimism in Perspectives on Somalia* (New York: Routledge, 2023). https://doi.org/10.4324/9781003283102.

Nelson, Zed, "Polar bears in shopping malls, fiberglass mountains, indoor rainforests: how humans got addicted to faking the natural world," *The Guardian*, July 24, 2025, https://www.theguardian.com/environment/2025/jul/24/humans-addicted-faking-natural-world-anthropocene-illusion-zed-nelson-aoe.

Noor, Dharna, "Climate advocates outrages at Trump administration plans to fast-track AI sector," *The Guardian*, July 23, 2025, https://www.theguardian.com/us-news/2025/jul/23/trump-ai-climate-change

Norton, Mary. *The Tame and the Wild: People and Animals after 1492* (Cambridge, USA: Harvard University Press, 2024).

O'Neil, Molly, "Food Porn," *Columbia Journalism Review*, September 1, 2003, https://www.cjr.org/from_the_archives/food-writing-cookbook.php

Painter, Luke E., Robert L. Beschta, and William J. Ripple, "Changing aspen stand structure following large carnivore restoration in Yellowstone," *Forest Ecology and Management* 594 (2025). https://doi.org/10.1016/j.foreco.2025.122941

Peterson, Christopher, *Bestial Traces: Race, Sexuality, Animality* (New York: Fordham University Press, 2012).

Polakovic, Gary, "Channel Island Woman's Bones May Rewrite History," *Los Angeles Times*, April 11, 1999, https://www.latimes.com/archives/la-xpm-1999-apr-11-mn-26401-story.html

Pope Alexander VI, "Demarcation Bull Granting Spain Possession of Lands Discovered by Columbus," *Gilder Lehrman Institute of American History*, https://www.gilderlehrman.org/sites/default/files/inline-pdfs/T-04093.pdf

Presner, Todd, David Shepard, and Yoh Kawano, *Hypercities: Thick Mapping in the Digital Humanities* (Cambridge, MA: Harvard University Press, 2014).

Raleigh, Sir Walter, *The Discovery of Guiana*, Project Gutenberg Etext; No. 2272 (Champaign, IL: Project Gutenberg, 2006), https://www.gutenberg.org/files/2272/2272-h/2272-h.htm

Redazione Roma, "Satnam Singh, morto dopo essere stato scaricato ferito davanti casa," *La Repubblica*, July 20, 2024, https://roma.repubblica.it/cronaca/2024/06/20/news/morte_satnam_singh_latina_lovato-423262714/

Richardson, Miles, "How we Lost Touch with Nature," *Finding Nature*, August 11, 2025, https://findingnature.org.uk/2025/08/11/model/

Robinson, Kylie, "OpenAI Is Poised to Become the Most Valuable Startup Ever. Should It Be?" *Wired*, August 19, 2025, https://www.wired.com/story/openai-valuation-500-billion-skepticism/#:~:text=OpenAI%20Is%20Poised%20to%20Become,Italiano%20

%E6%97%A5%E6%9C%AC%E8%AA%9E%20%E7%B9%81%E9%AB%94%E4%B8%AD%E6%96%87

Rosenwald, Mike, "Queen Arawelo," *Retropod*, April 10, 2019, https://www.washingtonpost.com/podcasts/retropod/queen-arawelo/?tid=aud_rsslink

Rothfels, Nigel, *Savages and Beasts: The Birth of the Modern Zoo* (Baltimore: The Johns Hopkins University Press, 2002).

Roy, Arundhati, "Approaching Gridlock: Free Speech and Failing Democracy," *The Wire*, March 24, 2023, https://thewire.in/politics/approaching-gridlock-free-speech-and-failing-democracy

Roy, Sutirtho, "Can The Non-Human Subaltern Speak? Addressing Injustice Through Parakeets, Penguins and Blue Macaws," *Studia Ubb Philologia* LXVII. 2 (2022), 227–50. https://doi.org/10.24193/subbphilo.2022.2.13.

Said, Edward, *Orientalism* (New York: Vintage Books, 1979).

Said, Edward, *Culture and Imperialism* (New York: Vintage Books, 1994).

Sarfatti, Margherita G, *The Life of Benito Mussolini*, trans. Frederic White (New York: Frederick A. Stokes Company, 1927).

Satia, Priya, "Stanford scholar explains the history of India's partition, its ongoing effects today," interview by Alex Shashkevich, *StanfordReport*, March 8, 2019, https://news.stanford.edu/stories/2019/03/partition-1947-continues-haunt-india-pakistan-stanford-scholar-says

Seymour, Nicole, *Strange Natures: Futurity, Empathy, and the Queer Ecological Imagination* (Champaign, IL: University of Illinois Press, 2013).

Sharif, Abusaleh and Ataulla Khan, "What is the Shadow Value of an Illegal Indian Immigrant in the US?" *The Wire*, October 24, 2024, https://thewire.in/world/indian-dunki-migration-us-shadow-value

Singh, Sandeep and Elena Cavallone, "India to Italy: a trail of broken dreams, loss and abuse," *Context*, November 1, 2024, https://www.context.news/money-power-people/india-to-italy-a-trail-of-broken-dreams-loss-and-abuse

Spiegel, Marjorie M., *The Dreaded Comparison: Human and Animal* (New York, NY: Mirror Books, 1996).

Staiger, Janet, *Interpreting Films: Studies in the Historical Reception of American Cinema* (Princeton, NJ: Princeton University Press, 1992).

Steger, Manfred, "Searching for *Satya* through *Ahimsa*: Gandhi's Challenge to Western Discourses of Power," *Constellations* 13.3 (2006): 332–353. https://doi.org/10.1111/j.1467-8675.2006.00405.x

TallBear, Kim and Angela Willey, "Critical Relationality: Queer, Indigenous, and Multispecies Belonging Beyond Settler Sex & Nature," *Imaginations* 10.1 (2019).

TallBear, Kim, "An Indigenous Reflection on Working beyond the Human/Not Human," *GLQ: A Journal of Lesbian and Gay Studies* 21.2–3 (2015): 230–35.

TallBear, Kim, "Standing With and Speaking as Faith: A Feminist-Indigenous Approach to Inquiry," *Journal of Research Practice* 10.2 (2014): 1–7.

Tanveer, Tuba, Nixon Mathur and Rakesh Sarwal, "Catastrophic impact of 1947 partition of India on people's health," *Journal of Family Medicine and Primary Care* 13.2 (February 2024): 401–8.

Tasty, "Mindy Kaling and Hawa Hassen Make Spicy Somali Pasta," *YouTube*, September 8, 2021, https://www.youtube.com/watch?v=PMqj2BSW0Yk

Taylor, Ashley P, "Why Do Parrots Talk?" *Audobon*, August 6, 2015, https://www.audubon.org/news/why-do-parrots-talk#:~:text=Parrots%20are%20vocal%20learners%2C%20meaning,until%20his%20death%20in%202007

Terranova, Tiziana, "Free Labor: Producing Culture for the Digital Economy," *Social Text* 18.2 (2000) 33–58. https://web.mit.edu/schock/www/docs/18.2terranova.pdf

Timbrook, Jan and Chris Chapman, *Chumash Ethnobotany: Plant Knowledge Among the Chumash People of Southern California*, 2nd ed. (Santa Barbara: Santa Barbara Museum of Natural History, 2007).

Todd, Zoe, "An Indigenous Feminist's Take On The Ontological Turn: 'Ontology' Is Just Another Word For Colonialism," *Journal of Historical Sociology* 29.1 (2016): 4–22. https://doi.org/10.1111/johs.12124

Tola, Miriam, "Intersectionality and the Environmental Humanities: Notes on Elisions and Encounters," *AboutGender* 11.22 (2022).

van Dooren, Thom, Eben Kirksey, and Ursula Münster, "Multispecies Studies: Cultivating Arts of Attentiveness," *Environmental Humanities* 8.1 (2016): 1–23. https://doi.org/10.1215/22011919-3527695

van Dooren, Thom, "Care," *Environmental Humanities* 5.1 (2014): 291–94. https://doi.org/10.1215/22011919-3615541

van Dooren, Thom, "Multispecies Storytelling in Catastrophic Times," *Thom van Dooren*, June 24, 2025, https://www.thomvandooren.org/2025/06/24/multispecies-storytelling-in-catastrophic-times/

Van Duzer, Chet, "Horror vacui," *Stanford University Libraries*, https://exhibits.stanford.edu/blrcc/feature/horror-vacui

Vandersommers, Daniel, *Entangled Encounters at the National Zoo: Stories from the Animal Archive* (Lawrence, KS: University of Kansas Press, 2023).

Venditti, Bruno, "Visualizing the Critical Metals in a Smartphone," *Visual Capitalist*, August 25, 2021, https://www.visualcapitalist.com/visualizing-the-critical-metals-in-a-smartphone/

Vespucci, Amerigo, *Mundus Novus: Letter to Lorenzo Pietro di Medici*, trans. George Tyler Northup (Princeton, NJ: Princeton University, Press, 1916), https://encyclopedia virginia.org/primary-documents/mundus-novus-1503/

Waldman, Devra and Asher Ghertner, "The enclaved body: Crises of personhood and the embodied geographies of urban gating," *Progress in Human Geography* 47.2 (2023): 280–97. https://doi.org/10.1177/03091325221150792

Wang, Xiaowei, *Blockchain Chicken Farm and Other Stories of Tech in China's Countryside* (New York: Farrar, Straus and Giroux, 2020).

Whyte, Kyle, "Against Crisis Epistemology" in Brendan Hokowhitu et al. (eds), *Routledge Handbook of Critical Indigenous Studies*, 1st edn (New York: Routledge, 2020), 52–64.

Wilson, Mark, "Smartphones Are Killing The Planet Faster Than Anyone Expected," *Fast Company*, March 27, 2018, https://www.fastcompany.com/90165365/smartphones-are-wrecking-the-planet-faster-than-anyone-expected

Wilson, Shawn, *Research Is Ceremony: Indigenous Research Methods* (Halifax: Fernwood Publishing, 2008).

Windchief, Sweeney and Timothy San Pedro, *Applying Indigenous Research Methods* (New York: Routledge, 2019).

Winfield, Nicole, "Vatican formally rejects 'Doctrine of Discovery' after Indigenous calls," *PBS News*, May 30, 2023, https://www.pbs.org/newshour/politics/vatican-formally-rejects-doctrine-of-discovery-after-Indigenous-calls#:~:text=Those%20decrees%20underpin%20the%20%E2%80%9CDoctrine,because%20they%20%E2%80%9Cdiscovered%E2%80%9D%20it

Wolfe, Cary, *Animal Rites: American Culture, the Discourse of Species, and Posthumanist Theory* (Chicago: University of Chicago Press, 2003).

Xiang, Chloe, "'He Would Still Be Here': Man Dies by Suicide After Talking with AI Chatbot, Widow Says," *Vice*, March 30, 2025, https://www.vice.com/en/article/man-dies-by-suicide-after-talking-with-ai-chatbot-widow-says/

Young, Robert J.C., *Colonial Desire: Hybridity in Theory, Culture and Race*, 1st edn (New York: Routledge, 1995).

Yusoff, Kathryn, "*A Billion Black Anthropocenes or None*: A New Book About Race and Geology," interview by Leah Kaplan, *Black Perspectives*, April 12, 2019, https://www.aaihs.org/a-billion-black-anthropocenes-or-none-a-new-book-about-race-and-geology/

Zavaleta, Erika, "The Emergence of Waterfowl Conservation Among Yup'ik Hunters in the Yukon-Kuskokwim Delta, Alaska," *Human Ecology* 27 (1999): 231–66. https://doi.org/10.1023/A:1018773211034

Zontek, Ken, *Buffalo Nation: American Indian Efforts to Restore the Bison* (Lincoln, NE: University of Nebraska Press, 2007).

Index

www.ingramcontent.com/pod-product-compliance
Ingram Content Group UK Ltd.
Pitfield, Milton Keynes, MK11 3LW, UK
UKHW041643031026
15587UKWH00001B/464

9 781804 132524